NONLINEAR SYSTEM ANALYSIS AND IDENTIFICATION FROM RANDOM DATA

NONLINEAR SYSTEM ANALYSIS AND IDENTIFICATION FROM RANDOM DATA

JULIUS S. BENDAT

A Wiley-Interscience Publication

JOHN WILEY & SONS

New York / Chichester / Brisbane / Toronto / Singapore

Library of Congress Cataloging in Publication Data:

Bendat, Julius S.
Nonlinear system analysis and identification from random data / Julius S. Bendat.
p. cm.
"A Wiley-Interscience publication."
Bibliography: p.
Includes index.
ISBN 0-471-60623-5
1. System analysis. 2. Nonlinear theories. 3. System identification. 4. Stochastic processes. I. Title.
QA402.B449 1990
003—dc20 89-16439
CIP

Printed in the United States of America

10 9 8 7 6 5 4 3 2 1

To my parents,
Benjamin and Frieda Bendat

PREFACE

This book deals with the analysis and identification of nonlinear systems from random data. Past work on these matters has been concerned mainly with topics of either (1) linear system analysis from random data or (2) nonlinear system analysis from deterministic data. Research studies from 1977 to date have given me more understanding and insight on the transformation of random data through many types of nonlinear systems. This book discusses various work done by others and contains my contributions to this field.

This book develops new standard nonlinear system analysis and identification techniques that extend and replace conventional linear system techniques. For single-input/single-output models with parallel linear, bilinear, and trilinear systems, formulas are derived that can predict the separate linear and nonlinear output spectra from measured input data and measured total output data. Practical techniques are outlined for modeling and analyzing a wide class of third-order polynomial nonlinear models consisting of linear systems in parallel with finite-memory square-law systems and finite-memory cubic systems. Optimum nonlinear system identification formulas are derived for bilinear and trilinear systems as well as for finite-memory square-law systems and finite-memory cubic systems. A general methodology is outlined for changing nonlinear models of parallel linear and nonlinear systems with correlated outputs into equivalent models of parallel linear and nonlinear systems with uncorrelated outputs. Useful procedures are shown for replacing important types of single-input/single-output nonlinear models with alternative multiple-input/single-output linear models, where input data can be Gaussian or non-Gaussian. Complicated Volterra functional methods for nonlinear system description are reviewed and compared with the new simpler methods.

Readers of this book are expected to be familiar with fundamental ideas of analyzing and identifying linear systems from random data as contained in two

Wiley-Interscience books that I coauthored with Allan Piersol, *Engineering Applications of Correlation and Spectral Analysis*, 1980, and *Random Data: Analysis and Measurement Procedures*, Second Edition, 1986. This background of notation, definitions, and procedures is assumed wherever needed. This book has been written for practicing engineers and scientists, students, and teachers who are concerned with extending past results by employing the techniques in this book to analyze and identify nonlinear systems from random data.

The emphasis here is on what can be learned from time-history records of stationary random data or transient random data measured at the input and output points of physical systems. When the nonlinear system properties are specified, procedures are shown for predicting the output data properties from measurement of the input data properties. When the nonlinear system properties are not specified, procedures are shown for identifying optimum nonlinear system properties from simultaneous measurement of the input data properties and the output data properties. Practical results are presented as functions of frequency using basic and advanced spectral density functions together with basic and advanced frequency response functions. These frequency-domain results are much more significant than any associated time-domain results and are also easier to compute and interpret. Statistical error analysis formulas are also derived in this book that can be used to design experiments and evaluate estimates obtained from the measured data.

The scope of this book is indicated in the contents. Chapters 1–4 involve older, known techniques. Chapter 1 reviews basic ideas of linear systems, random data, and spectral density functions. Chapter 2 discusses zero-memory nonlinear systems and gives many engineering examples. Relations are derived and illustrated for input/output probability density functions and correlation functions. Chapter 3 deals with theoretical properties of general bilinear and trilinear systems. Higher-order forms of multidimensional correlation and spectral density functions are defined. Chapter 4 develops multidimensional correlation and spectral input/output relationships for general third-order Volterra nonlinear models consisting of parallel linear, bilinear, and trilinear systems. Formulas are derived to identify optimum linear, bilinear, and trilinear system properties from measured input data and measured total output data. Other formulas show how to decompose the measured total output spectrum into its separate linear, bilinear, and trilinear components. These formulas are intricate to compute and difficult to interpret.

Chapters 5–7 involve new techniques. Chapter 5 concerns finite-memory square-law systems and finite-memory cubic systems that can be used to model many nonlinear physical situations via optimum third-order polynomial least-squares approximations. New, simpler, one-dimensional spectral input/output relations are derived and discussed in Chapter 5 that are special cases of the more general multidimensional spectral results in Chapters 3 and 4. Chapter 5 also shows how to solve many important nonlinear models using well-established multiple-input/single-output linear techniques. Chapter 6 derives new, useful, statistical error analysis formulas for estimates from measured data of various quantities needed to analyze the third-order polynomial nonlinear models in Chapter 5. Chapter 7 outlines a new, practical, recommended methodology for analyzing and identifying general nonlinear models of arbitrary linear systems in parallel with arbitrary nonlinear systems. A wide class of such nonlinear models is shown to be equivalent to two-input/single-output linear models. Formulas are stated for two engineering examples that represent nonlinear wave force models and nonlinear drift force models. Procedures are discussed for determining physical parameters in nonlinear differential equations of motion by appropriate analysis of measured data.

Material contained in this book originated from some of my nonlinear research work that was sponsored by Shell Internationale Petroleum Maatschappij (SIPM) in The Hague, Netherlands, under Jan Vugts, and by the Naval Civil Engineering Laboratory (NCEL) in Port Hueneme, California, under Paul Palo. These two men provided their expertise to guide and review the nonlinear results obtained for their projects. I thank Jan Vugts for starting me on these matters and for the support of SIPM from 1977 to 1982. I thank Paul Palo for his active technical participation with me and for suggesting many productive ideas on nonlinear system properties that are included in the text of this book. I gratefully acknowledge also the continuing support given to me since 1982 by NCEL and the Office of Naval Research. Part of the nonlinear research work for SIPM was performed by Allan Piersol and was enhanced by his efforts. Other nonlinear research work for NCEL to verify certain results in this book was conducted by Robert Coppolino of Measurement Analysis Corporation. My thanks are extended to Allan Piersol and Robert Coppolino for their contributions and engineering knowledge on practical ways to apply these techniques. I wish also to thank Carol Rosato and Phyllis Parris for their dedicated efforts and skill in typing this manuscript.

JULIUS S. BENDAT

Los Angeles, California
January 1990

CONTENTS

LIST OF FIGURES

NONLINEAR SYSTEM ANALYSIS AND IDENTIFICATION FROM RANDOM DATA

1

LINEAR SYSTEMS, RANDOM DATA, SPECTRAL DENSITY FUNCTIONS

This first chapter reviews how to classify systems and discusses basic properties for ideal physical systems to be linear, physically realizable, constant-parameter, and stable. The chapter also reviews how to classify random data as stationary, nonstationary, or transient, and discusses basic formulas for direct Fourier transform computation of spectral density functions for stationary random data and transient random data. These results represent background material in later chapters that deal with procedures for analyzing and identifying various types of linear and nonlinear systems when stationary random data or transient random data pass through these systems.

1.1 LINEAR SYSTEMS

Physical and engineering systems can be separated into two distinct classifications as linear or nonlinear, with parameters in these systems that are either constant-parameter or time-varying. This leads to the classifications shown in Figure 1.1. Linear system properties are discussed in references 1 and 2. Important types of nonlinear systems are discussed in this book.

A system H is a *linear* system if, for any inputs $x_1 = x_1(t)$ and $x_2 = x_2(t)$, and for any constants c_1, c_2,

$$H[c_1 x_1 + c_2 x_2] = c_1 H[x_1] + c_2 H[x_2] \tag{1.1}$$

This equation contains two properties required of linear systems:

Additive Property

$$H[x_1 + x_2] = H[x_1] + H[x_2] \tag{1.2}$$

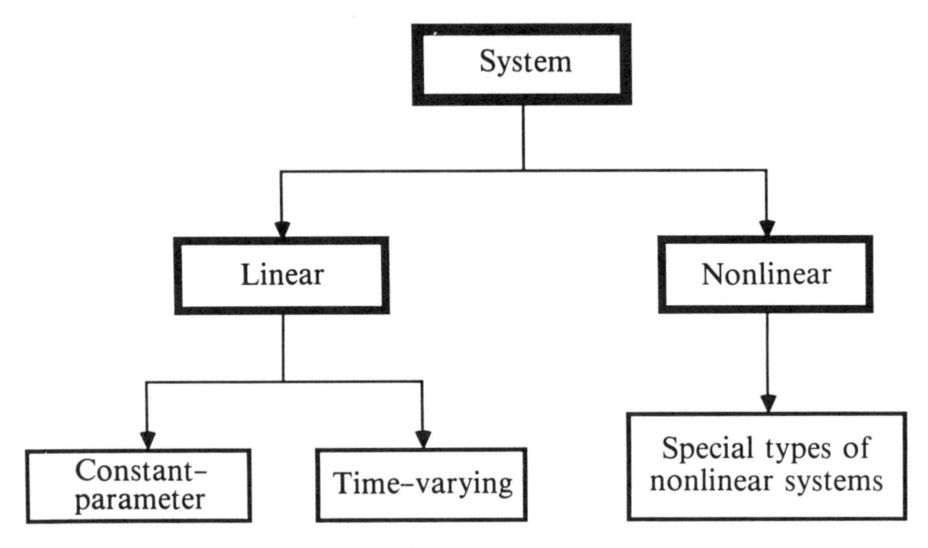

Figure 1.1 Classifications of systems.

Homogeneous Property

$$H[cx] = cH[x] \tag{1.3}$$

By superposition, for linear systems these two properties extend to any finite sum of input terms. The superposition property applies also to integrals that are limiting expressions of finite sums. Failure to satisfy either the additive or the homogeneous property defines a *nonlinear* system and provides a simple, basic way to detect nonlinear system operations.

Systems can be classified as constant-parameter or time-varying. A linear system or a nonlinear system is a *constant-parameter* system if the system response $y(t) = H[x(t)]$ due to an input $x(t)$ is independent of the time when $x(t)$ occurs. Stated another way, if $x(t)$ becomes $x(t + \tau)$, then $y(t)$ becomes $y(t + \tau)$, namely,

$$y(t + \tau) = H[x(t + \tau)] \qquad \text{for any } \tau \tag{1.4}$$

A constant-parameter system is governed by a differential-integral equation with constant coefficients. If the coefficients vary with time or if Eq. (1.4) is not satisfied, then the linear or nonlinear system is said to be *time-varying*. Unless stated otherwise, the systems considered in this book are constant-parameter systems.

Systems can be further classified as zero-memory, finite-memory, or infinite-memory. The idea of an *infinite memory* (infinite record lengths) is largely for mathematical convenience and gives negligible error if the system response due to all past inputs is known to be immeasurably small to inputs occurring in the far distant past and if long-time operation is involved. However, if these long-time effects do not die down, or if there is a starting switch in a system so that inputs are present only for a finite period after the switch is closed, then a definite restriction of *finite memory* (finite record lengths) must be imposed to agree with the physical situation. Data obtained experimentally always last only for finite record lengths, leading to statistical errors in estimates of desired theoretical quantities based upon infinite record lengths. A *zero-memory* system acts instantaneously on the input data without any memory operations.

For any linear system, random input data with a theoretical Gaussian probability density function describing its amplitude properties will produce random output data that will also have a theoretical Gaussian probability density function. However, for any nonlinear system, Gaussian input data will always produce non-Gaussian output data. Hence, another simple way to detect nonlinear system operations is to measure non-Gaussian output data when this output data is caused by input data that are known, assumed, or verified to be Gaussian. As will be developed in Chapter 2, the particular nature of the non-Gaussian output data can often help to identify the special type of nonlinear operation. Measured data, which can never be truly Gaussian over very wide amplitude ranges, should be judged to be Gaussian or non-Gaussian depending upon their properties over amplitude ranges of interest.

It should be noted that different nonlinear transformations may yield approximately similar non-Gaussian outputs so that precise nonlinear system identification usually requires other outside knowledge or additional analyses besides measurement of output probability density functions. Such matters are discussed in this book.

A system H is a *nonlinear* system if, for any inputs $x_1 = x_1(t)$ and $x_2 = x_2(t)$, and for any constant c, the system input/output relations are *not* additive and *not* homogeneous. Specifically,

$$H[x_1 + x_2] \neq H[x_1] + H[x_2] \tag{1.5}$$

$$H[cx] \neq cH[x] \tag{1.6}$$

These nonlinear operations will produce non-Gaussian output data when input data are Gaussian, whereas linear operations preserve the Gaussian structure. Non-Gaussian output data will also occur when input data are non-Gaussian, for example, sine waves. No true sine waves actually exist in nature, but they can be closely approximated.

Any physical system that must operate over very wide dynamic input ranges will ultimately exhibit nonlinear behavior because of saturation and clipping effects. Thus, physical systems should be judged to be linear or nonlinear depending upon their specific properties for desired types and levels of inputs. No general theory is available for the analysis of nonlinear systems as exists for linear systems. Instead, different techniques are required for special classifications of nonlinear systems such as those discussed in this book.

1.1.1 Weighting Functions

An ideal constant-parameter linear system can be defined mathematically using a real-valued *unit impulse response function* $h(\tau)$, also call the *weighting function*. This weighting function measures the response of the system at any time to a unit impulse delta function input applied a time τ before. Specifically, $h(\tau)$ measures the output at any time t due to an input applied at time $(t - \tau)$. For a real system to be *physically realizable*, the system cannot respond before the input occurs. Hence, one must have $h(\tau) = 0$ for $(t - \tau) > t$, corresponding to

$$h(\tau) = 0 \qquad \text{for} \quad \tau < 0 \tag{1.7}$$

This physically realizable weighting function is a special case of a *causal function*, defined as any real-valued function $f(t)$ that is zero for $t \leqslant 0$. Physically realizable systems, whether linear or nonlinear, are always causal.

Consider an arbitrary input $x(\tau)$ to a linear system satisfying Eq. (1.7) where the system operates over the past history of the input from $\tau = -\infty$, the infinite past, to $\tau = t$, the present time. The present value of $x(t)$ can be expressed by the integral

$$x(t) = \int_{-\infty}^{t+} x(\tau)\delta(t - \tau)\, d\tau \tag{1.8}$$

where $t+$ is an infinitesimal value above t and $\delta(\tau)$ is the usual symmetric unit impulse response delta function defined by

$$\begin{aligned} \delta(\tau) &= 0 \qquad \text{for} \quad \tau \neq 0, \quad \delta(0) = \infty \\ \int_{-\varepsilon}^{\varepsilon} \delta(\tau) &= 1 \qquad \text{for all} \quad \varepsilon > 0 \end{aligned} \tag{1.9}$$

Note that Eq. (1.8), where the integral is the limit of a sum of past delta functions $\delta(t - \tau)$, merely states that $x(t) = x(t)$.

Let the output of the linear system defined by a physically realizable $h(\tau)$ be denoted by $y(t)$. Now, by definition, an input $\delta(t - \tau)$ produces an output $h(t - \tau)$. Hence, by the superposition property for linear systems, the output $y(t)$ due to the input $x(t)$ of Eq. (1.8) becomes

$$y(t) = \int_{-\infty}^{t+} x(\tau)h(t - \tau)\, d\tau = \int_{0}^{\infty} h(\tau)x(t - \tau)\, d\tau \tag{1.10}$$

where the lower limit of zero is understood to be an infinitesimal value below zero. This is the well-known convolution integral expression that is given to describe ideal physically realizable constant-parameter linear systems in the time domain. See Figure 1.2.

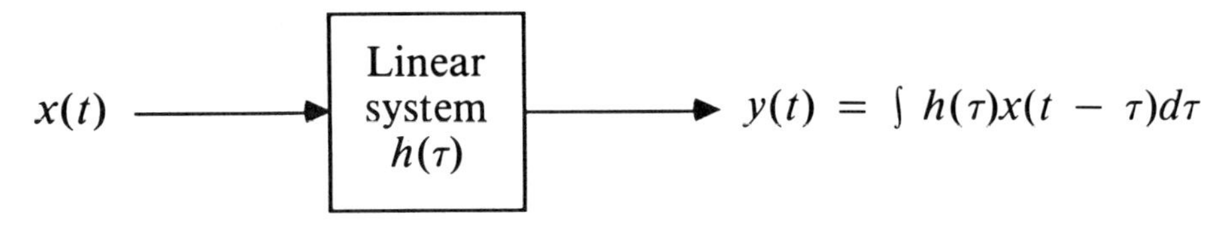

Figure 1.2 Convolution integral relation for ideal constant-parameter linear system.

For linear systems that are not physically realizable, Eq. (1.10) is replaced by

$$y(t) = \int_{-\infty}^{\infty} h(\tau)x(t - \tau)\, d\tau \tag{1.11}$$

Equation (1.11) includes Eq. (1.10) as a special case when $h(\tau) = 0$ for $\tau < 0$. It is convenient to use Eq. (1.11) even when Eq. (1.10) is true because of not having to change limits of integration when variables of integration are changed. In later derivations of various theoretical formulas, the theoretical limits of minus infinity to plus infinity will be understood when no limits are shown.

In place of Eq. (1.10), a time-varying linear system is defined by a two-parameter weighting function $h(\tau, t)$ that measures the response of the system at time t to a unit impulse function applied at time $(t - \tau)$. For physical realizability, one must have $h(\tau, t) = 0$ for $\tau < 0$. Thus, physically realizable time-varying linear systems are governed by the time-domain relation

$$y(t) = \int_{0}^{\infty} h(\tau, t)x(t - \tau)\, d\tau \tag{1.12}$$

When stationary random data pass through a constant-parameter linear system, the output data will also be stationary random data. However, if the system is time-varying, then the output data become nonstationary random data. To preserve stationarity, one must have a constant-parameter linear system where

$$h(\tau, t) = h(\tau) \qquad \text{for all } t \tag{1.13}$$

A physically realizable constant-parameter linear system is said to be *stable* if every possible bounded input produces a bounded output. A necessary and sufficient condition for a linear system to be stable is

$$\int_{-\infty}^{\infty} |h(\tau)|\, d\tau < \infty \tag{1.14}$$

In words, the weighting function $h(\tau)$ must be absolutely integrable over all τ.

TABLE 1.1 Properties of Ideal Physical Systems

System Property	Requirement	Benefit
Linear	Additive and homogeneous	Output stays Gaussian
Physically realizable	$h(\tau)=0 \quad \text{for } \tau<0$	Output follows input
Constant-parameter	$h(\tau, t)=h(\tau)$	Output stays stationary
Stable	$\int_{-\infty}^{\infty} \lvert h(\tau)\rvert \, d\tau<\infty$	Output stays bounded

Stability conditions are assumed for all linear and nonlinear systems considered in this book.

Table 1.1 displays the properties satisfied by ideal physical systems that are linear, physically realizable, constant-parameter, and stable. Many examples of such linear systems are discussed in references 1 and 2.

1.1.2 Frequency Response Functions

Fourier transforms of both sides of Eq. (1.10) or (1.11) give the familiar relation shown in Figure 1.3 that describes constant-parameter linear systems in the frequency domain, namely,

$$Y(f) = H(f)X(f) \tag{1.15}$$

where $X(f)$, $Y(f)$, and $H(f)$ are Fourier transforms of $x(t)$, $y(t)$ and $h(\tau)$, respectively, as defined by

$$X(f) = \int_{-\infty}^{\infty} x(t)e^{-j2\pi ft}\, dt \tag{1.16}$$

$$Y(f) = \int_{-\infty}^{\infty} y(t)e^{-j2\pi ft}\, dt \tag{1.17}$$

$$H(f) = \int_{-\infty}^{\infty} h(\tau)e^{-j2\pi f\tau}\, d\tau \tag{1.18}$$

A sufficient condition for $H(f)$ to exist is that the linear system be stable. The lower limit in Eq. (1.18) becomes zero for physically realizable (causal) systems

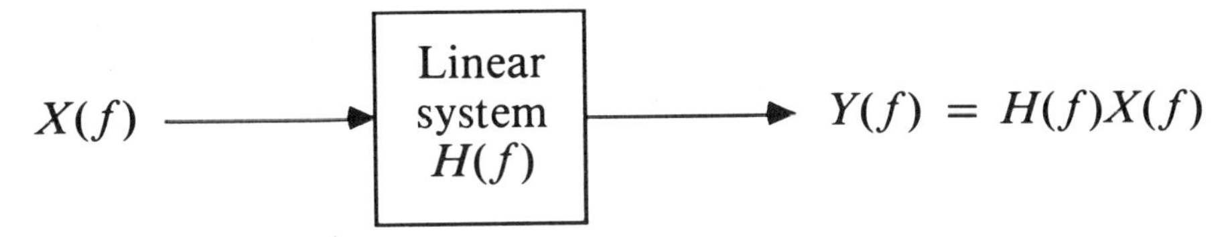

Figure 1.3 Fourier transform relation for ideal constant-parameter linear system.

where $h(\tau) = 0$ for $\tau < 0$. This complex-valued quantity $H(f)$ is called the *frequency response function* of the system. Note that $H(f)$ is defined for noncausal $h(\tau)$ by Eq. (1.18) when the lower limit can go to minus infinity. For linear systems, Eq. (1.15) shows that $Y(f)$ exists only at frequencies where $X(f)$ exists. This property is not true for nonlinear systems.

In terms of the frequency response function $H(f)$, a necessary and sufficient condition that a constant-parameter linear system be physically realizable and stable is that $H(f)$ should be bounded and analytic in the lower-half complex f-plane and on the real f-axis. All poles of $H(f)$ must be in the upper-half complex f-plane. When the complex-valued $H(f)$ is expressed in terms of a real part $H_R(f)$ and an imaginary part $H_I(f)$, a necessary and sufficient condition that a constant-parameter linear system be physically realizable (causal) is that the real part and the imaginary part are Hilbert transform pairs (reference 1).

The frequency response function $H(f)$ can be expressed in terms of a magnitude $|H(f)|$ and an associated phase angle $\phi(f)$ by the equation

$$H(f) = |H(f)|e^{-j\phi(f)} \tag{1.19}$$

The magnitude $|H(f)|$ is called the *gain factor* and the phase angle $\phi(f)$ is called the *phase factor*. The frequency response function can also be expressed in terms of a real part $H_R(f)$ and an imaginary part $H_I(f)$ by the equation

$$H(f) = H_R(f) - jH_I(f) \tag{1.20}$$

Equations (1.19) and (1.20) give the results

$$H_R(f) = |H(f)| \cos \phi(f) \tag{1.21}$$

$$H_I(f) = |H(f)| \sin \phi(f) \tag{1.22}$$

Also,

$$H(f) = [H_R^2(f) + H_I^2(f)]^{1/2} \tag{1.23}$$

$$\phi(f) = \tan^{-1}[H_I(f)/H_R(f)] \tag{1.24}$$

The use of $-j$ rather than $+j$ in Eq. (1.20) is preferred because it yields a $\phi(f)$ in Eq. (1.24) that is a positive instead of a negative arctangent, that is, it makes lag angles positive. This agrees with the physical requirement that it takes a *positive* time-delay for a signal to propagate from an input excitation point to an output response point, not a *negative* time-delay.

Many examples of gain and phase factors from different frequency response functions are discussed and pictured in references 1 and 2. These include theoretical results as well as practical results from measuring experimental data. In succeeding chapters, the notation $H(f)$ or equivalent will always represent a constant-parameter stable finite-memory linear system.

1.2 RANDOM DATA

Random data are any type of data occurring in physical phenomena or in engineering work that does not have an explicit mathematical formula to describe its properties. Instead, any particular time-history record represents only one record out of a collection of different time-history records that might have occurred, where the collection of records is called an *ensemble*. This ensemble produces a *random process* that can be defined by analyzing various statistical properties over the collection of records pictured in Figure 1.4. Basic classifications of random data are shown in Figure 1.5. Detailed discussions of each of these classifications and their properties are given in references 1 and 2.

Stationary random data are data whose ensemble-averaged statistical properties are invariant with respect to translations in time. For such data, ensemble-averaged mean values are the same at every time: correlation functions are not functions of two specific times but only of their time difference. *Ergodic data* are stationary random data where long-time averages on any arbitrary time-history record give results that are statistically equivalent to associated ensemble averages over a large collection of records. In practice, stationary random data will automatically be ergodic if there are no sine waves or other deterministic phenomena in the records. Throughout this book, unless stated otherwise, all stationary random data will be assumed to be ergodic. For such data, one long duration experiment is sufficient to obtain useful information. Theoretically, a stationary random record should never have a beginning or an end.

Nonstationary random data are data whose ensemble-averaged statistical properties change with translations in time. A special class of nonstationary data are called *transient random data* when there is a clearly defined beginning and end to the data. In general, techniques for analyzing stationary random data are not appropriate for analyzing nonstationary random data. An exception occurs for analyzing the spectral properties of transient random data, provided there exists an ensemble of similar transient records. Except for a scale factor denoting the time duration of the transient, the same procedures can be followed to obtain the "power" spectral density function of stationary random data versus the

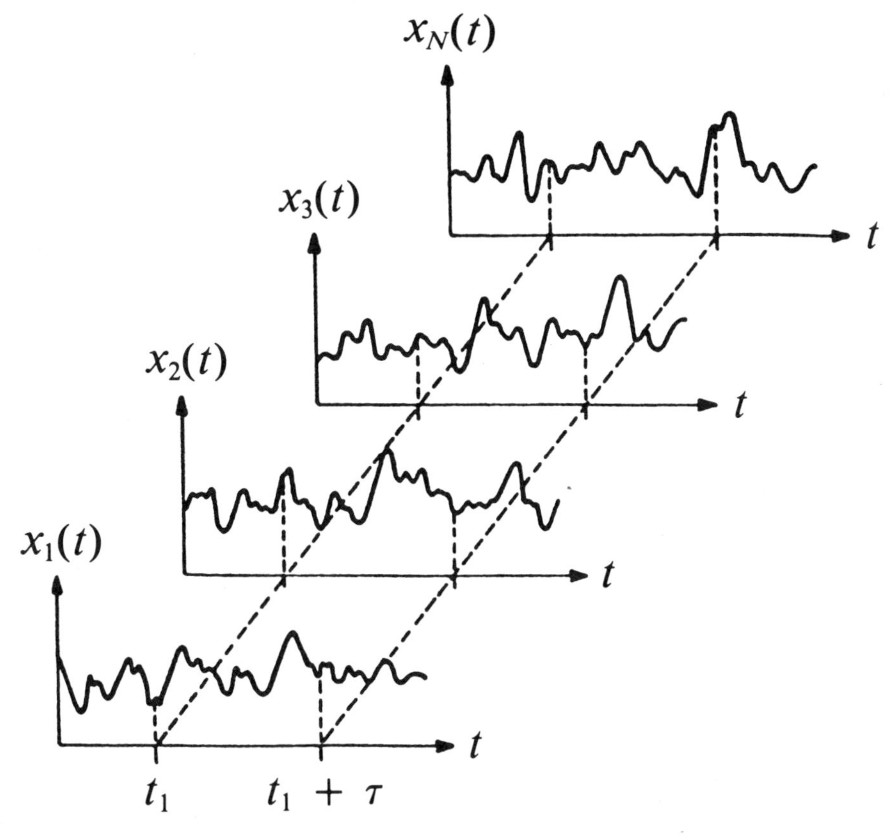

Figure 1.4 Ensemble of time-history records defining a random process.

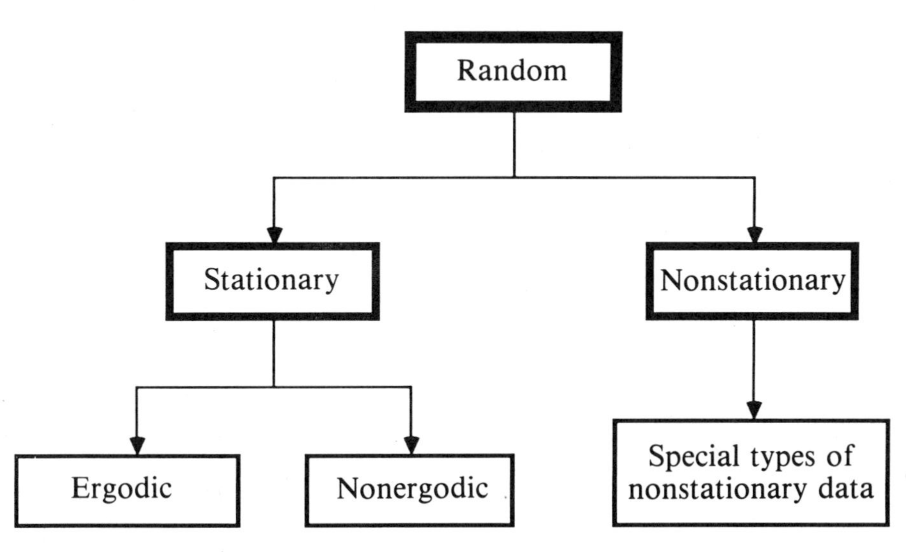

Figure 1.5 Classifications of random data.

"energy" spectral density function of transient random data. With stationary random data, only one long duration record is required to obtain desired results, since this one record can be subdivided into a collection of suitable shorter records to perform ensemble averages. With transient random data, it is never appropriate to subdivide the transient data into shorter records, since no shorter record is representative of the whole record. Instead, one must repeat the experiment over and over under similar conditions to obtain a collection of suitable records to perform ensemble averages.

When stationary random data pass through constant-parameter linear systems, the output data will also be stationary. Similarly, transient random data through constant-parameter linear systems produce transient output data. However, when stationary or transient random data pass through time-varying linear systems, the output data will be nonstationary. This book concerns only stationary or transient random data at the input and output locations. Special procedures for analyzing different types of nonstationary random data are in reference 1.

1.3 SPECTRAL DENSITY FUNCTIONS

Ideas from references 1 and 2 will now be reviewed for direct Fourier transform computation of basic spectral density functions (autospectra and cross-spectra) for either stationary random data or transient random data. Later formulas in this book will be concerned mostly with the spectral analysis of stationary random data passing through various combinations of linear and/or nonlinear systems. Only simple scale factor changes are required to have these results apply to transient random data. Special relations will also be developed for input/output probability density functions and input/output correlation functions, but the emphasis in this book will be devoted to input/output spectral density functions because such frequency-domain formulas are much simpler to compute and interpret than associated time-domain correlation formulas.

For stationary random data, the following expected values of Fourier transform quantities can define the two-sided autospectral density functions $S_{xx}(f)$ and $S_{yy}(f)$ and the two-sided cross-spectral density function $S_{xy}(f)$:

$$S_{xx}(f) = \frac{1}{T} E[|X(f)|^2] \tag{1.25}$$

$$S_{yy}(f) = \frac{1}{T} E[|Y(f)|^2] \tag{1.26}$$

$$S_{xy}(f) = \frac{1}{T} E[X^*(f)Y(f)] \tag{1.27}$$

In practice, $E[\ ]$ denotes an expected value ensemble average over the quantities inside the brackets. Here, the terms $X(f)$ and $Y(f)$ are finite Fourier transforms of length T for n_d distinct sample records, and the frequency f can theoretically be any value in $-\infty < f < \infty$. However, for digitized data, f is restricted to discrete values spaced $(1/T)$ apart in a bounded frequency range, and Fourier transform formulas become Fourier series formulas. The autospectral density functions $S_{xx}(f)$ and $S_{yy}(f)$ are positive, real-valued, even functions of f; the cross-spectral density function $S_{xy}(f)$ is a complex-valued function of f. If $x(t)$ and $y(t)$ are measured in volts, then $S_{xx}(f)$ and $S_{yy}(f)$ will have units of $(\text{volts})^2$ per hertz when t has units of seconds. Note that $S_{xy}(f)$ includes $S_{xx}(f)$ and $S_{yy}(f)$ as special cases, and that these quantities always exist for finite-length records.

For transient random data, the same type of expected value operation is required over n_d different sample transient records, but the scale factor $(1/T)$ should be omitted. One obtains the two-sided "energy" spectral density function results for transient random data defined by

$$\mathscr{S}_{xx}(f) = E[|X(f)|^2] \tag{1.28}$$

$$\mathscr{S}_{yy}(f) = E[|Y(f)|^2] \tag{1.29}$$

$$\mathscr{S}_{xy}(f) = E[X^*(f)Y(f)] \tag{1.30}$$

Note that

$$\mathscr{S}_{xy}(f) = TS_{xy}(f) \tag{1.31}$$

includes $\mathscr{S}_{xx}(f)$ and $\mathscr{S}_{yy}(f)$ as special cases.

For stationary random data, the one-sided measurable autospectral density functions and cross-spectral density function are zero for $f < 0$ and are defined for $f \geqslant 0$ by

$$G_{xx}(f) = \frac{2}{T} E[|X(f)|^2] \tag{1.32}$$

$$G_{yy}(f) = \frac{2}{T} E[|Y(f)|^2] \tag{1.33}$$

$$G_{xy}(f) = \frac{2}{T} E[X^*(f)Y(f)] \tag{1.34}$$

Note that

$$\begin{aligned} G_{xy}(f) &= 2S_{xy}(f), \quad f \geqslant 0 \\ &= 0, \qquad\qquad f < 0 \end{aligned} \tag{1.35}$$

includes $G_{xx}(f)$ and $G_{yy}(f)$ as special cases.

For transient random data, the one-sided measurable "energy" spectral density function results are zero for $f < 0$ and are defined for $f \geqslant 0$ by

$$\mathscr{G}_{xx}(f) = 2E[|X(f)|^2] \tag{1.36}$$

$$\mathscr{G}_{yy}(f) = 2E[|Y(f)|^2] \tag{1.37}$$

$$\mathscr{G}_{xy}(f) = 2E[X^*(f)Y(f)] \tag{1.38}$$

It follows that $\mathscr{G}_{xy}(f) = TG_{xy}(f)$ where

$$\begin{aligned} \mathscr{G}_{xy}(f) &= 2\mathscr{S}_{xy}(f), \quad f \geqslant 0 \\ &= 0, \qquad\qquad\quad f < 0 \end{aligned} \tag{1.39}$$

includes $\mathscr{G}_{xx}(f)$ and $\mathscr{G}_{yy}(f)$ as special cases.

For $f \geqslant 0$, these relations prove that

$$\frac{S_{xy}(f)}{S_{xx}(f)} = \frac{\mathscr{S}_{xy}(f)}{\mathscr{S}_{xx}(f)} = \frac{G_{xy}(f)}{G_{xx}(f)} = \frac{\mathscr{G}_{xy}(f)}{\mathscr{G}_{xx}(f)} \tag{1.40}$$

This is the reason why many useful linear system input/output results that involve these functions are valid regardless of whether the data are stationary random data or transient random data, and regardless of whether one computes two-sided or one-sided spectral quantities.

Table 1.2 shows the basic formulas of two- and one-sided spectral density functions for stationary and transient random data. This table demonstrates that only scale factor changes are required to move between these results. The material in this book for the most part will assume stationary random data and use two-sided spectra.

Many examples of autospectral density functions and cross-spectral density functions for both stationary and transient random data are discussed and pictured in references 1 and 2. Applications of these results to solve engineering problems of random data passing through linear systems are extensively treated in these references.

The *coherence function* $\gamma_{xy}^2(f)$ between two stationary random processes $x(t)$ and $y(t)$ is defined by

$$\gamma_{xy}^2(f) = \frac{|S_{xy}(f)|^2}{S_{xx}(f)S_{yy}(f)} = \frac{|G_{xy}(f)|^2}{G_{xx}(f)G_{yy}(f)} \tag{1.41}$$

As proved in references 1 and 2, this ordinary coherence function for all f is a nondimensional constant that is bounded between zero and one, namely,

$$0 \leqslant \gamma_{xy}^2(f) \leqslant 1 \tag{1.42}$$

The quantity $|\gamma_{xy}(f)|$ represents the positive square root of $\gamma_{xy}^2(f)$.

TABLE 1.2 Basic Spectral Density Functions

Stationary Random Data

Two-sided spectra

$$S_{xy}(f) = \frac{1}{T} E[X^*(f)Y(f)], \qquad -\infty < f < \infty$$

$$S_{xx}(f) = \frac{1}{T} E[X^*(f)X(f)] = \frac{1}{T} E[|X(f)|^2]$$

One-sided spectra

$$G_{xy}(f) = \frac{2}{T} E[X^*(f)Y(f)], \qquad f \geqslant 0 \text{ only}$$

$$G_{xx}(f) = \frac{2}{T} E[X^*(f)X(f)] = \frac{2}{T} E[|X(f)|^2]$$

Transient Random Data

Two-sided spectra

$$\mathscr{S}_{xy}(f) = E[X^*(f)Y(f)], \qquad -\infty < f < \infty$$

$$\mathscr{S}_{xx}(f) = E[X^*(f)X(f] = E[|X(f)|^2]$$

One-sided spectra

$$\mathscr{G}_{xy}(f) = 2E[X^*(f)Y(f)], \qquad f \geqslant 0 \text{ only}$$

$$\mathscr{G}_{xx}(f) = 2E[X^*(f)X(f)] = 2E[|X(f)|^2]$$

The coherence function of Eq. (1.41) is a measure of a possible linear relationship between $x(t)$ and $y(t)$. It does not indicate a causal relationship unless one has other physical grounds for knowing that, say, an input $x(t)$ produces an output $y(t)$. If a perfect linear relationship exists between $x(t)$ and $y(t)$ at some frequency f_o, then the coherence function $\gamma_{xy}^2(f_o)$ will be equal to unity at the frequency. If $x(t)$ and $y(t)$ are such that $S_{xy}(f_o) = 0$ at frequency f_o, then the coherence function will be zero at that frequency.

There are four main reasons in practice why a computed coherence function between a measured input $x(t)$ and a measured output $y(t)$ will differ from a value that was expected to be close to unity:

1. Extraneous noise in the input and output measurements.
2. Bias and random errors in spectral density function estimates.

3. The output $y(t)$ due in part to other inputs besides the measured $x(t)$.
4. Nonlinear system operations between $x(t)$ and $y(t)$.

If the first three possibilities are ruled out by good data acquisition, proper data processing, and physical understanding, it is reasonable to conclude that low coherence at particular frequencies is due to nonlinear system effects at these frequencies. Thus, the ordinary coherence function can provide a simple practical way to detect nonlinearity without identifying its precise nature.

For an easy reminder, Table 1.3 lists various assumptions common to later chapters unless stated otherwise.

This concludes Chapter 1.

TABLE 1.3 Assumptions in Book

1. Systems are constant-parameter stable systems.
2. Memory operations are of finite duration.
3. Data are stationary random ergodic data.
4. Missing integral limits are from $-\infty$ to $+\infty$.
5. Spectral density functions are two-sided spectra.
6. Input data are Gaussian with arbitrary spectral properties.

2

ZERO-MEMORY NONLINEAR SYSTEMS

This chapter begins with a guide to the overall nonlinear material contained in this book. General formulas are derived here to analyze the properties of stationary random data passing through zero-memory nonlinear systems. For any stationary random input data, Gaussian or non-Gaussian, this chapter shows how to determine the output probability density function from the input probability density function, and how this result can help identify different nonlinear systems. For Gaussian input data with a known autocorrelation function, it also shows how to determine the output autocorrelation function as well as the input/output cross-correlation function. Examples illustrating these results include data through nonlinear two-slope systems, dead-zone systems, clipped systems, smooth-limiter systems, square-law systems, cubic systems, hardening spring systems, and softening spring systems. The chapter concludes with material on how to describe any zero-memory nonlinear system by an optimum third-order polynomial least-squares approximation. Extensions of this optimum third-order polynomial will be used later in single-input/single-output nonlinear models discussed in Chapters 5–7.

2.1 NONLINEAR MATERIAL IN BOOK

To guide and motivate the reader regarding material in this book, an introduction will be given. The selected topics are based on previous results in the references and on new results developed here. The main types of nonlinear systems treated in this book (to be defined in this chapter and succeeding chapters) are

A. zero-memory nonlinear systems (Chapter 2),
B. bilinear and trilinear systems (Chapters 3 and 4),
C. square-law and cubic nonlinear systems (Chapters 5 and 6), and
D. parallel linear and nonlinear systems (Chapter 7).

Detailed mathematical procedures are explained for analyzing and identifying properties of these nonlinear systems from random data measured at input and output points. For many important engineering applications, new decomposition procedures are derived that are considerably simpler and more practical than previously used procedures. Where older procedures are still appropriate, many of them have been improved and made more understandable.

Higher-order systems beyond trilinear systems are not considered in this book because

(a) their effects are often negligible in practice,
(b) such higher-order systems involve multidimensional Volterra integrals that are difficult to compute and often require extremely large amounts of data to yield acceptable error analysis results, and
(c) other different procedures should be investigated when the practical procedures recommended in this book are not suitable.

Knowledge of material in this Chapter, 2, on zero-memory nonlinear systems is important for many engineering applications. The material in Chapters 3 and 4 is needed to know about past work involving complicated formulas that are often difficult to compute and interpret. The material in Chapters 5–7 shows how to replace a large class of single-input/single-output nonlinear systems with practical alternative multiple-input/single-output linear systems. Chapter 5 material shows also when and how to apply special single-frequency bispectral and trispectral density functions instead of using the multidimensional bispectral and trispectral density functions required in Chapters 3 and 4. Chapter 6 contains formulas to help design experiments and evaluate measured results. The last Chapter, 7, shows how to model and analyze general parallel linear and nonlinear systems, how to solve two important nonlinear engineering problems, and how to identify terms in proposed nonlinear differential equations of motion by using the techniques in this book.

For random data passing through either linear or nonlinear systems, there are always three distinct parts to be considered, namely, input data properties, system properties, and output data properties. From knowledge of any two of these three parts, one should try to compute the third part so that one can solve the following three problems:

I. *Output Prediction Problem.* To predict output data properties from knowledge of input data properties and system properties.

II. *System Identification Problem.* To identify system properties from knowledge of input data properties and output data properties.

III. *Input Determination Problem.* To determine input data properties from knowledge of output data properties and system properties.

With stationary or transient random data, to solve the above three problems, there exists relatively simple well-known input/output formulas for the passage of such data through general constant-parameter linear systems. These input/output formulas are extended in new standard ways for the various types of nonlinear systems treated in this book.

Special nonlinear system relationships and procedures are developed in this and succeeding chapters for many matters that include the following topics.

1. Practical ways to compute and interpret nonlinear system amplitude and frequency results for Gaussian and non-Gaussian data.
2. Input/output probability density functions.
3. Input/output correlation and spectral density functions.
4. Optimum identification of linear, bilinear, and trilinear systems.
5. Decomposition of output spectral density functions into linear and nonlinear components.
6. Determination of linear and nonlinear coherence functions.
7. Statistical error analysis criteria to design experiments and evaluate measured results.
8. Replacement of single-input/single-output nonlinear models by multiple-input/single-output linear models.
9. Applications to nonlinear wave force models and nonlinear drift force models.
10. Determination of physical parameters in nonlinear differential equations of motion by appropriate analysis of measured data.
11. Review of Volterra functional methods for nonlinear system description and comparisons with new simpler methods.
12. Establishment of new standard techniques for nonlinear system analysis and identification from random data that should be used to replace and extend conventional linear system techniques.

2.2 ZERO-MEMORY NONLINEAR SYSTEMS

A *zero-memory* nonlinear system is a system that acts instantaneously on present inputs in some nonlinear fashion. It does not weight past inputs due to "memory" operations as in convolution integrals for constant-parameter linear systems, where the present value of the output is a function of both present and past values of the input. The initial discussion will be restricted to such nonlinear systems where the output $y(t)$ at any time t is a single-valued nonlinear function of the input $x(t)$ at the *same* time t. Thus, as shown in Figure 2.1,

$$y(t) = g[x(t)] \tag{2.1}$$

where $g(x)$ is a single-valued nonlinear function of x. In succeeding material, the notation $g(x)$ will always represent a zero-memory nonlinear system operation on the input $x = x(t)$, but $g(x)$ is *not* a function of t.

From the definitions of a linear system in Section 1.1, when the system $g(x)$ is *nonlinear* it means that for any constants a_1, a_2 and any inputs $x_1 = x_1(t)$ and $x_2 = x_2(t)$,

$$g(a_1 x_1 + a_2 x_2) \neq a_1 g(x_1) + a_2 g(x_2) \tag{2.2}$$

either because $g(x)$ is not additive or because $g(x)$ is not homogeneous. For constant-parameter nonlinear systems,

1. when $x(t)$ is translated to $x(t + \tau)$, then $y(t) = g[x(t)]$ is translated to $y(t + \tau) = g[x(t + \tau)]$;
2. when $x(t)$ is a stationary random process, then $y(t) = g[x(t)]$ is also a stationary random process.

The input/output cross-correlation function $R_{xy}(\tau)$ and the output autocorrelation function $R_{yy}(\tau)$ are here functions only of τ as required for stationary random data, namely,

$$\begin{aligned} R_{xy}(\tau) &= E[x(t)y(t + \tau)] = E[x(t)g\{x(t + \tau)\}] \\ R_{yy}(\tau) &= E[y(t)y(t + \tau)] = E[g\{x(t)\}g\{x(t + \tau)\}] \end{aligned} \tag{2.3}$$

As usual, $E[\ ]$ denotes the expected value operation of the quantity inside the brackets.

Extensions of zero-memory nonlinear systems occur in practice that involve memory operations on past inputs. This can be modeled as described in the next section.

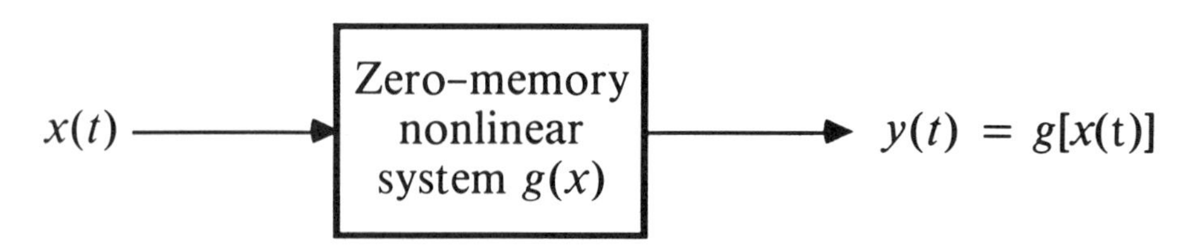

Figure 2.1 Zero-memory nonlinear system.

2.2.1 Finite-Memory Nonlinear Systems

Finite-memory nonlinear systems are defined by inserting constant-parameter linear systems after and/or before the zero-memory nonlinear systems using appropriate frequency response functions $A(f)$ and $B(f)$ as in Figure 2.2. Special physical cases of interest are when one of the linear systems can be considered to be unity so that memory operations occur only after or before the nonlinear operation but not in both places as in Figures 2.3 and 2.4.

Linear operations in Figure 2.2, as discussed in references 1 and 2, are appropriate for analyzing the passage of stationary random data from $x(t)$ to $u(t)$ and from $v(t)$ to $y(t)$. New results are required to determine how $v(t)$ is related to $u(t)$ for particular zero-memory nonlinear systems. In Figures 2.3 and 2.4, the passage of stationary random data from $z_1(t)$ to $y_1(t)$ and from $x(t)$ to $z_2(t)$ does not involve any new ideas. It is in the passage of $x(t)$ to $z_1(t)$ and from $z_2(t)$ to $y_2(t)$ through a particular zero-memory nonlinear system that new techniques are needed. Note also that even if $A(f) = B(f)$, and even if the same zero-memory nonlinear system is involved in Figures 2.3 and 2.4, the same input $x(t)$ will in general produce two different outputs $y_1(t) \neq y_2(t)$ that depend upon whether the linear systems follow or precede the zero-memory nonlinear system.

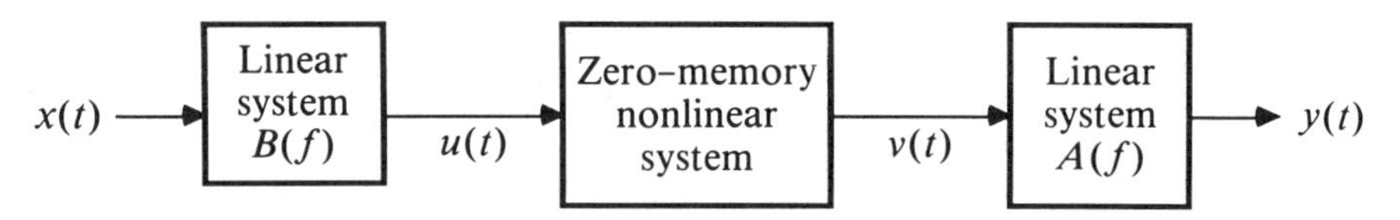

Figure 2.2 Finite-memory nonlinear system with linear systems before and after the zero-memory nonlinear system.

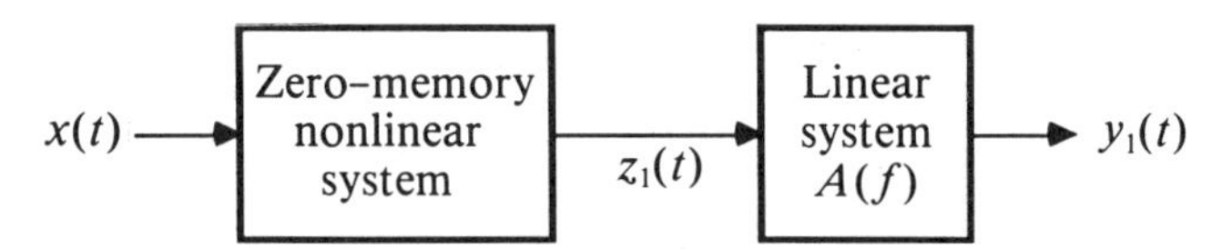

Figure 2.3 Finite-memory nonlinear system with a linear system after the zero-memory nonlinear system.

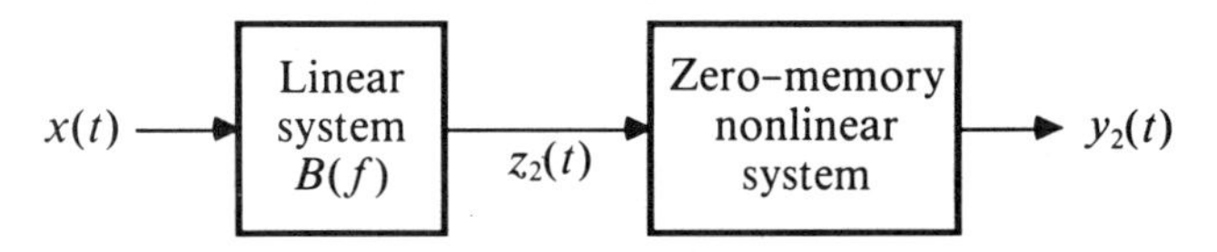

Figure 2.4 Finite-memory nonlinear systems with a linear system before the zero-memory nonlinear system.

Let the quantities $X(f)$, $U(f)$, $V(f)$, $Z(f)$, and $Y(f)$ be the Fourier transforms of $x(t)$, $u(t)$, $v(t)$, $z(t)$, and $y(t)$, respectively. In Figure 2.2, assume that $V(f)$ is some known nonlinear function of $U(f)$ as denoted by

$$V(f) = p[U(f)]$$

Then, the output in Figure 2.2 satisfies

$$Y(f) = A(f)V(f) = A(f)p[U(f)] = A(f)p[B(f)X(f)]$$

In Figure 2.3, assume that $Z_1(f)$ is some known nonlinear function of $X(f)$ as denoted by

$$Z_1(f) = p_1[X(f)]$$

In Figure 2.4, assume that $Y_2(f)$ is some other known nonlinear function of $Z_2(f)$ denoted by

$$Y_2(f) = p_2[Z_2(f)]$$

Now, the outputs in Figures 2.3 and 2.4 satisfy

$$Y_1(f) = A(f)Z_1(f) = A(f)p_1[X(f)]$$
$$Y_2(f) = p_2[Z_2(f)] = p_2[B(f)X(f)]$$

Finite-memory nonlinear systems will be used in Chapters 3–7. The remainder of this chapter concerns only zero-memory nonlinear systems.

2.2.2 Examples of Zero-Memory Nonlinear Systems

Example 1. Two-Slope Systems

$$\begin{aligned} y &= x, & |x| &\leqslant A \\ &= A + b(x - A), & x &\geqslant A \\ &= -A + b(x + A), & x &\leqslant -A \end{aligned}$$

Two cases can occur as sketched in Figure 2.5.

Example 2. Dead-Zone System

See Figure 2.6 and Equation (2.52).

Example 3. Clipped System

See Figure 2.7 and Equation (2.59).

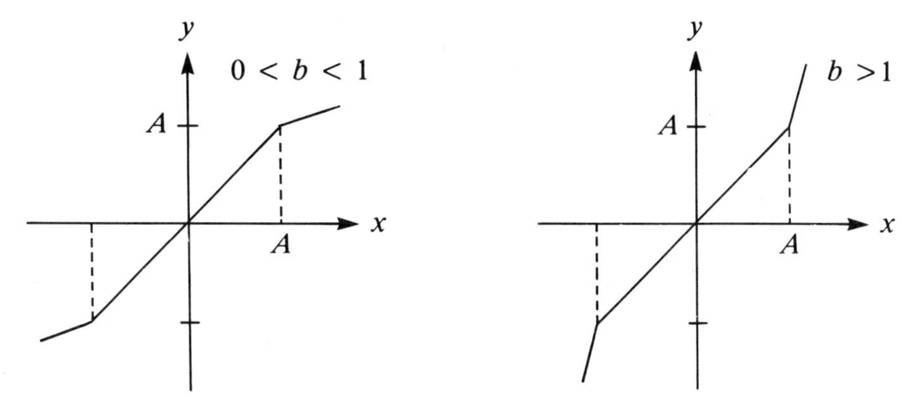

Figure 2.5 Two-slope systems.

Example 4. Hard-Clipped System

See Figure 2.8 and Equation (2.66).

Example 5. Square-Law System

See Figure 2.9 and Equation (2.101).

Example 6. Cubic System

See Figure 2.10 and Equation (2.120).

Example 7. Square-Law System with Sign

See Figure 2.11 and Equation (2.142).

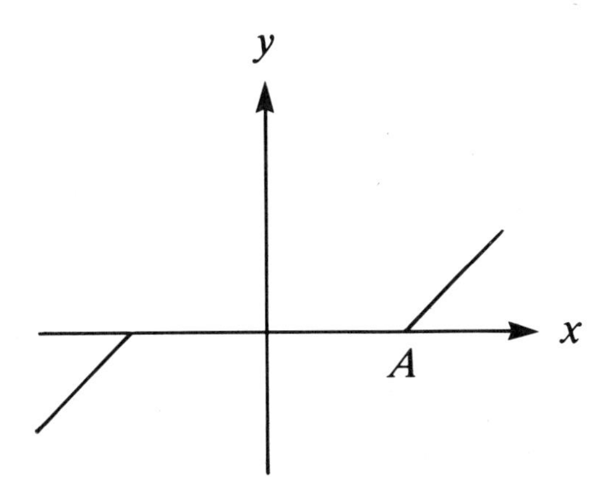

Figure 2.6 Dead-zone system.

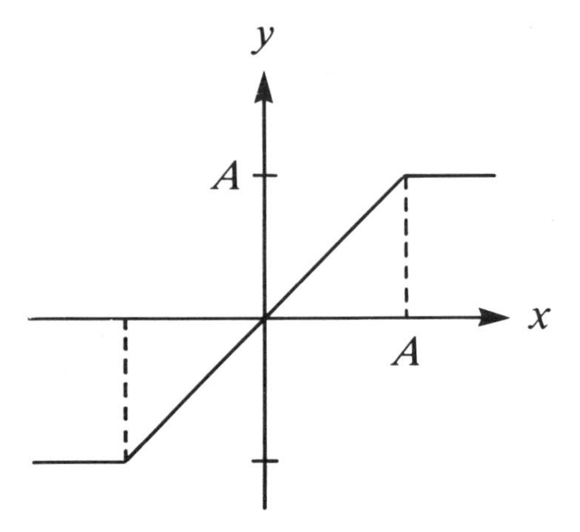

Figure 2.7 Clipped system.

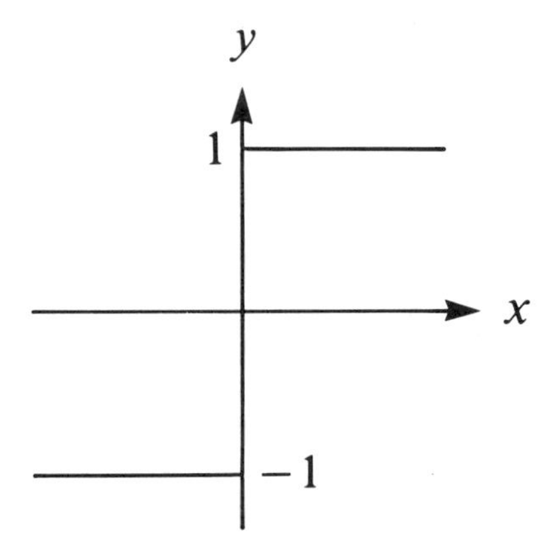

Figure 2.8 Hard-clipped system.

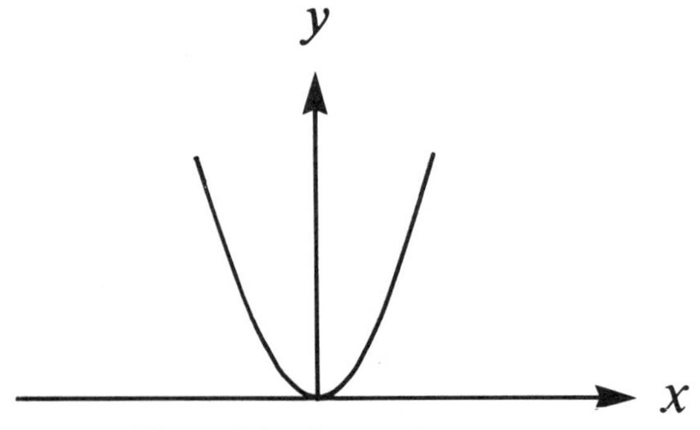

Figure 2.9 Square-law system.

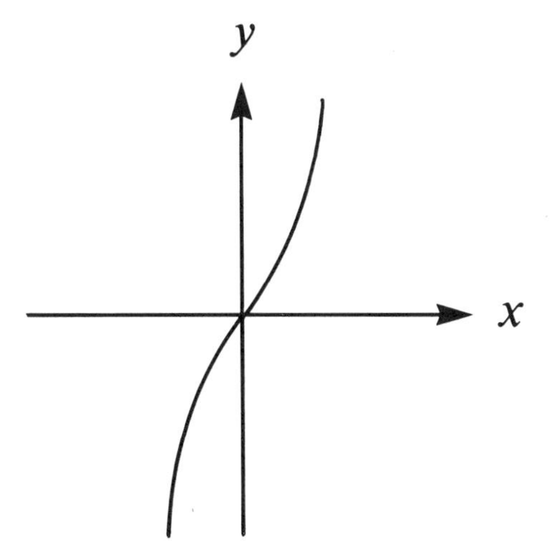

Figure 2.10 Cubic system.

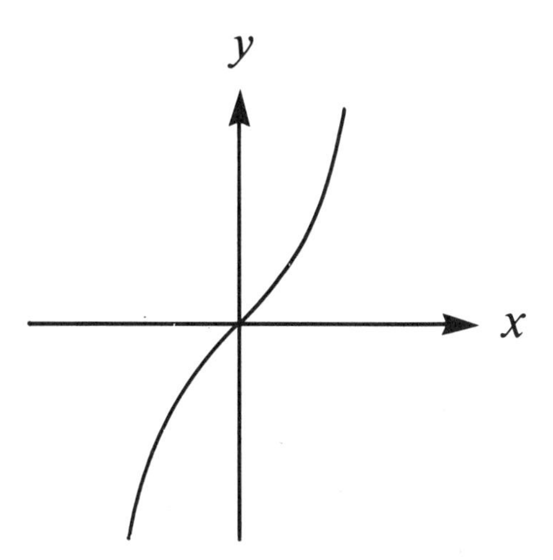

Figure 2.11 Square-law system with sign.

2.3 OUTPUT PROBABILITY DENSITY FUNCTION

An input record $x(t)$ is a representative member of a stationary Gaussian random process with zero mean value if, at any time t, the random variable $x = x(t)$ has the first-order Gaussian probability density function

$$p(x) = \frac{1}{\sigma_x\sqrt{2\pi}} \exp\left(\frac{-x^2}{2\sigma_x^2}\right) \tag{2.4}$$

The quantity

$$\sigma_x^2 = E[x^2] \tag{2.5}$$

is the variance of x when the mean value $\mu_x = E[x] = 0$. For the random variables $x_1 = x(t)$ and $x_2 = x(t + \tau)$ at any times t and τ, their joint probability density function is given by the second-order Gaussian form if

$$p(x_1, x_2) = \frac{1}{2\pi\sigma_x^2\sqrt{1-\rho^2}} \exp\left[\frac{-(x_1^2 - 2\rho x_1 x_2 + x_2^2)}{2\sigma_x^2(1-\rho^2)}\right] \tag{2.6}$$

where $\rho = \rho(\tau)$ is a function of the time-delay τ defined by

$$\rho = \frac{E[x_1 x_2]}{\sqrt{E[x_1^2]E[x_2^2)}} = \frac{E[x_1 x_2]}{\sigma_x^2} \tag{2.7}$$

This quantity ρ is the *correlation coefficient function* between x_1 and x_2. Note that $p(x_1, x_2) = p(x_1)p(x_2)$ when $\rho = 0$ for uncorrelated variables.

The autocorrelation function $R_{xx}(\tau)$ for a stationary random process is defined by

$$R_{xx}(\tau) = E[x(t)x(t + \tau)] = E[x_1 x_2] \tag{2.8}$$

At $\tau = 0$, this gives

$$R_{xx}(0) = E[x(t)x(t)] = E[x_1^2] = E[x_2^2] = \sigma_x^2 \tag{2.9}$$

Hence, the ρ of Eq. (2.7) is equivalent to

$$\rho(\tau) = \rho_{xx}(\tau) = \frac{R_{xx}(\tau)}{R_{xx}(0)} = \frac{R_{xx}(\tau)}{\sigma_x^2} \tag{2.10}$$

Moments for the nonlinear output data from Figure 2.1 can be obtained from knowledge of any $g(x)$ and any $p(x)$ as follows:

$$\begin{aligned} u_y = E[y] = E[g(x)] &= \int g(x)p(x)\,dx \\ E[y^n] = E\{[g(x)]^n\} &= \int [g(x)]^n p(x)\,dx \end{aligned} \tag{2.11}$$

The output data variance is given by

$$\sigma_y^2 = E[y^2] - (E[y])^2 = \psi_y^2 - u_y^2 \tag{2.12}$$

Consider the following important question. Suppose one knows the zero-memory nonlinear relationship $y = g(x)$ of Eq. (2.1). Assume it is not only single-valued but also one-to-one so that a given y corresponds to a unique x. Suppose one also knows the probability density function $p(x)$ for the input $x(t)$, which may or may not be of a Gaussian form. How can one determine the probability density function $p_2(y)$ for the output $y(t)$? The following theorem answers this question.

Theorem 1. Input/Output Probability Density Relation For any input data $x(t)$ with arbitrary probability density function $p(x)$ where $x(t)$ passes through a zero-memory nonlinear system to produce $y = g(x)$, which is single-valued and one-to-one, the output probability density function $p_2(y)$ for the output $y(t)$ satisfies the relation

$$p_2(y) = \frac{p(x)}{|dy/dx|} = \frac{p(x)}{|g'(x)|} \tag{2.13}$$

This theorem assumes that $(dy/dx) = g'(x)$ exists and is not equal to zero. The absolute value $|dy/dx|$ is required since $p_2(y)$ is nonnegative. The variable x on the right-hand side of Eq. (2.13) should be replaced by its equivalent y from $x = g^{-1}(y)$ to obtain $p_2(y)$ as a function of y.

Corollary If each value of $y = g(x)$ corresponds to n values of x that are equally likely, then

$$p_2(y) = \frac{np(x)}{|dy/dx|} = \frac{np(x)}{|g'(x)|} \tag{2.14}$$

Proof To prove Theorem 1, observe that for all $y_0 = g(x_0)$, the probability that x lies in $[x_0 - (\Delta x/2), x_0 + (\Delta x/2)]$ equals the probability that y lies in $[y_0 - (\Delta y/2), y_0 + (\Delta y/2)]$. Thus,

$$\begin{aligned} p(x_0)\,\Delta x &\approx \text{Prob}[x_0 - (\Delta x/2) < x \leqslant x_0 + (\Delta x/2)] \\ &= \text{Prob}[y_0 - (\Delta y/2) < y \leqslant y_0 + (\Delta y/2)] \approx p_2(y_0)\,\Delta y \end{aligned}$$

In the limit as $\Delta x \to 0$, this gives the desired result

$$p_2(y) = \frac{p(x)}{|dy/dx|}$$

The corollary of Eq. (2.14) comes by using $np(x_0)\,\Delta x \approx p_2(y_0)\,\Delta y$. This completes the proof.

Equations (2.13) and (2.14) are general results for arbitrary $p(x)$. Special results follow by using the Gaussian $p(x)$ of Eq. (2.4), together with the $g(x)$ of particular nonlinear system models. Note that

(1) when $p(x)$ is Gaussian and $g(x)$ is linear, then the resulting $p_2(y)$ will be Gaussian, and
(2) when $p(x)$ is Gaussian and $g(x)$ is nonlinear, then the resulting $p_2(y)$ will be non-Gaussian.

Hence, if output data are non-Gaussian when input data are Gaussian, then the system must be nonlinear. Analysis of input/output probability density functions thus provides a simple way to detect nonlinear system operations and can also be used to help identify their nature. Examples are discussed in Sections 2.6–2.8.

Identification of Zero-Memory Nonlinear System

When $y = g(x)$ is single-valued and one-to-one, it can be determined from measuring $p(x)$ and $p_2(y)$ by setting the probability distribution function $P_2(y)$ equal to the probability distribution function $P(x)$, namely, for any x_0 and its associated $y_0 = g(x_0)$,

$$P_2(y_0) = \int_{-\infty}^{y_0} p_2(y)\,dy = \int_{-\infty}^{x_0} p(x)\,dx = P(x_0) \tag{2.15}$$

The quantity $P_2(y_0)$ represents the probability that $y(t) \leqslant y_0$ and the quantity $P(x_0)$ represents the probability that $x(t) \leqslant x_0$. For each selected value of x_0 one should determine $P(x_0)$ from the measured $p(x)$, and then find the value of y_0 from the measured $p_2(y)$ such that $P_2(y_0) = P(x_0)$.

This procedure will also identify the form of a single-valued, one-to-one, nonlinear transformation $y = g(x)$ to produce a desired (non-Gaussian) output probability density function $p_2(y)$ from a given (Gaussian) input probability density function $p(x)$.

2.4 OUTPUT AUTOCORRELATION FUNCTION

Suppose one knows the nonlinear relationship $y = g(x)$ of Eq. (2.1), and suppose $x(t)$ is a Gaussian or non-Gaussian process. What is the predicted output autocorrelation function $R_{yy}(\tau)$? By definition from Eq. (2.3),

$$R_{yy}(\tau) = E[y(t)y(t+\tau)] = E[g\{x(t)\}g\{x(t+\tau)\}] \tag{2.16}$$

Let $x_1 = x(t)$, $x_2 = x(t+\tau)$, $y_1 = g(x_1)$, and $y_2 = g(x_2)$. Then

$$R_{yy}(\tau) = E[y_1 y_2] = \iint g(x_1)g(x_2)p(x_1, x_2)\, dx_1\, dx_2 \tag{2.17}$$

where $p(x_1, x_2)$ is the joint probability density function for x_1 and x_2. This expression is a general result for arbitrary $g(x)$ and arbitrary $p(x_1, x_2)$.

When $y(t)$ is measured, a simpler way to obtain $R_{yy}(\tau)$ is to compute it directly or via inverse Fourier transforms by

$$R_{yy}(\tau) = \mathscr{F}^{-1}[S_{yy}(f)] = \int S_{yy}(f)e^{j2\pi f\tau}\, df \tag{2.18}$$

where $S_{yy}(f)$ is the two-sided output autospectral density function of $y(t)$. With digital processing of data, special steps are required to avoid wrap-around errors and triangular weighting effects as explained in reference 1.

The following theorem is useful for predicting output autocorrelation functions after a given zero-memory nonlinear transformation when $y(t)$ is not measured. This theorem requires the input data to be Gaussian.

Theorem 2 (Price). Output Autocorrelation Relation For Gaussian input data $x(t)$ with known autocorrelation function $R_{xx}(\tau)$, where $x(t)$ passes through any zero-memory nonlinear system $y = g(x)$, the output autocorrelation function $R_{yy}(\tau)$ satisfies the relation

$$\frac{\partial R_{yy}(\tau)}{\partial R_{xx}(\tau)} = E[g'(x_1)g'(x_2)] \tag{2.19}$$

whenever $g'(x) = [dg(x)/dx]$ exists at $x_1 = x(t)$ and $x_2 = x(t + \tau)$.

Corollary Assuming higher-order derivatives $g^n(x) = [dg^n(x)/dx^n]$ exist,

$$\frac{\partial^n R_{yy}(\tau)}{\partial^n R_{xx}(\tau)} = E[g^n(x_1)g^n(x_2)] \tag{2.20}$$

Proof To prove Theorem 2, assume that $g(x)$ has a Fourier transform $Z(f)$ given by

$$Z(f) = \int g(x)e^{-j2\pi f x}\, dx$$

Then

$$g(x_1) = \int Z(f_1)e^{j2\pi f_1 x_1}\, df_1$$
$$g(x_2) = \int Z(f_2)e^{j2\pi f_2 x_2}\, df_2 \tag{2.21}$$

where $g(x)$ is the inverse Fourier transform of $Z(f)$. Now, substitute Eq. (2.21) into Eq. (2.16). This gives

$$R_{yy}(\tau) = \iiiint Z(f_1)Z(f_2)e^{j2\pi(f_1x_1 + f_2x_2)}p(x_1, x_2)\, dx_1\, dx_2\, df_1\, df_2$$
$$= \iint Z(f_1)Z(f_2)C_{x_1,x_2}(f_1, f_2)\, df_1\, df_2 \tag{2.22}$$

where

$$C_{x_1,x_2}(f_1, f_2) = \iint e^{j2\pi(f_1x_1 + f_2x_2)}p(x_1, x_2)\, dx_1\, dx_2 \tag{2.23}$$

defines the *joint characteristic function* of x_1 and x_2 at f_1 and f_2. Note that

$$C_{x_1,x_2}(f_1, f_2) = E[e^{j2\pi(f_1x_1 + f_2x_2)}] \tag{2.24}$$

From sources such as references 1 and 4 or by separate direct proof, it is known for Gaussian data that

$$C_{x_1,x_2}(f_1, f_2) = \exp[-2\pi^2\sigma_x^2(f_1^2 + 2\rho f_1 f_2 + f_2^2)] \tag{2.25}$$

Let the derivative of $g(x)$ with respect to x be denoted by $g'(x)$. Then from Eq. (2.21),

$$\begin{aligned} g'(x_1) &= \int (j2\pi f_1)Z(f_1)e^{j2\pi f_1x_1}\, df_1 \\ g'(x_2) &= \int (j2\pi f_2)Z(f_2)e^{j2\pi f_2x_2}\, df_2 \end{aligned} \tag{2.26}$$

Hence,

$$g'(x_1)g'(x_2) = \iint (-4\pi^2 f_1 f_2)Z(f_1)Z(f_2)e^{j2\pi(f_1x_1 + f_2x_2)}\, df_1\, df_2$$

Expected values of both sides yield

$$E[g'(x_1)g'(x_2)] = \iint (-4\pi^2 f_1 f_2)Z(f_1)Z(f_2)C_{x_1,x_2}(f_1, f_2)\, df_1\, df_2 \tag{2.27}$$

Consider now Eqs. (2.22) and (2.25). From Eq. (2.25), the partial derivative of $C_{x_1,x_2}(f_1, f_2)$ with respect to ρ yields the result

$$\frac{\partial C_{x_1,x_2}(f_1, f_2)}{\partial \rho} = (-4\pi^2\sigma_x^2 f_1 f_2)C_{x_1,x_2}(f_1, f_2) \tag{2.28}$$

From Eq. (2.22), one obtains the partial derivative

$$\frac{\partial R_{yy}(\tau)}{\partial \rho} = \int Z(f_1)Z(f_2)\frac{\partial C_{x_1,x_2}(f_1, f_2)}{\partial \rho}\, df_1\, df_2 \tag{2.29}$$

Substitution of Eq. (2.28) into Eq. (2.29) and comparison with Eq. (2.27) proves that

$$\frac{\partial R_{yy}(\tau)}{\partial \rho} = \sigma_x^2 E[g'(x_1)g'(x_2)] \tag{2.30}$$

From the definition of ρ in Eq. (2.10), one can replace Eq. (2.30) by the equivalent formula that proves the desired result:

$$\frac{\partial R_{yy}(\tau)}{\partial R_{xx}(\tau)} = E(g'(x_1)g'(x_2)] \tag{2.31}$$

The corollary of Eq. (2.20) follows by successive iteration procedures. Equations (2.30) and (2.31) are special cases of Price's theorem (reference 31). This useful result for Gaussian input data will be applied in later sections.

2.5 INPUT/OUTPUT CROSS-CORRELATION FUNCTION

Suppose one knows the nonlinear relationship $y = g(x)$ of Eq. (2.1), and suppose $x(t)$ is a Gaussian or non-Gaussian process. What is the predicted input/output cross-correlation function $R_{xy}(\tau)$? By definition from Eq. (2.3),

$$R_{xy}(\tau) = E[x(t)y(t + \tau)] = E[x(t)g\{x(t + \tau)\}] \tag{2.32}$$

Let $x_1 = x(t)$, $x_2 = x(t + \tau)$, and $y_2 = g(x_2)$. Then

$$R_{xy}(\tau) = E[x_1 y_2] = \iint x_1 g(x_2) p(x_1, x_2)\, dx_1\, dx_2 \tag{2.33}$$

where $p(x_1, x_2)$ is the joint probability density function for x_1 and x_2. This expression is a general result for arbitrary $g(x)$ and arbitrary $p(x_1, x_2)$.

When $x(t)$ and $y(t)$ are both measured, a simpler way to obtain $R_{xy}(\tau)$ is to compute it directly or via inverse Fourier transforms by

$$R_{xy}(\tau) = \mathscr{F}^{-1}[S_{xy}(f)] = \int S_{xy}(f)e^{j2\pi f\tau}\, df \tag{2.34}$$

where $S_{xy}(f)$ is the two-sided input/output cross-spectral density function between $x(t)$ and $y(t)$. As noted after Eq. (2.18), this requires special digital processing steps to avoid errors.

The following theorem is useful for predicting input/output cross-correlation functions after a given zero-memory nonlinear transformation when $y(t)$ is not measured. This theorem requires the input data to be Gaussian.

Theorem 3 (Bussgang). Input/Output Cross-Correlation Relation For Gaussian input data $x(t)$ with known autocorrelation function $R_{xx}(\tau)$, where $x(t)$ passes through any zero-memory nonlinear system $y = g(x)$, the input/output cross-correlation function $R_{xy}(\tau)$ satisfies the relation

$$R_{xy}(\tau) = \frac{R_{xx}(\tau)}{\sigma_x^2} \int xg(x)p(x)\,dx = \left(\frac{E[xg(x)]}{\sigma_x^2}\right) R_{xx}(\tau) \tag{2.35}$$

Corollary The input/output cross-spectral density function $S_{xy}(f)$ satisfies the relation

$$S_{xy}(f) = \left(\frac{E[xg(x)]}{\sigma_x^2}\right) S_{xx}(f) \tag{2.36}$$

and the optimum linear system $H_0(f)$ for this zero-memory nonlinear system $g(x)$ is given by

$$H_0(f) = \frac{S_{xy}(f)}{S_{xx}(f)} = \frac{E[xg(x)]}{\sigma_x^2} \tag{2.37}$$

Equation (2.36) follows directly from Eq. (2.35), and Eq. (2.37) from a basic result in references 1 and 2. Note that the optimum linear system here is independent of frequency because the $g(x)$ transformation has no memory. Note also that for nonlinear systems, the optimum linear system is a function of the input variance and will be different for different inputs.

Proof To prove Theorem 3, consider the inside integral of Eq. (2.33) with respect to x_1. This takes the form

$$\int x_1 p(x_1, x_2)\,dx_1 = \frac{1}{2\pi\sigma_x^2\sqrt{1-\rho^2}} \int x_1 \exp\left[\frac{-(x_1^2 - 2\rho x_1 x_2 + x_2^2)}{2\sigma_x^2(1-\rho^2)}\right] dx_1$$

when $p(x_1, x_2)$ is given by Eq. (2.6). The quantity

$$x_1^2 - 2\rho x_1 x_2 + x_2^2 = (x_1 - \rho x_2)^2 + (1-\rho)^2 x_2^2$$

Now

$$\int x_1 p(x_1, x_2)\,dx_1 = \frac{1}{2\pi\sigma_x^2\sqrt{1-\rho^2}} \int x_1 \exp\left[\frac{-(x_1 - \rho x_2)^2}{2\sigma_x^2(1-\rho^2)} - \frac{x_2^2}{2\sigma_2^2}\right] dx_1$$

Let

$$u = \frac{x_1 - \rho x_2}{\sqrt{1-\rho^2}}; \qquad du = \frac{dx_1}{\sqrt{1-\rho^2}}; \qquad x_1 = \rho x_2 + u\sqrt{1-\rho^2}$$

Then

$$\begin{aligned}\int x_1 p(x_1, x_2)\, dx_1 &= \frac{1}{2\pi\sigma_x^2}\int (\rho x_2 + u\sqrt{1-\rho^2}) \exp\left(\frac{-u^2}{2\sigma_x^2} - \frac{x_2^2}{2\sigma_x^2}\right) du \\ &= \frac{\rho x_2 \exp(-x_2^2/2\sigma_x^2)}{2\pi\sigma_x^2}\int \exp\left(\frac{-u^2}{2\sigma_x^2}\right) du\end{aligned}$$

where the integral involving u vanishes because u is part of an odd function between symmetric limits. The integral

$$\int \exp(-u^2/2\sigma_x^2)\, du = \sigma_x\sqrt{2\pi} \tag{2.38}$$

Thus,

$$\int x_1 p(x_1, x_2)\, dx_1 = \frac{\rho x_2 \exp(-x_2^2/2\sigma_x^2)}{\sigma_x\sqrt{2\pi}} \tag{2.39}$$

Finally, by substituting Eq. (2.39) into Eq. (2.33) and replacing ρ by $\rho_{xx}(\tau)$,

$$R_{xy}(\tau) = \frac{\rho_{xx}(\tau)}{\sigma_x\sqrt{2\pi}}\int x_2 g(x_2) \exp\left(\frac{-x_2^2}{2\sigma_x^2}\right) dx_2 \tag{2.40}$$

This proves the desired result

$$R_{xy}(\tau) = \rho_{xx}(\tau) E[xg(x)] = \left(\frac{E[xg(x)]}{\sigma_x^2}\right) R_{xx}(\tau) \tag{2.41}$$

Equations (2.40) and (2.41) are special cases of Bussgang's theorem (reference 13). This useful result for Gaussian input data will be applied in later sections.

Equation (2.41) shows that

1. $R_{xy}(\tau)$ is directly proportional to $R_{xx}(\tau)$ with the factor of proportionality given by $E[xg(x)]/\sigma_x^2$.
2. $R_{xy}(\tau) = 0$ whenever $g(x)$ is an even function of x, since $E[xg(x)]$ will then be the integral of an odd function between symmetric limits. In particular, $R_{xy}(\tau) = 0$ when $g(x) = x^2$, a square-law operation.

3. $S_{xy}(f)$, the two-sided Fourier transform of $R_{xy}(\tau)$, is directly proportional to $S_{xx}(f)$, the two-sided Fourier transform of $R_{xx}(\tau)$ with the factor of proportionality given by $E[xg(x)]/\sigma_x^2$. Hence, $S_{xy}(f)$ will be real-valued whenever $g(x)$ is real-valued, since $S_{xx}(f)$ is always a real-valued function of f. Also, the factor of proportionality here is the optimum linear system for this particular input $x(t)$ only.

2.6 EXAMPLES WITH DISCONTINUOUS DERIVATIVES

Examples of zero-memory nonlinear systems with discontinuous derivatives are discussed in this section. Other examples with continuous derivatives are discussed in Sections 2.7 and 2.8. All of these examples have engineering applications.

2.6.1 Two-Slope Systems

Two-slope systems as per Figure 2.5 are defined by the relation

$$\begin{aligned} y = g(x) &= x, & |x| &\leq A \\ &= A + b(x - A), & x &\geq A \\ &= -A + b(x + A), & x &\leq -A \end{aligned} \tag{2.42}$$

Note that

$$g(-x) = -g(x) \tag{2.43}$$

Thus, $g(x)$ is an odd function of x. Here,

$$\begin{aligned} \frac{dy}{dx} &= 1, & |x| &\leq A \\ &= b, & |x| &> A \end{aligned} \tag{2.44}$$

with

$$\begin{aligned} x = g^{-1}(y) &= y, & |y| &\leq A \\ &= A + \frac{(y - A)}{b}, & y &\geq A \\ &= -A + \frac{(y + A)}{b}, & y &\leq -A \end{aligned} \tag{2.45}$$

One should note that the derivative (dy/dx) is discontinuous at $|x| = A$. From Eq. (2.13), the output probability density function for $b \neq 0$ is now given by

$$
\begin{aligned}
p_2(y) &= p(x) = p(y), && |y| \leqslant A \\
&= \frac{p(A + [(y - A)/b])}{b}, && y > A \\
&= \frac{p(-A + [(y + A)/b])}{b}, && y < -A
\end{aligned} \tag{2.46}
$$

Without loss of generality, one can let $\sigma_x = 1$ so that

$$
p(x) = \frac{1}{\sqrt{2\pi}} \exp\left(\frac{-x^2}{2}\right) \tag{2.47}
$$

Now, the output probability density function

$$
\begin{aligned}
p_2(y) &= \frac{1}{\sqrt{2\pi}} \exp\left(\frac{-y^2}{2}\right), && |y| \leqslant A \\
&= \frac{1}{b\sqrt{2\pi}} \exp\left[\frac{-(A + [(y - A)/b])^2}{2}\right], && y > A \\
&= \frac{1}{b\sqrt{2\pi}} \exp\left[\frac{-(A - [(y + A)/b])^2}{2}\right], && y < -A
\end{aligned} \tag{2.48}
$$

Observe that immediately to the left and right of A,

$$
\begin{aligned}
p_2(A-) &= \frac{1}{\sqrt{2\pi}} \exp\left(\frac{-A^2}{2}\right) \\
p_2(A+) &= \frac{1}{b\sqrt{2\pi}} \exp\left(\frac{-A^2}{2}\right)
\end{aligned} \tag{2.49}
$$

Two distinct cases occur as follows:

$$
\begin{aligned}
&\text{if } 0 < b < 1, \quad \text{then } p_2(A+) > p_2(A-); \\
&\text{if } b > 1, \quad \text{then } p_2(A+) < p_2(A-).
\end{aligned}
$$

Two-slope softening spring systems can be represented by cases where $0 < b < 1$, and two-slope hardening spring systems can be represented by cases where $b > 1$ when x is the displacement and y is the force.

Figures 2.12 and 2.13 show the relationships when $0 < b < 1$ and when $b > 1$ between a Gaussian $p(x)$ input probability density function (PDF) and a non-Gaussian $p_2(y)$ output probability density function for a nonlinear two-slope system. In these figures, the probability value α is defined by the formula

$$
\alpha = \int_{-\infty}^{-A} p(x)\,dx = \int_{A}^{\infty} p(x)\,dx = \int_{A}^{\infty} p_2(y)\,dy = \int_{-\infty}^{-A} p_2(y)\,dy \tag{2.50}
$$

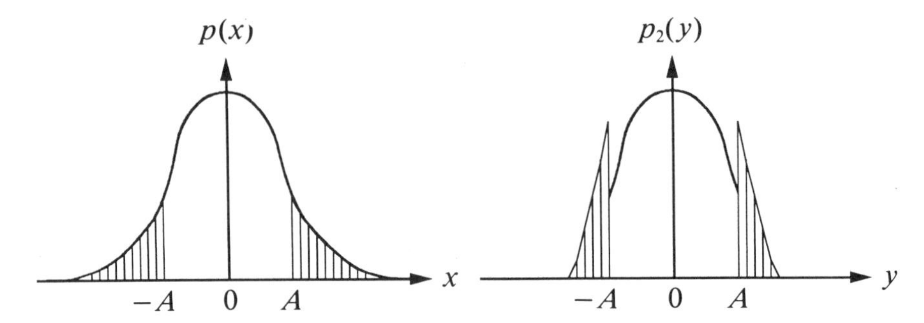

Figure 2.12 Input/output PDF for two-slope system when $0 < b < 1$.

representing the shaded probability areas. It follows that the probability of x falling inside the range from $-A$ to A is given by

$$1 - 2\alpha = \int_{-A}^{A} p(x)\,dx = \int_{-A}^{A} p_2(y)\,dy \tag{2.51}$$

Figure 2.12 shows that $p_2(y)$ is discontinuous at the value $y = A$ where $p_2(A+) > p_2(A-)$ as y approaches the value A from the right of A versus from the left of A, whereas Figure 2.13 shows that $p_2(A+) < p_2(A-)$ as y approaches the value A from the right of A versus from the left of A. Since Figures 2.12 and 2.13 both have the same probability of α that $y \geqslant A$, this requires that the tail in Figure 2.12 where $y \geqslant A$ should be much shorter than the tail in Figure 2.13 where $y \geqslant A$. Similar results apply to the tails in Figures 2.12 and 2.13 where $y \leqslant -A$ because of the discontinuity in $p_2(y)$ at the value $y = -A$.

2.6.2 Dead-Zone System

A dead-zone system as per Figure 2.6 is defined by the relation

$$\begin{aligned} y = g(x) &= 0, && |x| \leqslant A \\ &= b(x - A), && x \geqslant A \\ &= b(x + A), && x \leqslant -A \end{aligned} \tag{2.52}$$

Here, the discontinuous derivative

$$\begin{aligned} \frac{dy}{dx} &= 0, && |x| \leqslant A \\ &= b, && |x| > A \end{aligned} \tag{2.53}$$

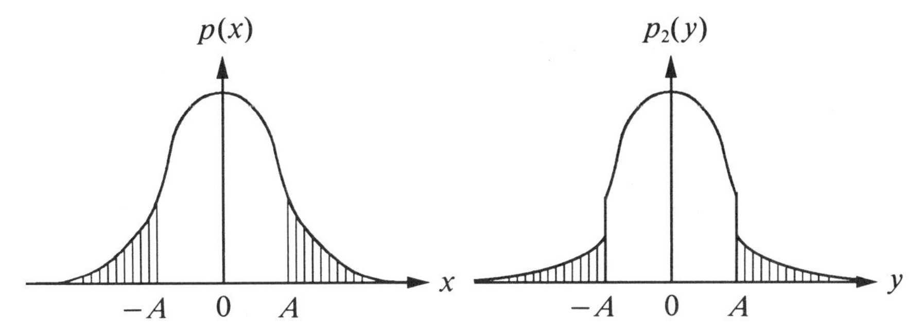

Figure 2.13 Input/output PDF for two-slope system when $b > 1$.

with

$$\begin{aligned} &x \text{ undefined}, && y = 0 \\ &x = A + \frac{y}{b}, && y > 0 \\ &\;\;= -A + \frac{y}{b}, && y < 0 \end{aligned} \tag{2.54}$$

As before, let

$$p(x) = \frac{1}{\sqrt{2\pi}} \exp\left(\frac{-x^2}{2}\right) \tag{2.55}$$

$$\alpha = \int_A^\infty p(x)\,dx \qquad 1 - 2\alpha = \int_{-A}^{A} p(x)\,dx \tag{2.56}$$

Now, the output probability density function

$$\begin{aligned} p_2(y) &= (1 - 2\alpha)\delta(y), && y = 0 \\ &= \frac{p(A + [y/b])}{b}, && y > 0 \\ &= \frac{p(-A + [y/b])}{b}, && y < 0 \end{aligned} \tag{2.57}$$

where $\delta(y)$ is the usual delta function. Observe that

$$\begin{aligned} &\int_{0+}^{\infty} p_2(y)\,dy = \int_{-\infty}^{0-} p_2(y)\,dy = \alpha \\ &\int_{0-}^{0+} p_2(y)\,dy = 1 - 2\alpha \end{aligned} \tag{2.58}$$

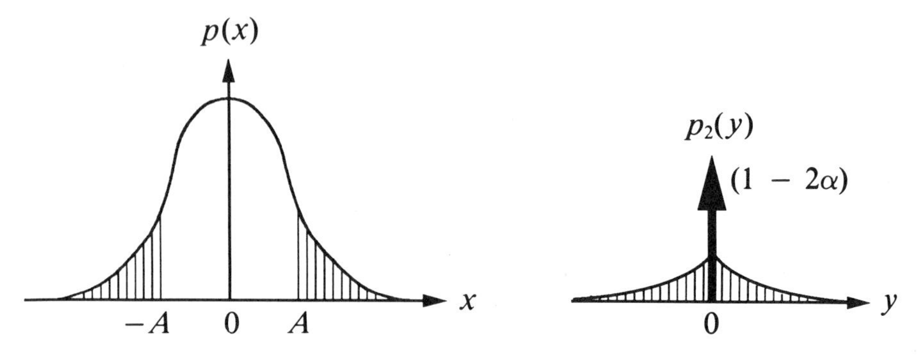

Figure 2.14 Input/output PDF for dead-zone system.

Figure 2.14 shows the relationship between a Gaussian input probability density function and a non-Gaussian output probability density function for a dead-zone system. The reason for the theoretical delta function at $y = 0$ is that all of the values of x inside the range $|x| \leqslant A$ are restricted to this single value. Thus,

$$\text{probability}[y(t) = 0] = 1 - 2\alpha$$

2.6.3 Clipped System

A clipped system as per Figure 2.7 is defined by the relation

$$\begin{aligned} y = g(x) &= x, \qquad |x| \leqslant A \\ &= A, \qquad x \geqslant A \\ &= -A, \qquad x \leqslant -A \end{aligned} \tag{2.59}$$

Here, the discontinuous derivative

$$\begin{aligned} \frac{dy}{dx} &= 1, \qquad |x| \leqslant A \\ &= 0, \qquad |x| > A \end{aligned} \tag{2.60}$$

with

$$\begin{aligned} &x = y, \qquad |y| \leqslant A \\ &x \text{ undefined}, \qquad |y| > A \end{aligned} \tag{2.61}$$

As per Eqs. (2.55) and (2.56), let

$$p(x) = \frac{1}{\sqrt{2\pi}} \exp\left(\frac{-x^2}{2}\right) \tag{2.62}$$

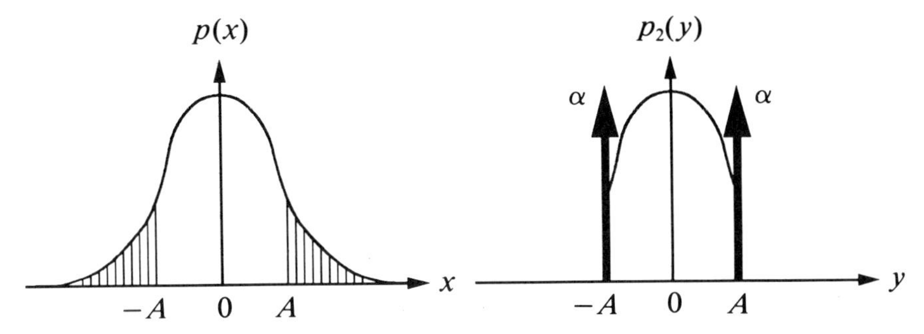

Figure 2.15 Input/output PDF for clipped system.

$$\alpha = \int_A^\infty p(x)\,dx, \qquad 1 - 2\alpha = \int_{-A}^{A} p(x)\,dx \tag{2.63}$$

The output probability density function becomes here

$$\begin{aligned} p_2(y) &= p(y), & |y| \leqslant A \\ &= \alpha\delta(y - A), & y = A \\ &= \alpha\delta(y + A), & y = -A \end{aligned} \tag{2.64}$$

where $\delta(y)$ is the usual delta function. Observe that

$$\begin{aligned} \int_{-A+}^{0} p_2(y)\,dy &= \int_0^{A-} p_2(y)\,dy = \frac{1}{2} - \alpha \\ \int_{-\infty}^{-A+} p_2(y)\,dy &= \int_{A-}^{\infty} p_2(y)\,dy = \alpha \end{aligned} \tag{2.65}$$

Figure 2.15 shows the relationship between a Gaussian input probability density function and a non-Gaussian output probability density function for a clipped system. The reason for the theoretical delta functions at $y = A$ and $y = -A$ is that all of the values of x where $x \geqslant A$ are restricted to the single value $y = A$, and all of the values of x where $x \leqslant -A$ are restricted to the single value $y = -A$. Thus,

$$\text{probability}\,[y(t) = -A] = \text{probability}[y(t) = A] = \alpha$$

2.6.4 Hard-Clipped System

A hard-clipped system as per Figure 2.8 is defined as an extreme form of a clipped system where

$$\begin{aligned} y = g(x) &= 1, \qquad && x > 0 \\ &= 0, && x = 0 \\ &= -1, && x < 0 \end{aligned} \tag{2.66}$$

The output probability density function becomes

$$\begin{aligned} p_2(y) &= \frac{1}{2}\,\delta(y - 1), \qquad && y = 1 \\ &= \frac{1}{2}\,\delta(y + 1), && y = -1 \\ &= 0 && \text{otherwise} \end{aligned} \tag{2.67}$$

where $\delta(y)$ is the usual delta function. The special nature of Eq. (2.67) is sketched in Figure 2.16. Here,

$$\text{probability}\ [y(t) = -1] = \text{probability}\ [y(t) = 1] = 0.50.$$

For the hard-clipped system, the discontinuous derivative

$$\begin{aligned} g'(x) &= 2\delta(x) \qquad && \text{at } x = 0 \\ &= 0 && \text{otherwise} \end{aligned} \tag{2.68}$$

The delta function occurs at the value $x = 0$, since there is a discontinuous jump of two units in $g(x)$ when $g(x)$ changes from -1 to $+1$ as x passes through zero.

Price's theorem of Eq. (2.30) and Bussgang's theorem of Eq. (2.35) will now be applied to obtain the output autocorrelation function and the input/output cross-correlation function for the hard-clipped system. Similar results were not

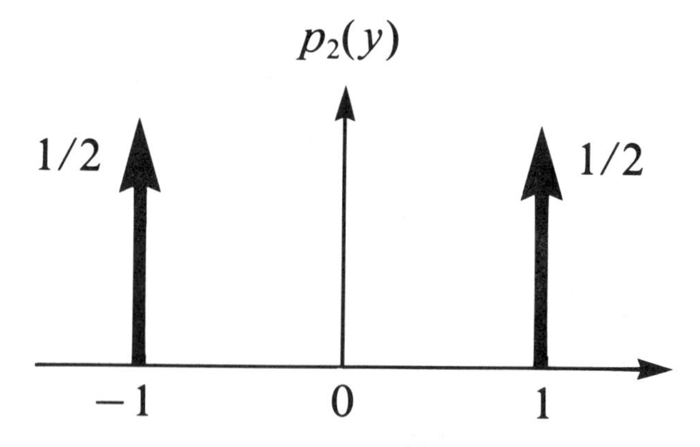

Figure 2.16 Output PDF for hard-clipped system.

carried out for previous systems because of the difficulty of deriving closed-form answers. If desired, however, numerical approximations can be obtained.

Output Autocorrelation Function for Hard-Clipped System

From Eq. (2.30), where $\rho = \rho_{xx}(\tau)$,

$$\frac{\partial R_{yy}(\tau)}{\partial \rho} = \sigma_x^2 E[g'(x_1)g'(x_2)]$$

$$= \sigma_x^2 \iint g'(x_1)g'(x_2)p(x_1, x_2)\, dx_1\, dx_2 \tag{2.69}$$

Substitutions from Eqs. (2.6) and (2.68) give

$$\frac{\partial R_{yy}(\tau)}{\partial \rho} = \frac{1}{2\pi\sqrt{1-\rho^2}} \iint 4\delta(x_1)\delta(x_2) \exp\left\{\frac{-[x_1^2 - 2\rho x_1 x_2 + x_2^2]}{2\sigma_x^2(1-\rho^2)}\right\} dx_1\, dx_2$$

$$= \frac{2}{\pi\sqrt{1-\rho^2}} \iint \delta(x_1)\delta(x_2)\, dx_1\, dx_2 = \frac{2}{\pi\sqrt{1-\rho^2}} \tag{2.70}$$

Hence,

$$R_{yy}(\tau) = \frac{2}{\pi} \int \frac{d\rho}{\sqrt{1-\rho^2}} + C = \frac{2}{\pi} \sin^{-1}(\rho) + C \tag{2.71}$$

when C is a constant of integration.

Consider now Eq. (2.17). When $\rho = 0$, this reduces to

$$R_{yy}(\tau) = \iint g(x_1)g(x_2)p(x_1)p(x_2)\, dx_1\, dx_2$$

$$= \left[\int g(x)p(x)\, dx\right]^2 \tag{2.72}$$

For cases where $g(x)$ is an odd function of x, as is true for the $g(x)$ of Eq. (2.66), one obtains

$$R_{yy}(\tau) = 0 \qquad \text{for } \rho = 0 \tag{2.73}$$

Thus, the constant $C = 0$ in Eq. (2.71). This proves that

$$R_{yy}(\tau) = \frac{2}{\pi} \sin^{-1}[\rho_{xx}(\tau)] = \frac{2}{\pi} \sin^{-1}\left[\frac{R_{xx}(\tau)}{\sigma_x^2}\right] \tag{2.74}$$

This famous result is known as the *arc-sine law* for the autocorrelation function of hard-clipped Gaussian signals. Measurement of $R_{yy}(\tau)$ enables one to estimate $\rho_{xx}(\tau)$ using the relation

$$\rho_{xx}(\tau) = \sin[(\pi/2)R_{yy}(\tau)] \tag{2.75}$$

One can also determine σ_x^2 from measurement of $x(t)$ to obtain

$$R_{xx}(\tau) = \sigma_x^2 \rho_{xx}(\tau) \tag{2.76}$$

As expected, note also that $R_{yy}(0) = 1$ here.

Input/Output Cross-Correlation Function for Hard-Clipped System

From Eq. (2.35), the $g(x)$ of Eq. (2.66) gives

$$\begin{aligned} R_{xy}(\tau) &= \frac{R_{xx}(\tau)}{\sigma_x^3\sqrt{2\pi}}\left[\int_{-\infty}^{0}(-u)e^{-u^2/2\sigma_x^2}\,du + \int_{0}^{\infty} ue^{-u^2/2\sigma_x^2}\,du\right] \\ &= \frac{2R_{xx}(\tau)}{\sigma_x^3\sqrt{2\pi}}\int_0^{\infty} ue^{-u^2/2\sigma_x^2}\,du = \left(\frac{\sqrt{2}}{\sigma_x\sqrt{\pi}}\right)R_{xx}(\tau) \end{aligned} \tag{2.77}$$

The associated input/output cross-spectrum formula is

$$S_{xy}(f) = \left(\frac{\sqrt{2}}{\sigma_x\sqrt{\pi}}\right)S_{xx}(f) \tag{2.78}$$

2.6.5 Nonsymmetric Hard-Clipped System

Suppose the hard-clipped system is nonsymmetric as defined by

$$\begin{aligned} y = g(x) &= 0, \qquad x < 0 \\ &= 0.5, \qquad x = 0 \\ &= 1, \qquad x > 0 \end{aligned} \tag{2.79}$$

The discontinuous derivative $g'(x)$ is now

$$\begin{aligned} g'(x) &= \delta(x) \qquad \text{at } x = 0 \\ &= 0 \qquad \text{for } x \neq 0 \end{aligned} \tag{2.80}$$

Price's theorem gives

$$\frac{\partial R_{yy}(\tau)}{\partial \rho} = \frac{1}{2\pi\sqrt{1-\rho^2}} \tag{2.81}$$

leading to

$$R_{yy}(\tau) = \left(\frac{1}{2\pi}\right) \sin^{-1}\left[\frac{R_{xx}(\tau)}{\sigma_x^2}\right] + C \tag{2.82}$$

At $\tau = 0$, $R_{xx}(0) = \sigma_x^2$ and $R_{yy}(0) = (1/2)$ here so that

$$R_{yy}(0) = \left(\frac{1}{2\pi}\right) \sin^{-1}(1) + C = \frac{1}{4} + C = \frac{1}{2} \tag{2.83}$$

Thus, the constant C in Eq. (2.82) has the value $C = (1/4)$ for this nonsymmetric hard-clipped system. Results in Eqs. (2.82) and (2.74) should be compared.

2.6.6 Smooth-Limiter System

Consider the nonlinear nonsymmetric smooth-limiter system sketched in Figure 2.17, where

$$\begin{aligned} y = g(x) &= \sqrt{\frac{2}{\pi}} \int_0^x \exp\left(\frac{-t^2}{2}\right) dt, \qquad x \geqslant 0 \\ &= 0, \qquad\qquad\qquad\qquad\qquad\quad x \leqslant 0 \end{aligned} \tag{2.84}$$

Equation (2.84) is the equation for the standardized normal probability integral.

For this nonlinear transformation, special values are $g(0) = 0$, $g(1) = 0.683$, $g(2) = 0.954$, $g(3) = 0.997$, and $g(\infty) = 1.000$. Tables are available that give $g(x)$ as a function of x, which also can be used to find x for any $g(x)$ in range $0 < g(x) < 1$.

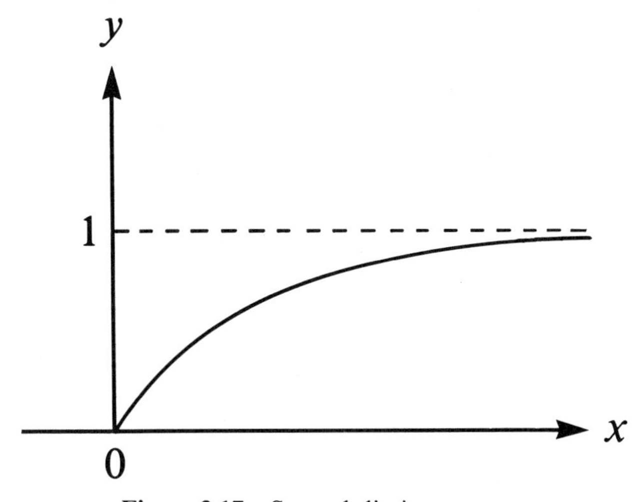

Figure 2.17 Smooth-limiter system.

The inverse relations are

$$\begin{aligned} x = g^{-1}(y) &= \text{any value of } x \leqslant 0 \text{ for } y = 0 \\ &= \text{unique value of } x \text{ in range } 0 < x < \infty \text{ for } 0 < y < 1 \\ &= \text{undefined for } y < 0 \text{ or } y > 1 \end{aligned}$$

The derivative relations are

$$\begin{aligned} \frac{dy}{dx} = g'(x) &= \sqrt{\frac{2}{\pi}} \exp\left(\frac{-x^2}{2}\right), \qquad x > 0 \\ &= 0, \qquad x < 0 \end{aligned} \tag{2.85}$$

Here, the derivative is discontinuous at $x = 0$.

Assume that $p(x)$, the input probability density function, is Gaussian with mean zero and standard deviation σ, where

$$p(x) = \frac{1}{\sigma\sqrt{2\pi}} \exp\left(\frac{-x^2}{2\sigma^2}\right) \tag{2.86}$$

All of the values of $x \leqslant 0$ are associated with the single value of $y = 0$. Hence, $p_2(y)$, the output probability density function, is such that

$$p_2(y) = \frac{1}{2}\delta(y), \qquad \text{at } y = 0 \tag{2.87}$$

where the factor (1/2) occurs because the probability is (1/2) that $x \leqslant 0$. There are no values of x where $y < 0$ or $y > 1$. Hence,

$$p_2(y) = 0 \qquad \text{for } y < 0 \text{ or } y > 1 \tag{2.88}$$

In the range where $0 < y < 1$, the result is

$$p_2(y) = \frac{p(x)}{|dy/dx|} = \sqrt{\frac{\pi}{2}}\, p(x) \exp\left(\frac{x^2}{2}\right) \tag{2.89}$$

where the value of x on the right-hand side should be replaced by its equivalent y from $x = g^{-1}(y)$. As y approaches the value 0 from the inside where $y > 0$, the value of x also approaches zero so that

$$p_2(0+) = \sqrt{\frac{\pi}{2}}\, p(0) = \frac{1}{2\sigma} \tag{2.90}$$

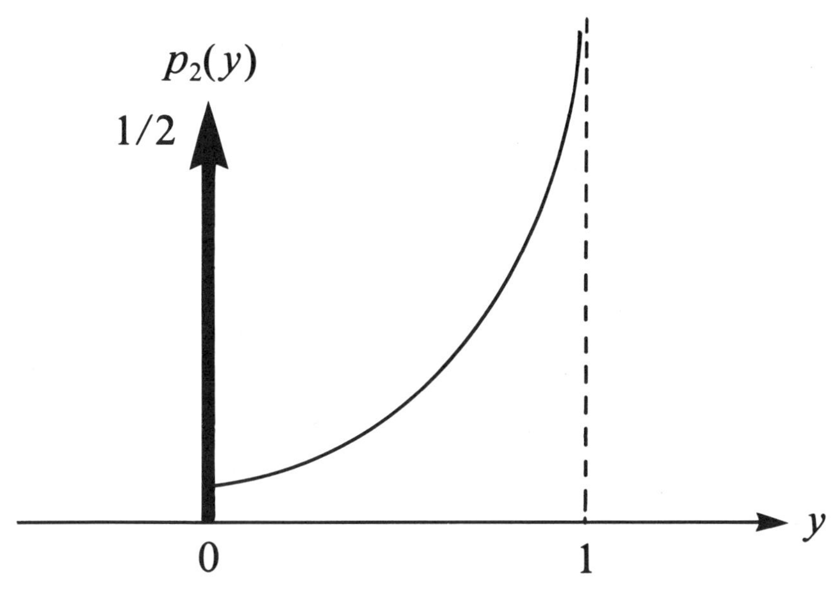

Figure 2.18 Output PDF for smooth-limiter system.

As y approaches the value 1 from the inside where $y < 1$, the value of x approaches infinity so that

$$p_2(1-) = \infty \tag{2.91}$$

A typical shape for $p_2(y)$ is sketched in Figure 2.18. The area under the curve from $0+$ to $1-$ is (1/2). Here,

$$\text{probability } [y(t) = 0] = 0.50$$

$$\text{probability } [y(t) < 0 \text{ or } y(t) > 1] = 0$$

$$\text{probability } [0 < y(t) < 1] = 0.50$$

Input/Output Cross-Correlation Function for Smooth-Limiter System

By Bussgang's theorem, Eq. (2.35), the input/output cross-correlation function $R_{xy}(\tau)$ for Gaussian input data satisfies the equation

$$R_{xy}(\tau) = \frac{R_{xx}(\tau)}{\sigma^2} \int_0^\infty x g(x) p(x)\, dx \tag{2.92}$$

where, from Eqs. (2.84) and (2.86),

$$g(x) = \sqrt{\frac{2}{\pi}} \int_0^x \exp\left(\frac{-t^2}{2}\right) dt, \qquad x \geqslant 0 \tag{2.93}$$

$$p(x) = \frac{1}{\sigma\sqrt{2\pi}} \exp\left(\frac{-x^2}{2\sigma^2}\right) \tag{2.94}$$

Integration by parts formula gives

$$\int u\,dv = uv - \int v\,du$$

Let

$$u = g(x), \qquad dv = xp(x)\,dx$$

Then

$$du = g'(x)\,dx = \sqrt{\frac{2}{\pi}} \exp\left(\frac{-x^2}{2}\right) dx, \qquad v = -\sigma^2 p(x)$$

At $x = 0$ and $x = \infty$, the product $uv = 0$. Hence,

$$\begin{aligned} \int_0^\infty xg(x)p(x)\,dx &= \sigma^2 \int_0^\infty p(x)g'(x)\,dx \\ &= \frac{\sigma}{\pi} \int_0^\infty \exp\left[\frac{-x^2}{2\sigma^2} - \frac{x^2}{2}\right] dx = \frac{\sigma}{\pi} \int_0^\infty \exp\left(\frac{-x^2}{2\sigma_1^2}\right) dx \end{aligned} \tag{2.95}$$

where σ_1^2 is defined by

$$\sigma_1^2 = \frac{\sigma^2}{1 + \sigma^2} \tag{2.96}$$

The integrand is now in the form of a first-order Gaussian probability density function for $x \geqslant 0$, satisfying

$$\frac{1}{\sigma_1\sqrt{2\pi}} \int_0^\infty \exp\left(\frac{-x^2}{2\sigma_1^2}\right) dx = \frac{1}{2} \tag{2.97}$$

Hence,

$$\int_0^\infty xg(x)p(x)\,dx = \frac{\sigma\sigma_1}{\sqrt{2\pi}} = \frac{\sigma^2}{\sqrt{2\pi(1 + \sigma^2)}} \tag{2.98}$$

and

$$R_{xy}(\tau) = \frac{R_{xx}(\tau)}{\sqrt{2\pi(1 + \sigma^2)}} \tag{2.99}$$

The associated input/output cross-spectrum formula is

$$S_{xy}(f) = \frac{S_{xx}(f)}{\sqrt{2\pi(1 + \sigma^2)}} \tag{2.100}$$

2.6.7 Summary of Results

Sections 2.6.1–2.6.6 produce the following results for the non-Gaussian output probability density function (PDF) compared to a Gaussian PDF, when zero mean value Gaussian input data pass through nonlinear systems.

1. *Two-Slope System* (*Figure 2.12*). For cases where the slope decreases, the output PDF has a discontinuity that produces shorter tails. This also occurs from softening spring and clipped systems.
2. *Two-Slope System* (*Figure 2.13*). For cases where the slope increases, the output PDF has a discontinuity that produces longer tails. This also occurs from hardening spring systems.
3. *Dead-Zone System* (*Figure 2.14*). A delta function effect will appear in the output PDF at output values where the dead-zone effect occurs, with similar tails on both sides of the peak value.
4. *Clipped-System* (*Figure 2.15*). A delta function effect will appear in the output PDF at output values where the clipping effect occurs, with a small tail (theoretically zero) for values above a positive clipped value or below a negative clipped value.
5. *Hard-Clipped System* (*Figure 2.16*). Two delta function effects will appear in the output PDF at the two output values where the clipping effect occurs. Similar appearing results can be due to the presence of a strong sine wave in the output data, so one should investigate whether the output data contains such sine waves to distinguish these two possibilities. Sine wave PDF's are discussed in references 1 and 2.
6. *Smooth-Limiter System* (*Figure 2.18*). This output PDF exists only for positive values, with a minimum near $y = 0$ and a very high value where the limit occurs, completely different from a Gaussian PDF and from symmetric results due to the presence of sine waves. It is similar to one-sided limiter results on sine waves.
7. For Gaussian input data, Sections 2.6.4 and 2.6.6 contain other formulas that predict the output autocorrelation function and input/output cross-correlation functions for the examples of hard-clipped and smooth-limiter systems.

2.7 SQUARE-LAW AND CUBIC SYSTEMS

Three types of zero-memory nonlinear systems will now be discussed representing square-law, cubic, and square-law with sign operations. Such nonlinear systems can occur alone or in parallel with linear operations. From Section 2.9, by least-square procedures, combinations of linear, square-law, and cubic systems can often be obtained to yield useful approximations to many other types of nonlinear systems that occur in practice. To obtain closed-form answers, input data will be assumed to be a zero mean value Gaussian stationary random process.

2.7.1 Square-Law System

A nonlinear square-law system as per Figure 2.9 is defined by the relation

$$y = g(x) = x^2 \tag{2.101}$$

Here, the continuous derivative

$$\frac{dy}{dx} = g'(x) = 2x \qquad \text{and} \qquad x = \pm\sqrt{y} \tag{2.102}$$

Thus, the inverse function $x = g^{-1}(y)$ is double-valued. From Eq. (2.14), the output probability density function

$$p_2(y) = \frac{2p(x)}{|2x|} = \frac{p(\sqrt{y})}{\sqrt{y}}, \qquad y > 0 \tag{2.103}$$

with $p_2(y) = 0$ for $y < 0$ and $p_2(0) = \infty$. If $p(x)$ has the Gaussian form of Eq. (2.4), then

$$p(\sqrt{y}) = \frac{1}{\sigma_x\sqrt{2\pi}} \exp\left(\frac{-y}{2\sigma_x^2}\right), \qquad y > 0 \tag{2.104}$$

Hence,

$$p_2(y) = \frac{1}{\sigma_x\sqrt{2\pi y}} \exp\left(\frac{-y}{2\sigma_x^2}\right), \qquad y > 0 \tag{2.105}$$

This is a *chi-square probability density function* with one degree of freedom as sketched in Figure 2.19.

For zero mean value Gaussian input data through a square-law system, the output moments are

$$E[y^n] = E[x^{2n}] = (1)(3)(5)\cdots(2n-1)\sigma_x^{2n} \tag{2.106}$$

As special cases,

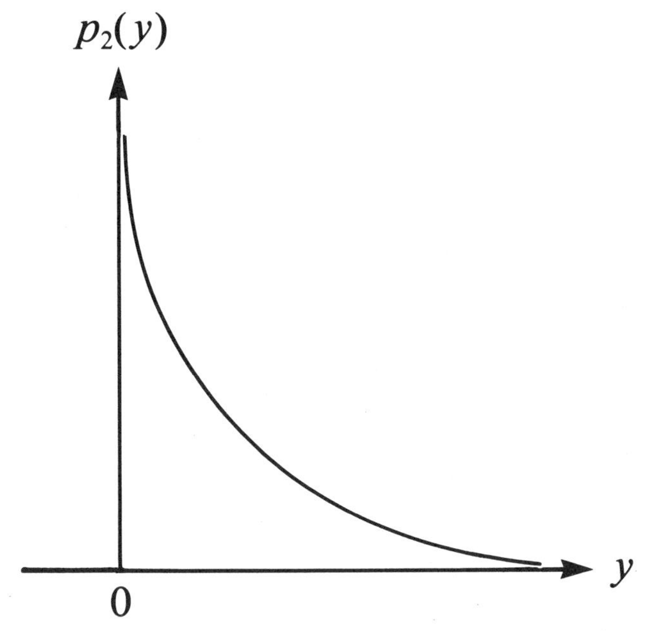

Figure 2.19 Output chi-square PDF for Gaussian data through a square-law system.

Mean value

$$\mu_y = E[y] = E[x^2] = \sigma_x^2 \tag{2.107}$$

Mean square value

$$\psi_y^2 = R_{yy}(0) = E[y^2] = E[x^4] = 3\sigma_x^4 \quad (2.(108)$$

Variance

$$\sigma_y^2 = E[y^2] - (E[y])^2 = 2\sigma_x^4 \tag{2.109}$$

Note that $\mu_y \neq 0$ even though $\mu_x = 0$. Note also that all odd-order moments of the zero mean value Gaussian input data are zero.

The fourth-order moment of the zero mean value Gaussian inputs reduces to the following products of second-order moments, known as the *Fourth-Order Moment Relation* (reference 1):

$$E[x_1x_2x_3x_4] = E[x_1x_2]E[x_3x_4] + E[x_1x_3]E[x_2x_4] + E[x_1x_4]E[x_2x_3] \tag{2.110}$$

Higher even-order moments can similarly be reduced.

Output Autocorrelation Function for Square-Law System

From Eq. (2.110), for the zero mean value Gaussian input data, the output autocorrelation function for a square-law system where $y_1 = y(t)$ and $y_2 = y(t + \tau)$ is

$$
\begin{aligned}
R_{yy}(\tau) &= E[y_1 y_2] = E[x_1^2 x_2^2] = E[x_1 x_1 x_2 x_2] \\
&= E[x_1 x_1]E[x_2 x_2] + 2E[x_1 x_2]E[x_1 x_2] \\
&= \sigma_x^4 + 2R_{xx}^2(\tau) \qquad (2.111)
\end{aligned}
$$

This same result can also be derived from Price's theorem of Eq. (2.19) using

$$\frac{\partial R_{yy}(\tau)}{\partial R_{xx}(\tau)} = E[(2x_1)(2x_2)] = 4R_{xx}(\tau) \qquad (2.112)$$

Now

$$R_{yy}(\tau) = 2R_{xx}^2(\tau) + C \qquad (2.113)$$

and

$$R_{yy}(0) = 2R_{xx}^2(0) + C = 2\sigma_x^4 + C \qquad (2.114)$$

From separate considerations of Eq. (2.85),

$$R_{yy}(0) = E[y^2] = 3\sigma_x^4 \qquad (2.115)$$

Hence, $C = \sigma_x^4$ and

$$R_{yy}(\tau) = \sigma_x^4 + 2R_{xx}^2(\tau) \qquad (2.116)$$

in agreement with Eq. (2.111).

Input/Output Cross-Correlation Function for Square-Law System

The input/output cross-correlation function for a Gaussian input through a square-law system between $x_1 = x(t)$ and $y_2 = y(t + \tau)$ will be zero, since all odd-order moments are zero for zero mean value Gaussian data. Thus,

$$R_{xy}(\tau) = E[x_1 y_2] = E[x_1 x_2^2] = 0 \qquad (2.117)$$

This same result also follows from Bussgang's theorem of Eq. (2.35), since $g(x)$ is an even function of x. The input/output two-sided cross-spectral density function for a Gaussian input through a square-law system is also zero, namely,

$$S_{xy}(f) = 0 \tag{2.118}$$

Thus, usual linear analysis does not recognize square-law effects. This very important result enables one to analyze situations like Figure 2.20 representing a linear system $H(f)$ in parallel with a square-law system. Calculation of $S_{xx}(f)$ and $S_{xy}(f)$ will give $H(f)$ using $S_{xy}(f) = H(f)S_{xx}(f)$, since $S_{xy_2}(f) = 0$. Detection and analysis of the square-law system requires new linear techniques or more advanced calculations involving special bispectral density functions to be developed in Chapters 4 and 5. A spectral function of $x(t)$ and $y(t)$ is desired which will be zero for the linear system and nonzero for the square-law system.

Output Autospectral Density Function for Square-Law System

From Eq. (2.116), the output autospectral density function for the Gaussian input through a square-law system is

$$S_{yy}(f) = \sigma_x^4\delta(f) + 2\int_{-\infty}^{\infty} S_{xx}(u)S_{xx}(f-u)\,du \tag{2.119}$$

where $\delta(f)$ is the usual delta function, while $S_{xx}(f)$ and $S_{yy}(f)$ are two-sided autospectral density functions obtained from the Fourier transforms of $R_{xx}(\tau)$ and $R_{yy}(\tau)$, respectively. For $f \neq 0$, the output spectrum $S_{yy}(f)$ does not include the delta function.

2.7.2 Cubic System

A nonlinear cubic system as per Figure 2.10 is defined by the relation

$$y = g(x) = x^3 \tag{2.120}$$

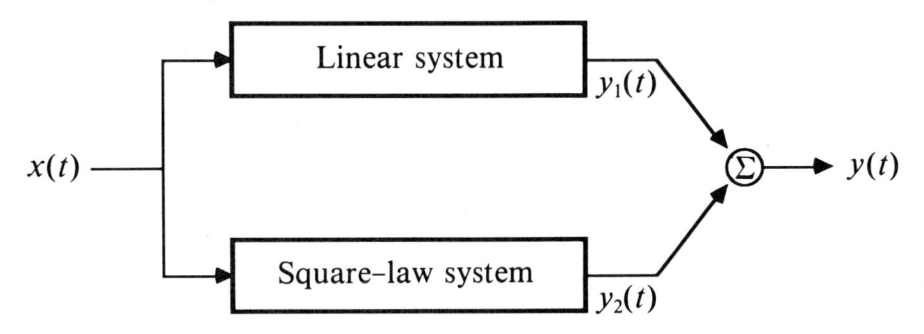

Figure 2.20 Parallel linear and square-law systems.

Here, $x = y^{1/3}$ and the continuous derivative

$$\frac{dy}{dx} = g'(x) = 3x^2 = 3y^{2/3} \tag{2.121}$$

From Eq. (2.13), the output probability density function is

$$p_2(y) = \frac{p(x)}{|3x^2|} = \frac{p(y^{1/3})}{3y^{2/3}} \tag{2.122}$$

If $p(x)$ has the Gaussian form of Eq. (2.4), then

$$p(y^{1/3}) = \frac{1}{\sigma_x\sqrt{2\pi}} \exp\left(\frac{-y^{2/3}}{2\sigma_x^2}\right) \tag{2.123}$$

Hence,

$$p_2(y) = \frac{1}{3\sigma_x y^{2/3}\sqrt{2\pi}} \exp\left(\frac{-y^{2/3}}{2\sigma_x^2}\right) \tag{2.124}$$

Note that $p_2(-y) = p_2(y)$ and $p_2(0) = \infty$ as sketched in Figure 2.21.

For zero mean value Gaussian input data through a cubic system, the output moments are

$$\mu_y = E[y] = E[x^3] = 0 \tag{2.125}$$

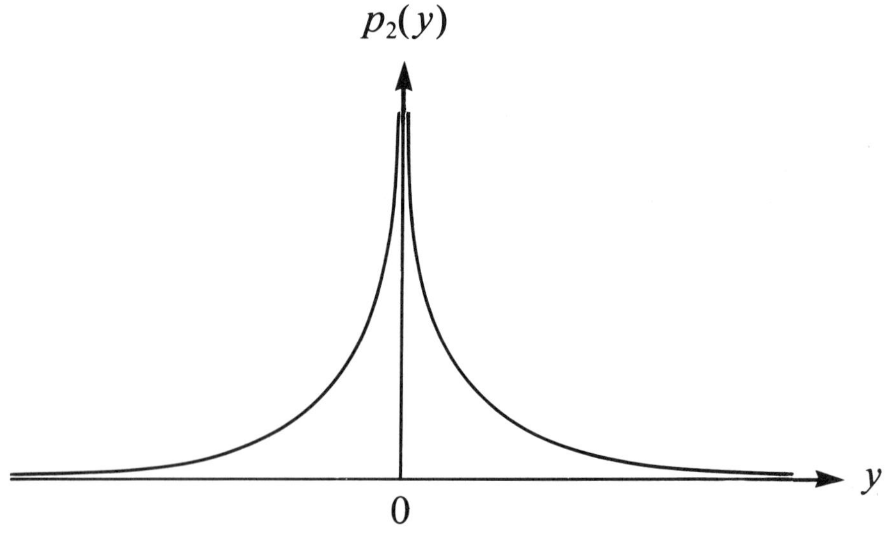

Figure 2.21 Output PDF for Gaussian data through a cubic system.

$$R_{yy}(0) = E[y^2] = E[x^6] = 15\sigma_x^6 \tag{2.126}$$

Note that $\mu_y = 0$ here when $\mu_x = 0$.

The sixth-order moment of the zero mean value Gaussian inputs satisfies the relation (reference 1)

$$\begin{aligned} E[x_1x_2x_3x_4x_5x_6] = {} & E[x_1x_2]E[x_3x_4x_5x_6] \\ & + E[x_1x_3]E[x_2x_4x_5x_6] + E[x_1x_4]E[x_2x_3x_5x_6] \\ & + E[x_1x_5]E[x_2x_3x_4x_6] + E[x_1x_6]E[x_2x_3x_4x_5] \end{aligned} \tag{2.127}$$

Output Autocorrelation Function for Cubic System

The output autocorrelation function for the Gaussian input through a cubic system is given by

$$\begin{aligned} R_{yy}(\tau) &= E[y_1y_2] = E[x_1^3x_2^3] = E[x_1x_1x_1x_2x_2x_2] \\ &= 2E[x_1x_1]E[x_1x_2x_2x_2] + 3E[x_1x_2]E[x_1x_1x_2x_2] \\ &= 2\sigma_x^2[3\sigma_x^2R_{xx}(\tau)] + 3R_{xx}(\tau)[\sigma_x^4 + 2R_{xx}^2(\tau)] \\ &= 9\sigma_x^4R_{xx}(\tau) + 6R_{xx}^3(\tau) \end{aligned} \tag{2.128}$$

This same result can also be derived from Price's theorem of Eq. (2.19) using

$$\begin{aligned} \frac{\partial R_{yy}(\tau)}{\partial R_{xx}(\tau)} &= E[(3x_1^2)(3x_2^2)] = 9E[x_1^2x_2^2] \\ &= 9\sigma_x^4 + 18R_{xx}^2(\tau) \end{aligned} \tag{2.129}$$

Now

$$R_{yy}(\tau) = 9\sigma_x^4R_{xx}(\tau) + 6R_{xx}^3(\tau) + C \tag{2.130}$$

and

$$R_{yy}(0) = 15\sigma_x^6 + C \tag{2.131}$$

But $R_{yy}(0) = 15\sigma_x^6$ from Eq. (2.126). Hence, $C = 0$ and

$$R_{yy}(\tau) = 9\sigma_x^4R_{xx}(\tau) + 6R_{xx}^3(\tau) \tag{2.132}$$

in agreement with Eq. (2.128).

Input/Output Cross-Correlation Function for Cubic System

The input/output cross-correlation function for the Gaussian input through a cubic system is given by

$$R_{xy}(\tau) = E[x_1 y_2] = E[x_1 x_2^3] = 3\sigma_x^2 R_{xx}(\tau) \tag{2.133}$$

This same result also follows from Bussgang's theorem of Eq. (2.35) using

$$R_{xy}(\tau) = \frac{R_{xx}(\tau)}{\sigma_x^2} E[x^4] = 3\sigma_x^2 R_{xx}(\tau) \tag{2.134}$$

Equations (2.133) and (2.134) show that the input/output cross-spectral density function for the Gaussian input through a cubic system is

$$S_{xy}(f) = 3\sigma_x^2 S_{xx}(f) \tag{2.135}$$

These relations should be compared with Eqs. (2.117) and (2.118) for the square-law system.

Output Autospectral Density Function for Cubic System

From Eq (2.128) or (2.132), the output autospectral density function for the Gaussian input through a cubic system is

$$S_{yy}(f) = 9\sigma_x^4 S_{xx}(f) + 6 \iint S_{xx}(u) S_{xx}(v-u) S_{xx}(f-v)\, du\, dv \tag{2.136}$$

In Eq. (2.135), the output $y(t) = x^3(t)$ is correlated with the input $x(t)$. For later applications, it is desirable to have a different output $y_c(t)$ that will be uncorrelated with $x(t)$. This can be done as follows. Let

$$y_c(t) = x^3(t) - 3\sigma_x^2 x(t) \tag{2.137}$$

Then instead of Eqs. (2.135) and (2.136), one will obtain

$$S_{xy_c}(f) = 0 \tag{2.138}$$

$$S_{y_c y_c}(f) = 6 \iint S_{xx}(u) S_{xx}(v-u) S_{xx}(f-v)\, du\, dv \tag{2.139}$$

The corresponding correlation functions become

$$R_{xy_c}(\tau) = 0 \tag{2.140}$$

$$R_{y_c y_c}(\tau) = 6 R_{xx}^3(\tau) \tag{2.141}$$

2.7.3 Square-Law System with Sign

Square-law systems with sign as per Figure 2.11 are of great importance in fluid dynamics models involving "dynamic pressure," defined as $q = (1/2)\rho v|v|$ where ρ is the fluid density and v is the fluid velocity. This quantity q is a key parameter in the description of boundary-layer pressures and drag forces due to the fluid flow over structures.

A nonlinear square-law system with sign is defined by the relation

$$y = g(x) = x|x| \tag{2.142}$$

Hence,

$$\begin{aligned} y &= x^2, & x > 0 \\ &= -x^2, & x < 0 \end{aligned} \tag{2.143}$$

The inverse relation

$$\begin{aligned} x &= +\sqrt{y}, & y > 0 \\ &= -\sqrt{-y}, & y < 0 \end{aligned} \tag{2.144}$$

Also, the continuous derivative

$$\begin{aligned} \frac{dy}{dx} = g'(x) &= 2x, & x > 0 \\ &= -2x, & x < 0 \end{aligned} \tag{2.145}$$

From Eq. (2.13), the output probability density function

$$\begin{aligned} p_2(y) = \frac{p(x)}{|2x|} &= \frac{p(\sqrt{y})}{2\sqrt{y}}, & y > 0 \\ &= \frac{p(\sqrt{-y})}{2\sqrt{-y}}, & y < 0 \end{aligned} \tag{2.146}$$

Thus, $p_2(-y) = p_2(y)$ and $p_2(0) = \infty$ with $p_2(y)$ as sketched in Figure 2.22 if $p(x)$ has the Gaussian form of Eq. (2.4).

For zero mean value Gaussian input data through a square-law system with sign, the output moments

$$\mu_y = E[y] = E[x|x|] = 0 \tag{2.147}$$

$$R_{yy}(0) = E[y^2] = E[x^4] = 3\sigma_x^4 \tag{2.148}$$

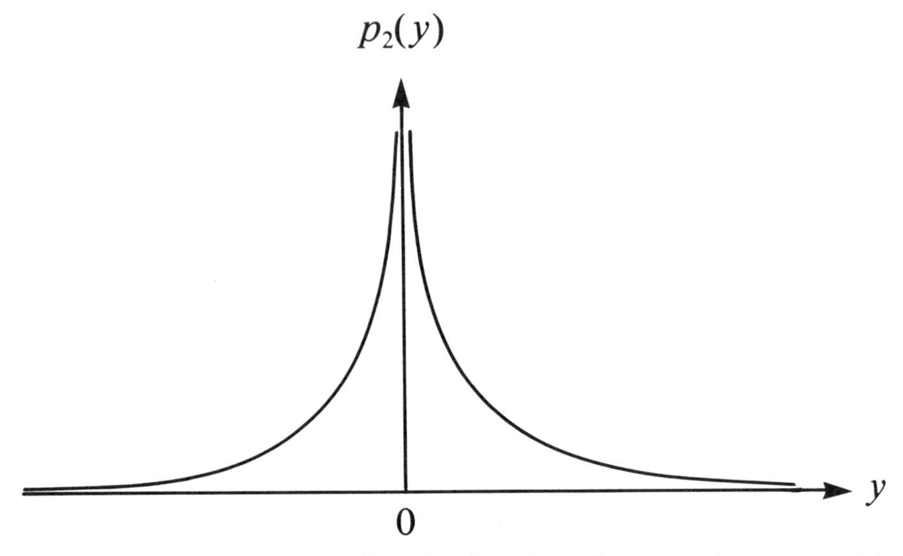

Figure 2.22 Output PDF for Gaussian data through a square-law system with sign.

Note that the mean value here is the same as for a cubic system but the mean square value is the same as for a square-law system.

For the square-law system with sign,

$$E[xy] = E[x^2|x|] = E[-x^3]_{x<0} + E[x^3]_{x>0}$$

$$= 2E[x^3]_{x>0} = 2\int_0^{\infty} x^3 p(x)\,dx = (2\sigma_x^3)\sqrt{(2/\pi)} \tag{2.149}$$

using the $p(x)$ of Eq. (2.4). Hence, from Bussgang's theorem of Eq. (2.35), the input/output cross-correlation function for a Gaussian input through a square-law system with sign is

$$R_{xy}(\tau) = 2\sigma_x\sqrt{(2/\pi)}\, R_{xx}(\tau) \tag{2.150}$$

The associated input/output cross-spectrum is

$$S_{xy}(f) = 2\sigma_x\sqrt{(2/\pi)}\, S_{xx}(f) \tag{2.151}$$

The output autocorrelation function for a Gaussian input through a square-law system with sign is given by

$$R_{yy}(\tau) = E[y_1 y_2] = E[(x_1|x_1|)(x_2|x_2|)] \tag{2.152}$$

A useful approximation for this result, derived later in Eq. (2.201), is

$$R_{yy}(\tau) \approx (2.55\sigma_x^2)R_{xx}(\tau) + (0.42/\sigma_x^2)R_{xx}^3(\tau) \tag{2.153}$$

The associated output autospectrum is

$$S_{yy}(f) \approx (2.55\sigma_x^2) S_{xx}(f) + (0.42/\sigma_x^2) \iint S_{xx}(u) S_{xx}(v-u) S_{xx}(f-v)\, du\, dv \quad (2.154)$$

2.7.4 Summary of Results

Sections 2.7.1–2.7.3 produce the following results for the square-law system, the cubic system, and the square-law system with sign when zero mean value Gaussian input data pass through these systems.

1. *Square-Law System.* The output PDF (Figure 2.19) exists only for positive values of y with a very high value near $y = 0$ and with a long tail, as described by a chi-square PDF with one degree of freedom that is completely different from a Gaussian PDF.
2. *Cubic System.* The output PDF (Figure 2.21) will be very sharply peaked near $y = 0$ compared to a Gaussian PDF. Similar measured output PDF can occur for dead-zone systems (Figure 2.16).
3. *Square-Law System with Sign.* This gives an output PDF (Figure 2.22) that appears the same as for a cubic system output PDF (Figure 2.21). Other analysis is required to distinguish these two situations.

For these three systems, from knowledge of $R_{xx}(\tau)$ for $\tau \geqslant 0$ where $R_{xx}(0) = \sigma_x^2$, the predicted results for $E[y]$, $E[y^2]$, $R_{xy}(\tau)$, and $R_{yy}(\tau)$ are listed in Table 2.1 for easy comparison.

TABLE 2.1 Square-Law, Cubic, and Square-Law with Sign Systems

	System		
	Square-Law	Cubic	Square-Law with Sign
Relation	$y = x^2$	$y = x^3$	$y = x\|x\|$
$E[y]$	σ_x^2	0	0
$E[y^2]$	$3\sigma_x^4$	$15\sigma_x^6$	$3\sigma_x^4$
$R_{xy}(\tau)$	0	$3\sigma_x^2 R_{xx}(\tau)$	$2\sigma_x\sqrt{(2/\pi)}\, R_{xx}(\tau)$
$R_{yy}(\tau)$	$\sigma_x^4 + 2R_{xx}^2(\tau)$	$9\sigma_x^4 R_{xx}(\tau) + 6R_{xx}^3(\tau)$	$(2.55\sigma_x^2)R_{xx}(\tau) + (0.42/\sigma_x^2)R_{xx}^3(\tau)$

2.8 HARDENING AND SOFTENING SPRING SYSTEMS

Output probability density functions will now be derived for the force output $y(t)$ versus the displacement input $x(t)$ in four models that can represent continuously hardening and softening spring systems. Related alternative output probability density functions follow for the displacement output versus the force input when the roles of $x(t)$ and $y(t)$ are reversed.

2.8.1 Hardening Spring System

For Gaussian input data governed by Eq. (2.4) with $\sigma_x = 1.0$, a continuously hardening spring system can be defined for suitable x and a by the single-valued, one-to-one, zero-memory nonlinear function

$$y = g(x) = \begin{cases} x + ax^2, & x \geqslant 0 \\ x - ax^2, & x \leqslant 0 \end{cases} \tag{2.155}$$

This special $g(x)$ is an odd function of x. Note that the hardening spring system of Eq. (2.155) is equivalent to having a linear system in parallel with a square-law system with sign, since

$$y = x + ax|x| \tag{2.156}$$

Equation (2.155) is sketched in Figure 2.23 for $a = 0.14$ and $|x| \leqslant 3.0$.

For the assumed Gaussian displacement input data,

$$E[x] = 0, \qquad E[x^2] = 1.0, \qquad E[x^3] = 0, \qquad E[x^4] = 3.0 \tag{2.157}$$

For the hardening spring force output data of Eq. (2.155),

$$E[y] = 0, \qquad E[y^2] \approx 1.51, \qquad E[y^3] = 0, \qquad E[y^4] \approx 9.16 \tag{2.158}$$

Thus, the hardening spring causes large kurtosis where $E[y^4] > E[x^4]$. Note also that $\sigma_y \approx 1.23$.

The inverse function $x = g^{-1}(y)$ is given here by

$$\begin{aligned} x &= (-1 + \sqrt{1 + 4ay})/2a, && y > 0 \\ &= (1 - \sqrt{1 - 4ay})/2a, && y < 0 \\ &= 0, && y = 0 \end{aligned} \tag{2.159}$$

The derivative function

$$\begin{aligned} \frac{dy}{dx} &= 1 + 2ax, && x \geqslant 0 \\ &= 1 - 2ax, && x \leqslant 0 \end{aligned} \tag{2.160}$$

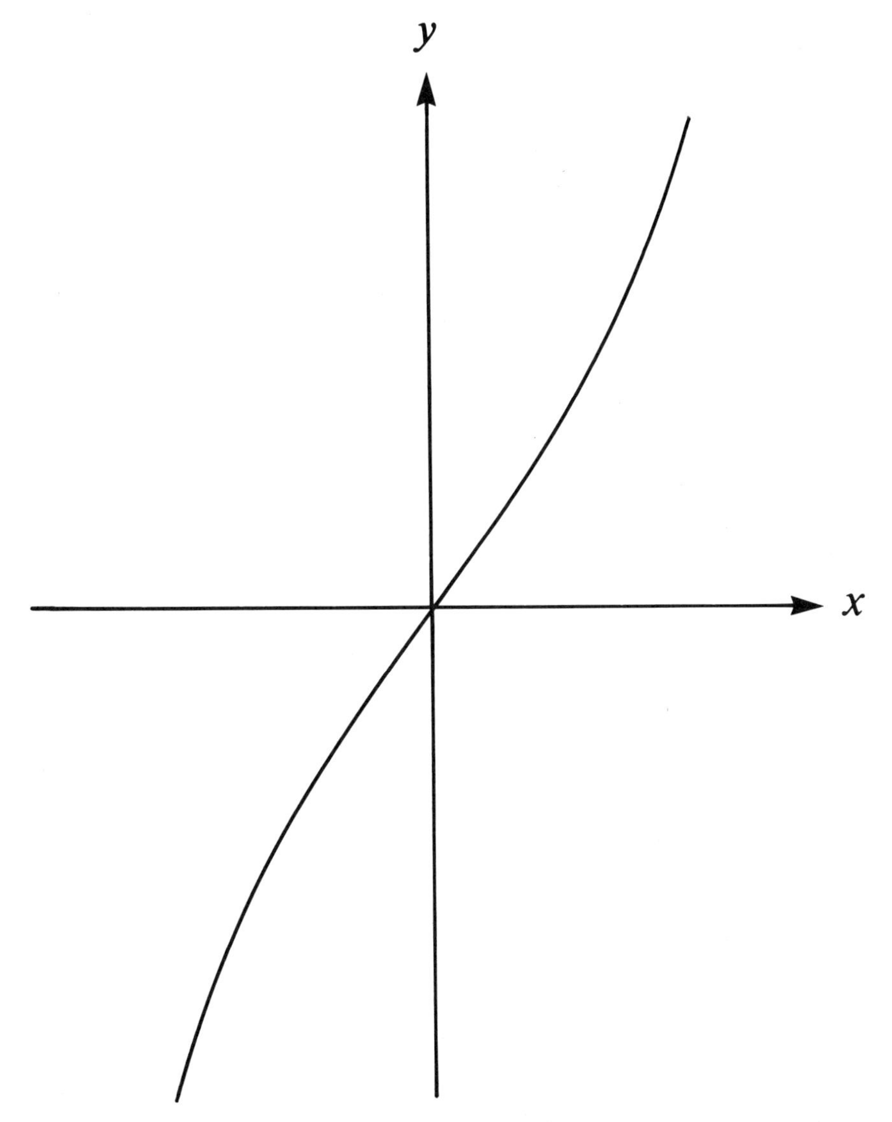

Figure 2.23 Hardening spring system.

Substituting Eq. (2.159) into Eq. (2.160) yields

$$\begin{aligned} \frac{dy}{dx} &= \sqrt{1 + 4ay}, \qquad y > 0 \\ &= \sqrt{1 - 4ay}, \qquad y < 0 \\ &= 1, \qquad y = 0 \end{aligned} \tag{2.161}$$

Hence, by Eq. (2.13), $p_2(0) = p(0)$ with

$$p_2(y) = \frac{p[(-1+\sqrt{1+4ay})/2a]}{\sqrt{1+4ay}}, \qquad y > 0$$
$$= \frac{p[(1-\sqrt{1-4ay})/2a]}{\sqrt{1-4ay}}, \qquad y < 0 \tag{2.162}$$

This formula shows that the output $p_2(y)$ is an even function of y.

Equation (2.162) is sketched in Figure 2.24 for $a = 0.14$ and $|y| \leqslant 3.5$. Figure 2.24 demonstrates that the PDF for a hardening spring, compared with a Gaussian PDF, will have a sharper peak in the vicinity of $y = 0$ and will have longer tails. The Gaussian PDF is drawn with dashed lines.

2.8.2 Softening Spring System

For Gaussian input data governed by Eq. (2.4) with $\sigma_x = 1.0$, a continuously softening spring system can be defined for suitable x and a by the single-valued

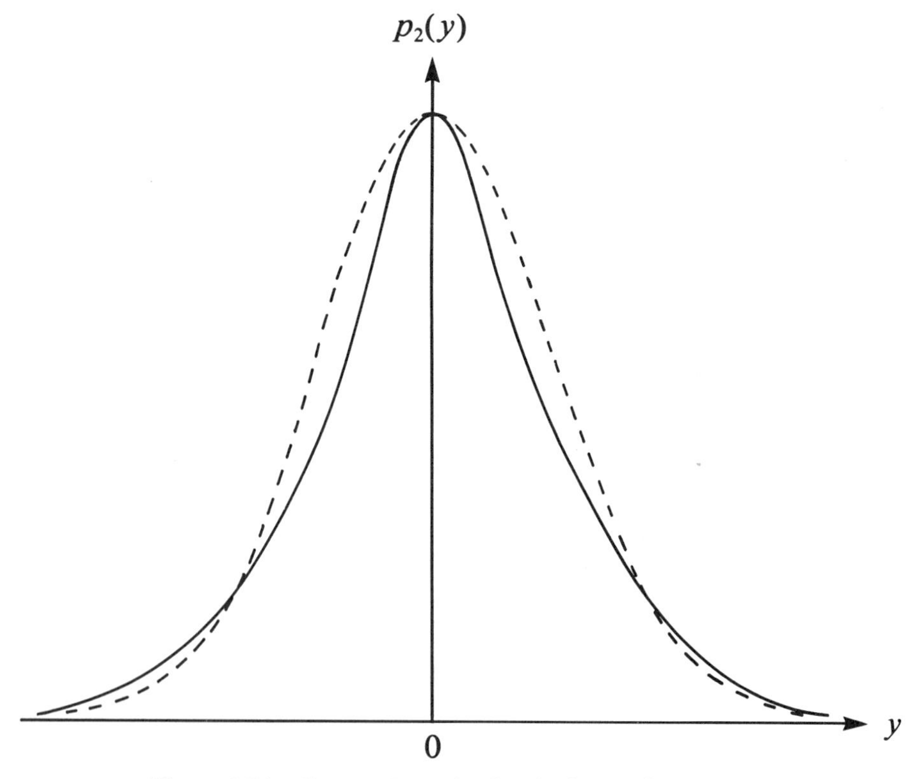

Figure 2.24 Output PDF for hardening spring system.

one-to-one zero-memory nonlinear function

$$y = g(x) = \begin{cases} x - ax^2, & x \geqslant 0 \\ x + ax^2, & x \leqslant 0 \end{cases} \tag{2.163}$$

As in Eq. (2.155), this different $g(x)$ is an odd function of x. Note that this softening spring system consists of a linear system in parallel with a square-law system with sign, since

$$y = x - ax|x| \tag{2.164}$$

Equation (2.163) is sketched in Figure 2.25 for $a = 0.14$ and $|x| \leqslant 3.0$.

For the assumed Gaussian displacement input data,

$$E[x] = 0, \qquad E[x^2] = 1.0, \qquad E[x^3] = 0, \qquad E[x^4] = 3.0 \tag{2.165}$$

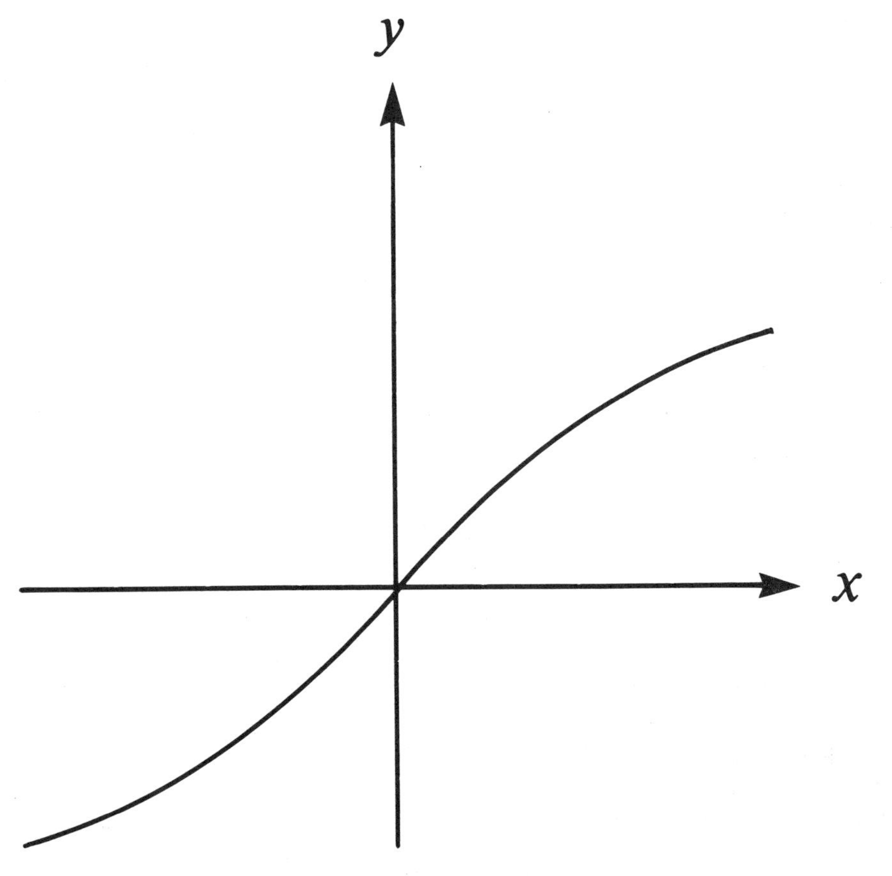

Figure 2.25 Softening spring system.

For the softening spring force output data of Eq. (2.163),

$$E[y] = 0, \qquad E[y^2] \approx 0.61, \qquad E[y^3] = 0, \qquad E[y^4] \approx 1.17 \tag{2.166}$$

Thus, the softening spring causes small kurtosis where $E[y^4] < E[x^4]$. Note here that $\sigma_y \approx 0.78$.

The derivative function

$$\begin{aligned} \frac{dy}{dx} &= 1 - 2ax, \qquad x \geqslant 0 \\ &= 1 + 2ax, \qquad x \leqslant 0 \end{aligned} \tag{2.167}$$

In particular, for $x > 0$ and $a = 0.14$,

$$\frac{dy}{dx} = 0 \quad \text{when} \quad x = \frac{1}{2a} \approx 3.57$$

with a maximum value of y given by

$$y_{\text{max}} = \frac{1}{4a} \approx 1.785$$

By restricting $|x| \leqslant 3.5$, it follows that the inverse function $x = g^{-1}(y)$ will be single-valued. This inverse function is given here by

$$\begin{aligned} x &= (1 - \sqrt{1 - 4ay})/2a, \qquad y > 0 \\ &= (-1 + \sqrt{1 + 4ay})/2a, \qquad y < 0 \\ &= 0, \qquad y = 0 \end{aligned} \tag{2.168}$$

Now, by Eq. (2.13), $p_2(0) = p(0)$ with

$$\begin{aligned} p_2(y) &= \frac{p[(1 - \sqrt{1 - 4ay})/2a]}{\sqrt{1 - 4ay}}, \qquad y > 0 \\ &= \frac{p[(-1 + \sqrt{1 + 4ay})/2a]}{\sqrt{1 + 4ay}}, \qquad y < 0 \end{aligned} \tag{2.169}$$

As before, observe that this output $p_2(y)$ is an even function of y.

Equation (2.169) is sketched in Figure 2.26 for $a = 0.14$ and $|y| \leqslant 3.5$. Figure 2.26 demonstrates that the PDF for a softening spring, compared with a Gaussian PDF, will have a flatter peak in the vicinity of $y = 0$ and will have shorter tails. Note also the two humps immediately to the right and left of $y = 0$. The Gaussian PDF is drawn with dashed lines.

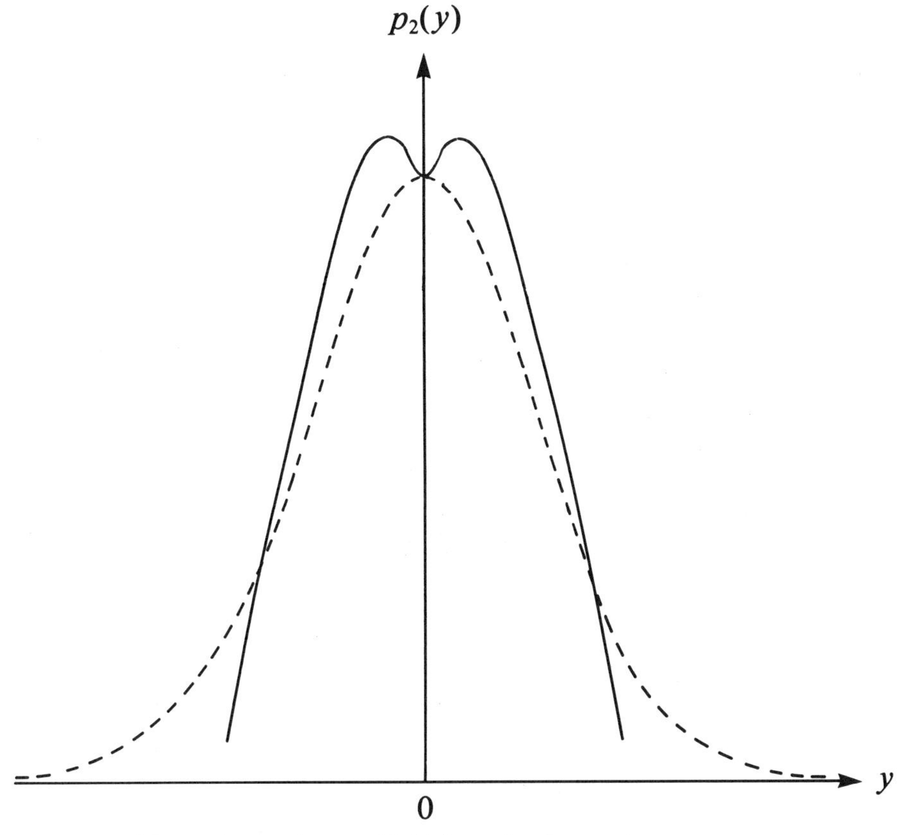

Figure 2.26 Output PDF for softening spring system.

2.8.3 Hardening/Softening Spring System

A continuously hardening/softening spring system with hardening for $x > 0$ and softening for $x < 0$ can be defined by combining the results from Sections 2.8.1 and 2.8.2 as follows:

$$y = g(x) = x + ax^2 \tag{2.170}$$

This function is neither even nor odd. Equation (2.170) is sketched in Figure 2.27 for $a = 0.14$ and $|x| \leqslant 3.0$. Note that this particular nonlinear transformation has the shape of a catenary.

For this case, the output PDF is given by the relations $p_2(0) = p(0)$ with

$$p_2(y) = \frac{p[(-1 + \sqrt{1 + 4ay})/2a]}{\sqrt{1 + 4ay}} \qquad \text{for all } y \neq 0 \tag{2.171}$$

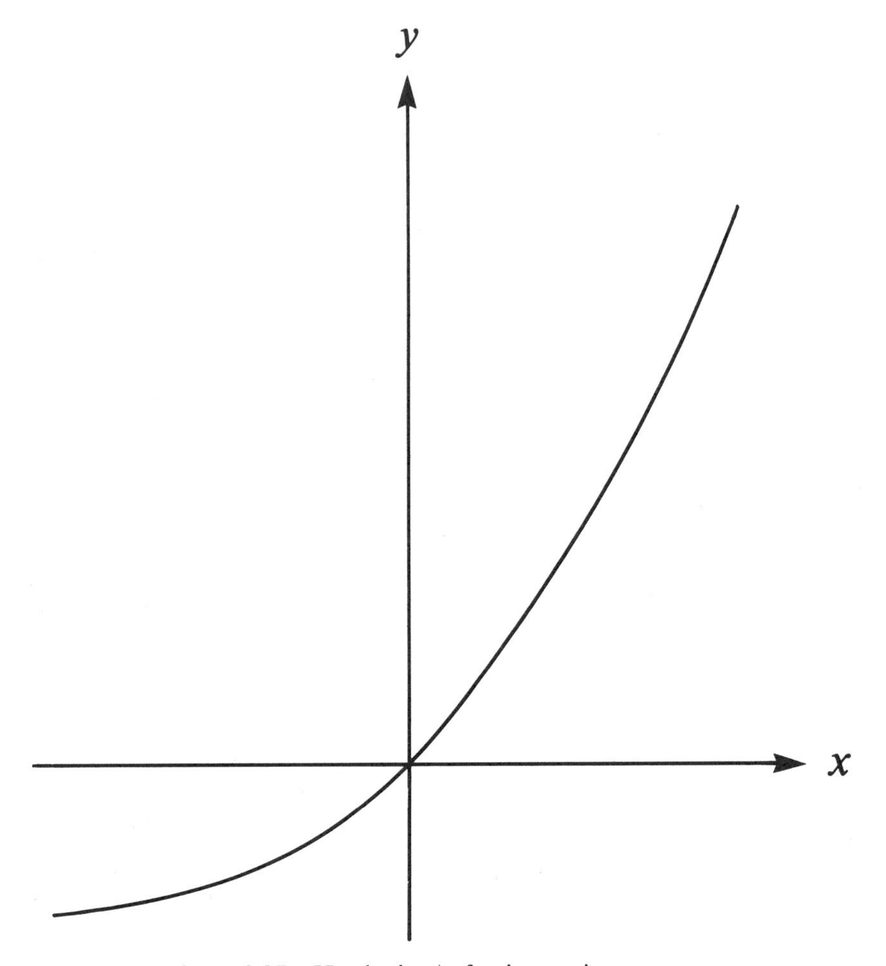

Figure 2.27 Hardening/softening spring system.

This is sketched in Figure 2.28 for $a = 0.14$ and $|y| \leqslant 3.5$ by combining previous results from Figures 2.24 and 2.26. Note that Figure 2.28 demonstrates that the output PDF for a hardening/softening spring system as defined here, compared with a Gaussian PDF, will be skewed to the left. The Gaussian PDF is drawn with dashed lines.

2.8.4 Softening/Hardening Spring System

A continuously softening/hardening spring system, with softening for $x > 0$ and hardening for $x < 0$, can be defined by combining the results from Sections 2.8.1 and 2.8.2 as follows:

$$y = g(x) = x - ax^2 \tag{2.172}$$

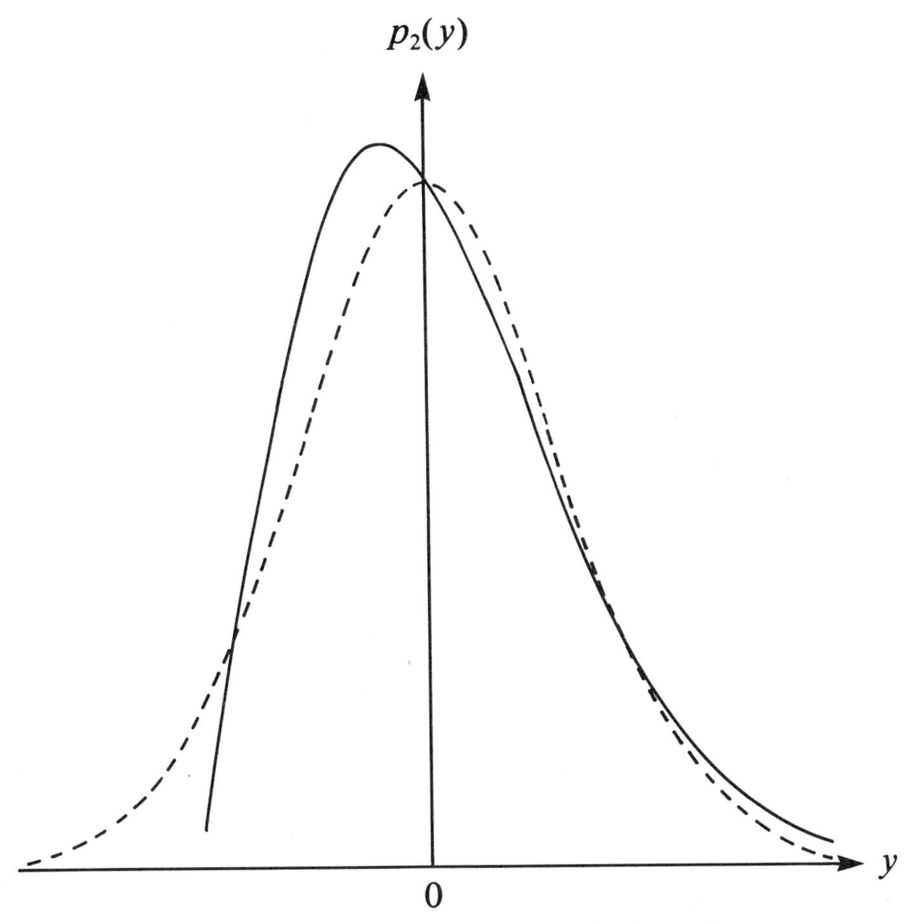

Figure 2.28 Output PDF for hardening/softening spring system.

This function is neither even nor odd. Equation (2.172) is sketched in Figure 2.29 for $a = 0.14$ and $|x| \leqslant 3.0$.

For this case, the output PDF is now given by the relations $p_2(0) = p(0)$ with

$$p_2(y) = \frac{p[(1 - \sqrt{1 - 4ay})/2a]}{\sqrt{1 - 4ay}} \qquad \text{for all } y \neq 0 \tag{2.173}$$

This is sketched in Figure 2.30 for $a = 0.14$ and $|y| \leqslant 3.5$ by combining in the reverse order from Figure 2.28 the previous results of Figures 2.24 and 2.26. In particular, note that Figure 2.30 demonstrates that the output PDF for a softening/hardening spring as defined here, compared with a Gaussian PDF, will be skewed to the right. The Gaussian PDF is drawn with dashed lines.

Figures 2.26, 2.28, 2.30, and 2.32 show the output PDF for unstandardized output variables. However, most computer programs standardize the output

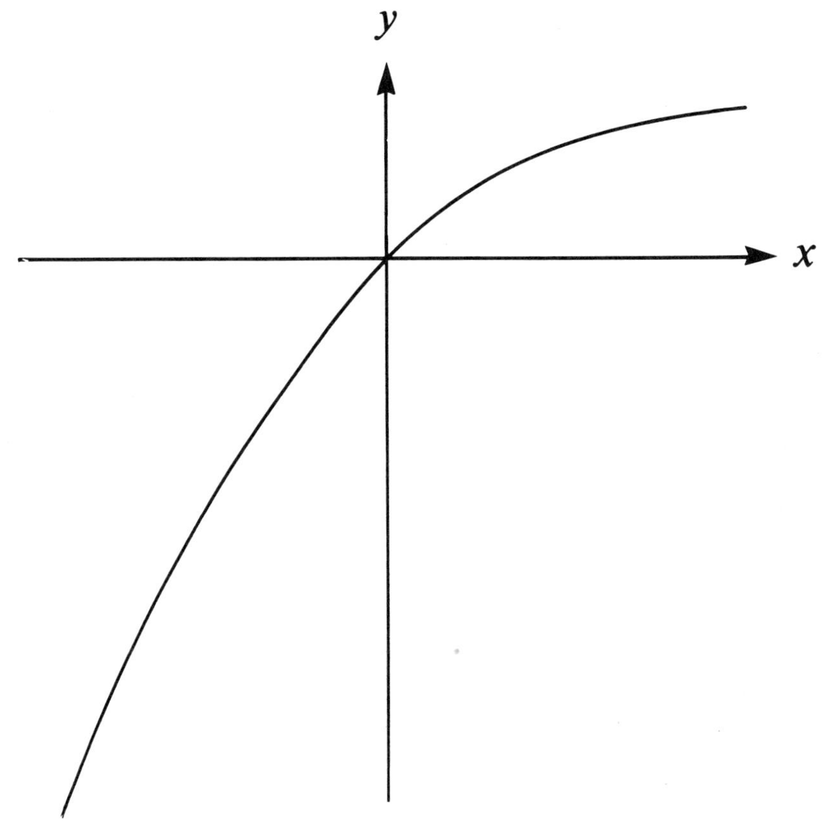

Figure 2.29 Softening/hardening spring system.

variable before computing its PDF. This produces different results as discussed in the next section.

2.8.5 Standardized Output Variable

Suppose the output variable $y(t)$ with probability density function $p_2(y)$, mean value μ_y, and standard deviation σ_y is replaced by a standardized output variable $z(t)$ with probability density function $p_3(z)$, zero mean value $\mu_z = 0$, and unit standard deviation $\sigma_z = 1.0$. The transformation is

$$z(t) = \frac{y(t) - \mu_y}{\sigma_y} \tag{2.174}$$

Now, by the probability density relation of Eqs. (2.13),

$$p_3(z) = \frac{p_2(y)}{|dz/dy|} = \sigma_y p_2(y) = \sigma_y p_2(\mu_y + \sigma_y z) \tag{2.175}$$

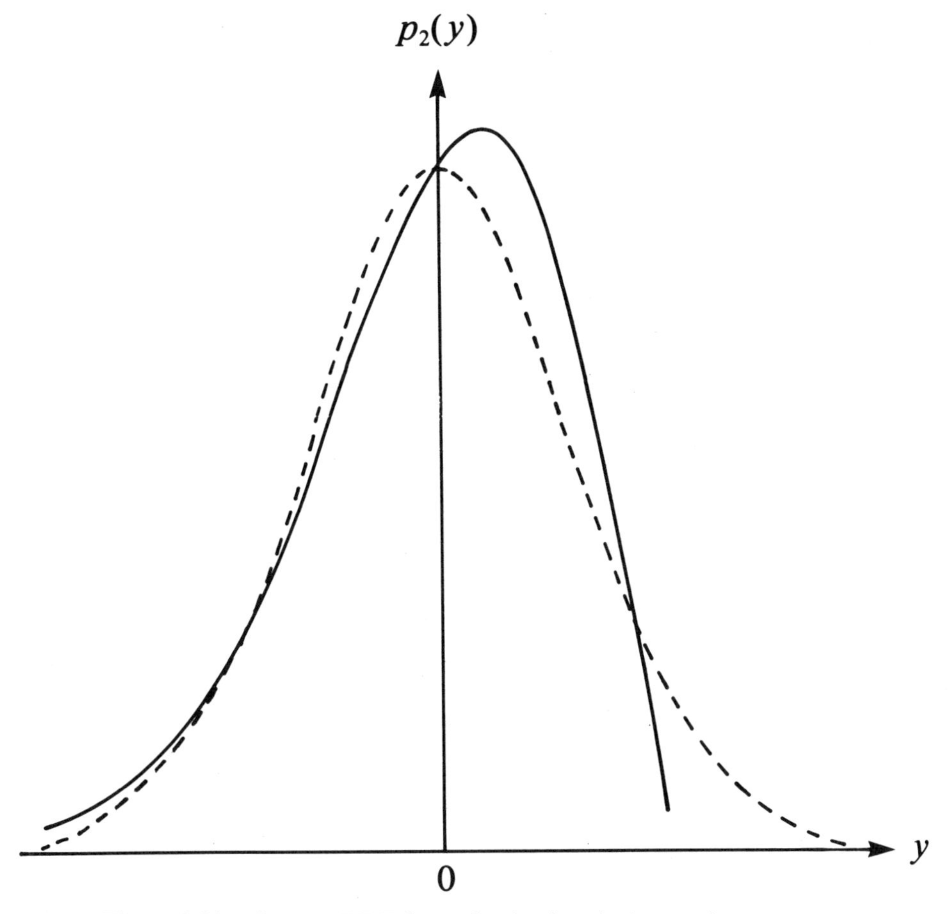

Figure 2.30 Output PDF for softening/hardening spring system.

This is a general result. When $\mu_y = 0$, one obtains the special formula

$$p_3(z) = \sigma_y p_2(\sigma_y z) \tag{2.176}$$

Consider the continuously hardening and softening spring systems in Figures 2.23 and 2.25 where $\mu_y = 0$. For these cases, at the value $z = 0$,

$$p_3(0) = \sigma_y p_2(0) \tag{2.177}$$

It follows that

$$\begin{aligned} p_3(0) &> p_2(0) \quad \text{when } \sigma_y > 1 \\ p_3(0) &< p_2(0) \quad \text{when } \sigma_y < 1 \end{aligned} \tag{2.178}$$

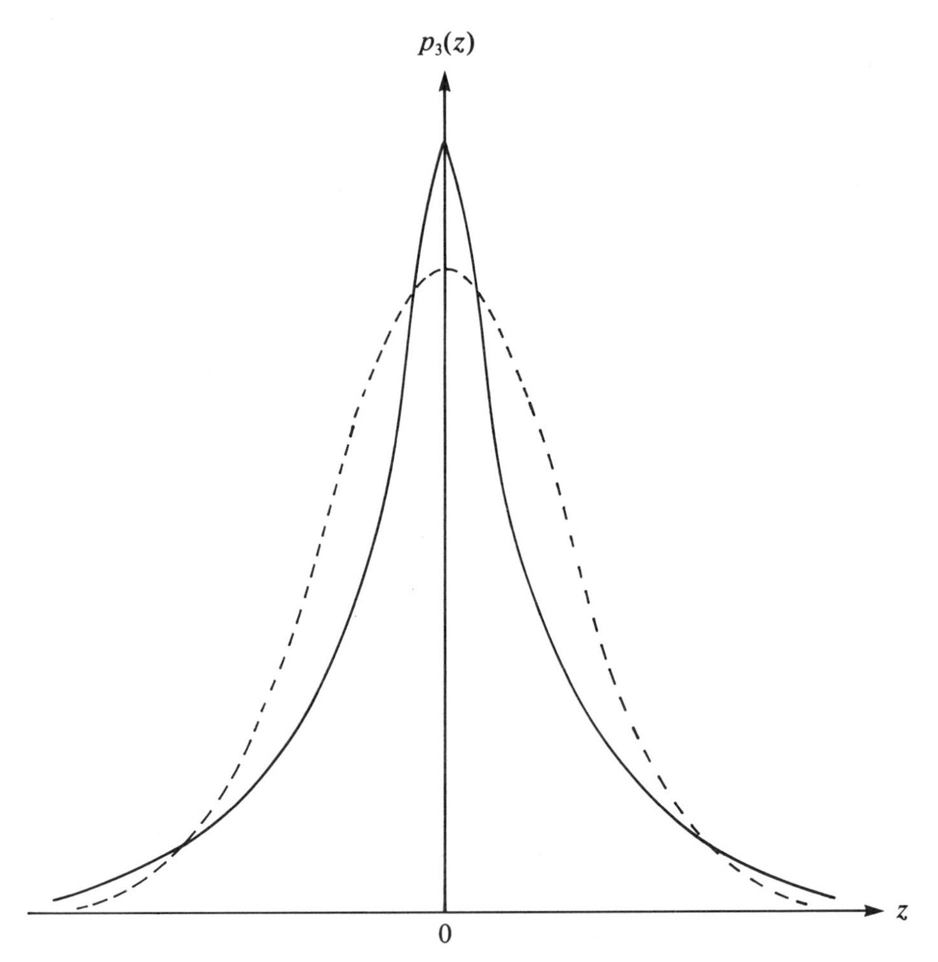

Figure 2.31 Output PDF for hardening spring system with standardized output variable.

Thus, for the hardening spring system where Eq. (2.158) gives $\sigma_y \approx 1.23$, the output PDF of $p_3(z)$ versus z will be more peaked than shown in Figure 2.24 with $p_3(0) \approx 1.23p_2(0)$. Conversely, for the softening spring system where Eq. (2.166) gives $\sigma_y \approx 0.78$, the output PDF of $p_3(z)$ versus z will be flatter than shown in Figure 2.26 with $p_3(0) \approx 0.78p_2(0)$. Figures 2.31 and 2.32 based upon Eq. (2.176) are plots of $p_3(z)$ versus z for the hardening and softening spring systems where $|z| \leqslant 3.0$. The Gaussian PDF is drawn with dashed lines.

2.8.6 Summary of Results

This chapter proves that when Gaussian input data pass through different types of zero-memory nonlinear systems, the output probability density function (PDF) will be non-Gaussian and depend on the type of nonlinearity. The form of

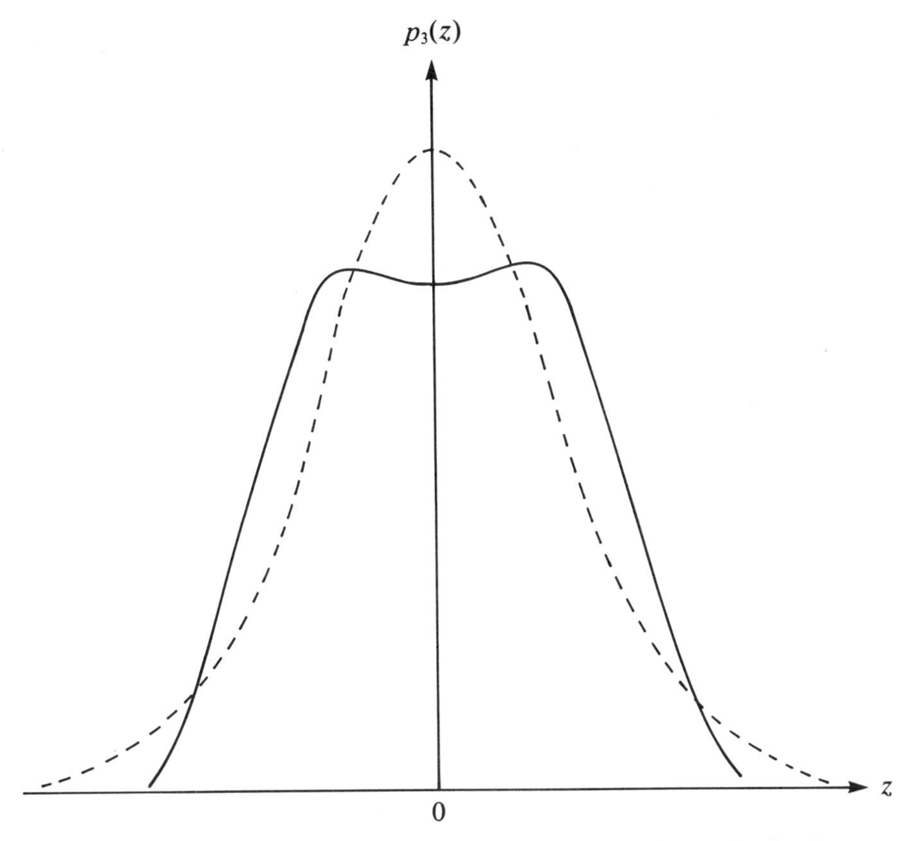

Figure 2.32 Output PDF for softening spring system with standardized output variable.

the output PDF can help determine the type of nonlinear operation that is occurring.

Sections 2.8.1–2.8.5 produce the following results for the non-Gaussian output PDF compared to a Gaussian PDF when zero mean value Gaussian input data pass through hardening and softening spring systems.

1. *Hardening Spring System* (*Figures 2.23, 2.24, 2.31*). The output PDF will have a sharper peak in the vicinity of $y = 0$, longer tails, and higher kurtosis values.
2. *Softening Spring System* (*Figures 2.25, 2.26, 2.32*). The output PDF will have a flatter peak in the vicinity of $y = 0$, shorter tails, and lower kurtosis values.
3. *Hardening/Softening Spring System* (*Figures 2.27, 2.28*). The output PDF will be skewed to the left.
4. *Softening/Hardening Spring System* (*Figures 2.29, 2.30*). The output PDF will be skewed to the right.

5. The plots shown in Figures 2.23, 2.25, 2.27, and 2.29 are for the force output $y(t)$ versus the displacement input $x(t)$. These same plots can also be used to represent a displacement output versus a force input by interchanging $x(t)$ with $y(t)$ and replacing "hardening" with "softening." The plots shown in Figures 2.24, 2.26, 2.28, and 2.30 are output probability density functions for the force output to a Gaussian displacement input. When the roles of $x(t)$ and $y(t)$ are reversed, then these same particular plots can represent the output probability density functions for the displacement output to a Gaussian force input for systems where "hardening" and "softening" are reversed. For example, for a Gaussian force input $x(t)$, Figure 2.24 is the plot of the output probability density function $p_2(y)$ for the displacement output $y(t)$ in a "softening" spring system, and Figure 2.26 is the plot of the output probability density function $p_2(y)$ for the displacement output $y(t)$ in a "hardening" spring system.

2.9 THIRD-ORDER POLYNOMIAL LEAST-SQUARES APPROXIMATION

The replacement of zero-memory nonlinear systems by an optimum third-order polynomial least-squares approximation will now be derived. Suppose $y = g(x)$ represents a zero-memory nonlinear system. What is the least-squares approximation to $y = g(x)$ by the *third-order* polynomial

$$v = a_1 x + a_2 x^2 + a_3 x^3 \tag{2.179}$$

under the assumption that the input data x follow a zero mean value Gaussian distribution? To be specific, what should be the choices of the coefficients a_1, a_2, a_3 so as to minimize the mean square error Q defined by

$$Q = E[(y - v)^2] = E[(y - a_1 x - a_2 x^2 - a_3 x^3)^2] \tag{2.180}$$

over all possible choices of these coefficients?

For any y, the assumptions on x give

$$\begin{aligned} Q = {} & E[y^2] - 2a_1 E[xy] - 2a_2 E[x^2 y] - 2a_3 E[x^3 y] \\ & + a_1^2 E[x^2] + 2a_1 a_3 E[x^4] + a_2^2 E[x^4] + a_3^2 E[x^6] \end{aligned} \tag{2.181}$$

using the fact that $E[x] = E[x^3] = E[x^5] = 0$. The setting of partial derivatives of Q with respect to a_1, a_2, and a_3 equal to zero shows

$$\begin{aligned} \frac{\partial Q}{\partial a_1} &= -2E[xy] + 2a_1 E[x^2] + 2a_3 E[x^4] = 0 \\ \frac{\partial Q}{\partial a_2} &= -2E[x^2 y] + 2a_2 E[x^4] = 0 \\ \frac{\partial Q}{\partial a_3} &= -2E[x^3 y] + 2a_1 E[x^4] + 2a_3 E[x^6] = 0 \end{aligned} \tag{2.182}$$

For zero mean value Gaussian data, the expected values $E[x^2] = \sigma_x^2$, $E[x^4] = 3\sigma_x^4$, and $E[x^6] = 15\sigma_x^6$ now give the following set of equations to be solved for a_1, a_2, and a_3:

$$\begin{aligned} a_1\sigma_x^2 + 3a_3\sigma_x^4 &= E[xy] \\ 3a_2\sigma_x^4 &= E[x^2y] \\ 3a_1\sigma_x^4 + 15a_3\sigma_x^6 &= E[x^3y] \end{aligned} \tag{2.183}$$

Hence, for any $y = g(x)$, solving Eq. (2.183) shows that one obtains the v of Eq. (2.179) by choosing

$$a_1 = \frac{15\sigma_x^6 E[xy] - 3\sigma_x^4 E[x^3y]}{6\sigma_x^8} \tag{2.184}$$

$$a_2 = \frac{E[x^2y]}{3\sigma_x^4} \tag{2.185}$$

$$a_3 = \frac{\sigma_x^2 E[x^3y] - 3\sigma_x^4 E[xy]}{6\sigma_x^8} \tag{2.186}$$

These three equations are important results for many engineering applications. When substituted into Eq. (2.179), they show how an arbitrary zero-memory nonlinear system can be replaced by the optimum third-order polynomial of Figure 2.33, consisting of a linear operation x in parallel with a

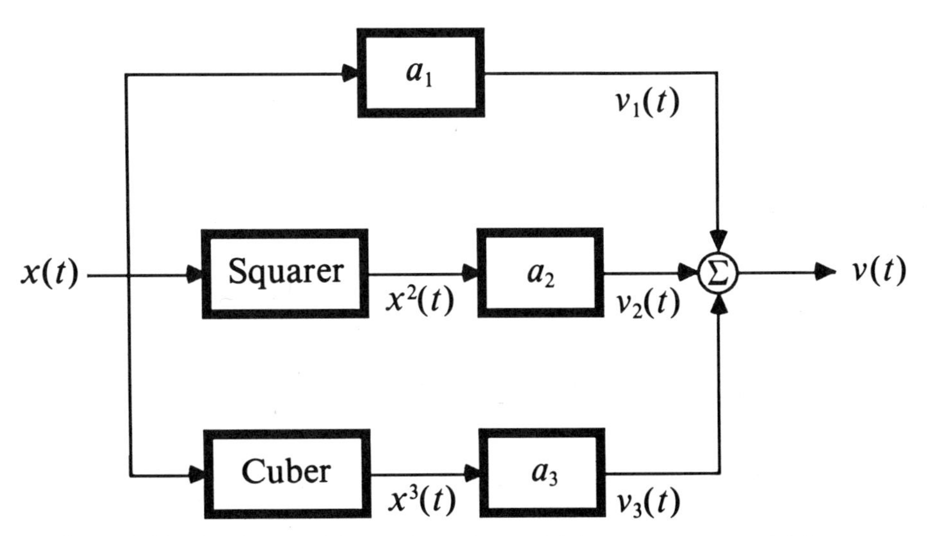

Figure 2.33 Optimum third-order polynomial least-squares model for a zero-memory nonlinear system.

squaring operation x^2 and a cubing operation x^3. This model will be applied further in Chapters 5–7.

These optimum third-order results should be compared with a *linear* least-squares approximation to y given by

$$v = b_1 x \tag{2.187}$$

Here, one minimizes the mean square error Q defined by

$$\begin{aligned} Q &= E[(y - v)^2] = E[(y - b_1 x)^2] \\ &= E[y^2] - 2b_1 E[xy] + b_1^2 E[x^2] \end{aligned}$$

Now

$$\frac{\partial Q}{\partial b_1} = -2E[xy] + 2b_1 E[x^2] = 0$$

shows that for any $y = g(x)$,

$$b_1 = \frac{E[xy]}{E[x^2]} = \frac{E[xy]}{\sigma_x^2} \tag{2.188}$$

This value of b_1 is *not* the same as the a_1 in Eq. (2.184), and Eq. (2.187) is inferior to Eq. (2.179). Note that b_1 is the same as the factor of proportionality in Bussgang's theorem, Eq. (2.35). Similarly, other approximations involving subsets of the terms in Eq. (2.179) will, in general, give poorer results than Eq. (2.179).

Application to Square-Law System with Sign

Consider application of these matters to the case of a *square-law system with sign* where

$$y = x|x| \tag{2.189}$$

For this particular y, the Gaussian assumptions on x yield

$$\begin{aligned} E[xy] &= E[x^2|x|] = 2E[x^3]_{x>0} = 2\sigma_x^3\sqrt{(2/\pi)} \\ E[x^2 y] &= E[x^3|x|] = 0 \\ E[x^3 y] &= E[x^4|x|] = 2E[x^5]_{x>0} = 8\sigma_x^5\sqrt{(2/\pi)} \end{aligned} \tag{2.190}$$

Hence, from Eq. (2.188),

$$b_1 = 2\sigma_x\sqrt{(2/\pi)} \tag{2.191}$$

so that the *linear* least-squares approximation to $x|x|$ by Eq. (2.187) is

$$v = 2\sigma_x\sqrt{(2/\pi)}\,x \tag{2.192}$$

However, from Eqs. (2.184)–(2.186), one obtains

$$a_1 = \sigma_x\sqrt{(2/\pi)} \tag{2.193}$$

$$a_2 = 0 \tag{2.194}$$

$$a_3 = \sqrt{(2/\pi)}/3\sigma_x \tag{2.195}$$

Hence, the *third-order* polynomial least-squares approximation to $x|x|$ by Eq. (2.179) is

$$v = [\sigma_x\sqrt{(2/\pi)}]x + [\sqrt{(2/\pi)}/3\sigma_x]x^3 \tag{2.196}$$

This useful result shows how to approximate $x|x|$ by a combination of a linear system in parallel with a cubic system. This gives a much better representation for $x|x|$ than the simpler linear result of Eq. (2.192) as will now be demonstrated by computing $R_{xv}(\tau)$ and $R_{vv}(\tau)$.

From Eq. (2.179) for third-order approximations, the input/output cross-correlation function for the assumed x, using $x_1 = x(t)$ and $x_2 = x(t + \tau)$, is, in general,

$$\begin{aligned} R_{xv}(\tau) &= a_1 E[x_1 x_2] + a_3 E[x_1 x_2^3] \\ &= a_1 R_{xx}(\tau) + 3a_3\sigma_x^2 R_{xx}(\tau) \end{aligned} \tag{2.197}$$

For the example of a squarc-law system with sign, where a_1 and a_3 satisfy Eqs. (2.193) and (2.195), the special result is

$$R_{xv}(\tau) = 2\sigma_x\sqrt{(2/\pi)}\,R_{xx}(\tau) \tag{2.198}$$

in agreement with the known correct result of Eq. (2.150). The linear approximations of Eqs. (2.187) and (2.192) also give

$$R_{xv}(\tau) = b_1 R_{xx}(\tau) = 2\sigma_x\sqrt{(2/\pi)}\,R_{xx}(\tau) \tag{2.199}$$

Thus, for the square-law system with sign, $R_{xv}(\tau)$ is the same for third-order and linear approximations. This is the expected result from Bussgang's theorem of Eq. (2.35).

From Eq. (2.179) for third-order approximations, the output autocorrelation function for the assumed x, using $x_1 = x(t)$ and $x_2 = x(t + \tau)$, is, in general,

$$
\begin{aligned}
R_{vv}(\tau) &= a_1^2 E[x_1 x_1] + a_1 a_3 E[x_1 x_2^3] + a_1 a_3 E[x_1^3 x_2] \\
&\quad + a_2^2 E[x_1^2 x_2^2] + a_3^2 E[x_1^3 x_2^3] \\
&= a_1^2 R_{xx}(\tau) + 3a_1 a_3 \sigma_x^2 R_{xx}(\tau) + 3a_1 a_3 \sigma_x^2 R_{xx}(\tau) \\
&\quad + a_2^2[\sigma_x^4 + 2R_{xx}^2(\tau)] + a_3^2[9\sigma_x^4 R_{xx}(\tau) + 6R_{xx}^3(\tau)]
\end{aligned}
\tag{2.200}
$$

For the example of a square-law system with sign, from Eqs. (2.193)–(2.195) one obtains

$$R_{vv}(\tau) = [8\sigma_x^2/\pi]R_{xx}(\tau) + [4/3\pi\sigma_x^2]R_{xx}^3(\tau) \tag{2.201}$$

with

$$R_{vv}(0) = [28/3\pi]\sigma_x^4 \approx 2.971\sigma_x^4 \tag{2.202}$$

This differs by less than 1% from the known correct result of Eq. (2.148) that $R_{yy}(0) = 3\sigma_x^4$. For the linear approximation of Eqs. (2.187) and (2.192) one obtains

$$R_{vv}(\tau) = b_1^2 R_{xx}(\tau) = [8\sigma_x^2/\pi]R_{xx}(\tau) \tag{2.203}$$

with

$$R_{vv}(0) = [8/\pi]\sigma_x^4 \approx 2.546\sigma_x^4 \tag{2.204}$$

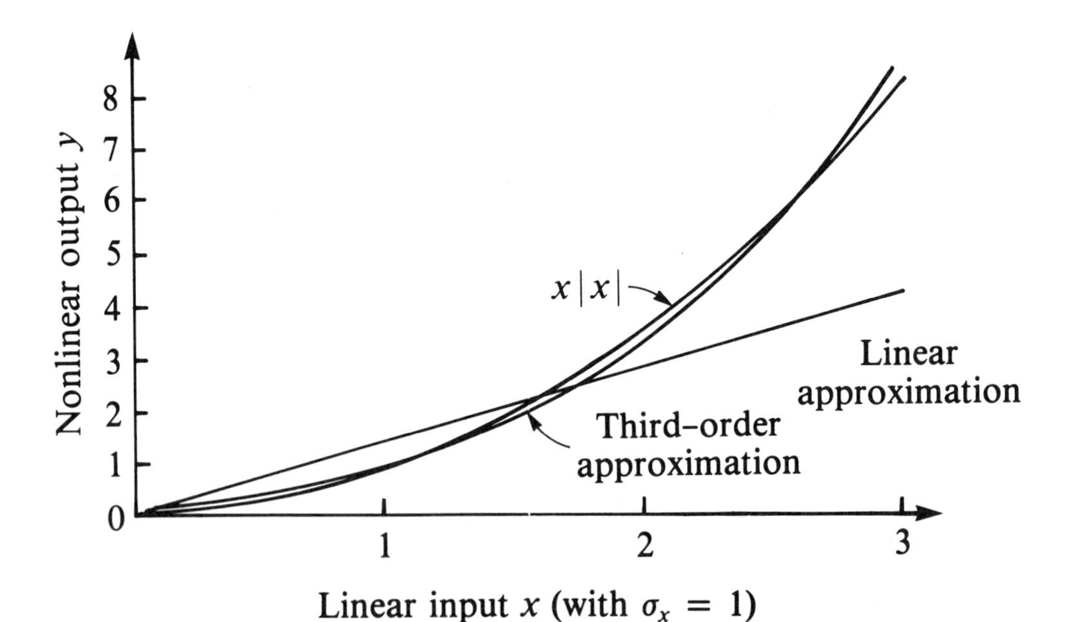

Figure 2.34 Least-squares approximation of $x|x|$.

This result is significantly different than the correct $R_{yy}(0) = 3\sigma_x^4$. The conclusion is that one should use the third-order polynomial approximation of Eq. (2.196) in preference to the linear approximation of Eq. (2.192) if one desires to analyze a square-law system with sign by such least-squares approximations. Figure 2.34 shows graphically the differences between these two approximations and the desired curve of $x|x|$ for values of x in the range $0 \leqslant x \leqslant 3\sigma_x$.

This concludes Chapter 2.

3

BILINEAR AND TRILINEAR SYSTEMS

This chapter concerns the theoretical properties of general nonlinear systems known as bilinear (second-order) and trilinear (third-order) systems, representing functional Volterra extensions of well-known properties for linear (first-order) systems. This chapter shows how to determine the multidimensional second- and third-order weighting and frequency response functions that describe these systems. Examples illustrating these results include finite-memory nonlinear systems represented by zero-memory square-law systems and cubic systems that are preceded or followed by linear systems. The chapter contains material on how to synthesize a time-varying linear system by a constant-parameter bilinear system. Also discussed are multidimensional higher-order cross-correlation and cross-spectral density functions that will be used in Chapter 4 to analyze and identify nonlinear system input/output properties of stationary random data passing through bilinear and trilinear systems. The material in Chapters 3 and 4 is needed to understand when and how these complicated techniques should be applied, and to appreciate the benefits when they can be replaced by the new, simpler, practical techniques in Chapters 5–7.

3.1 FUNCTIONAL REPRESENTATION OF NONLINEAR SYSTEMS

The output of a nonlinear system can often be expressed as a "power series with memory" by using a Volterra functional representation as discussed in references 17, 26, 35, 38, and 40. By this representation, the output $y(t)$ due to an input $x(t)$ is given by a sum of Volterra functionals as follows:

$$y(t) = y_0 + y_1(t) + y_2(t) + y_3(t) + \cdots \tag{3.1}$$

where

$$\begin{aligned} y_0 &= \text{constant} \\ y_1(t) &= \text{linear output} \\ y_2(t) &= \text{bilinear output} \\ y_3(t) &= \text{trilinear output} \end{aligned}$$

and so on. The mean value $\bar{y} = E[y(t)]$ satisfies

$$\bar{y} = y_0 + \bar{y}_1 + \bar{y}_2 + \bar{y}_3 + \cdots \tag{3.2}$$

Now

$$y(t) - \bar{y} = [y_1(t) - \bar{y}_1] + [y_2(t) - \bar{y}_2] + [y_3(t) - \bar{y}_3] + \cdots \tag{3.3}$$

In the frequency domain, by taking Fourier transforms of both sides of Eq. (3.3), one obtains

$$Y(f) = Y_1(f) + Y_2(f) + Y_3(f) + \cdots \tag{3.4}$$

where

$$Y(f) = \mathscr{F}[y(t) - \bar{y}] = \int_{-\infty}^{\infty} [y(t) - \bar{y}]e^{-j2\pi ft}\,dt \tag{3.5}$$

with

$$\begin{aligned} Y_1(f) &= \mathscr{F}[y_1(t) - \bar{y}_1] \\ Y_2(f) &= \mathscr{F}[y_2(t) - \bar{y}_2] \end{aligned} \tag{3.6}$$

and so on.

General ways will be developed in Section 3.1 to describe the separate quantities $y_1(t)$, $y_2(t)$, and $y_3(t)$ in the time domain, and their corresponding $Y_1(f)$, $Y_2(f)$, and $Y_3(f)$ in the frequency domain. This will be followed in Section 3.2 by examples deemed to be of physical importance. Without loss of generality, it will be assumed henceforth that $y_0 = 0$. Otherwise, results obtained here apply to $[y(t) - y_0]$ rather than to $y(t)$ by itself. Figure 3.1 illustrates the situation for third-order nonlinear input/output models where $y(t)$ does not go beyond $y_3(t)$. This is a single-input/single-output nonlinear model between the input $x(t)$ and the output $y(t)$.

In further analysis, to obtain closed-form answers, assume that $x(t)$ is a stationary (ergodic) random process such that any n random variables $x_i = x(t_i)$, $i = 1, 2, \ldots, n$, follow an n-dimensional Gaussian (normal) distribution with zero

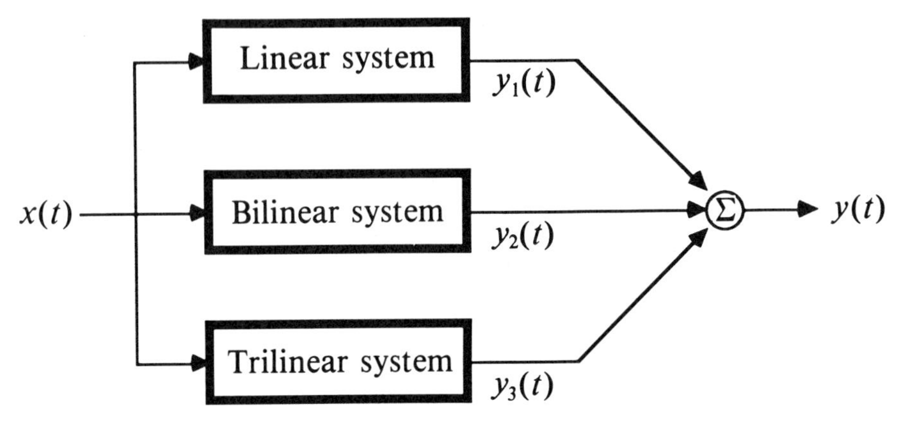

Figure 3.1 Third-order nonlinear input/output model.

mean values. Even-order moments of such random variables then reduce to products of second-order moments, and all odd-order moments are zero. It follows that the mean values

$$\bar{y}_1 = E[y_1(t)] = 0, \qquad \bar{y}_3 = E[y_3(t)] = 0, \qquad \bar{y}_2 = E[y_2(t)] \neq 0 \tag{3.7}$$

where $E[\ ]$ denotes an appropriate expected value operation.

Definitions and properties will now be discussed for linear, bilinear, and trilinear systems.

3.1.1 Linear Systems

By definition, as per Eq. (1.1), L_1 is a *linear system operator* with respect to inputs $\{x(t)\}$ if for all inputs $x_1(t)$, $x_2(t)$ and all constants a, b

$$L_1[ax_1(t) + bx_2(t)] = aL_1[x_1(t)] + bL_1[x_2(t)] \tag{3.8}$$

The operator L_1 has the additive and homogeneous properties of Eqs. (1.2) and (1.3).

$$\text{Additive} \qquad L_1[x_1(t) + x_2(t)] = L_1[x_1(t)] + L_1[x_2(t)]$$
$$\text{Homogeneous} \qquad L_1[ax(t)] = aL_1[x(t)] \tag{3.9}$$

From basic theory, let $h_1(\tau)$ be the response of the linear system L_1 at time τ to a delta function input $\delta(\tau)$, namely,

$$h_1(\tau) = L_1[\delta(\tau)] \tag{3.10}$$

The quantity $h_1(\tau)$ is the *linear weighting function* of the system or the *unit impulse response function*. It is also called the *first-order time-domain kernel.*

From properties of the delta function, any arbitrary input $x(t)$ can be described by the integral

$$x(t) = \int_{-\infty}^{\infty} x(t-\tau)\delta(\tau)\,d\tau \tag{3.11}$$

By letting $d\tau = \Delta\tau$ and $\tau = \tau_n = n\Delta\tau$ for all integers n, this can be approximated by the summation

$$x(t) \approx \sum_n x(t-\tau_n)\delta(\tau_n)\Delta\tau \tag{3.12}$$

Now, from properties of L_1 and Eq. (3.10), it follows that

$$L_1[x(t)] \approx \sum_n x(t-\tau_n)h_1(\tau_n)\Delta\tau$$

where

$$h_1(\tau_n) = L_1[\delta(\tau_n)]$$

Hence, in the limit as $\Delta\tau \to 0$, for any input $x(t)$, one obtains the linear system output response of Eq. (1.11):

$$y_1(t) = L_1[x(t)] = \int_{-\infty}^{\infty} x(t-\tau)h_1(\tau)\,d\tau \tag{3.13}$$

From the assumption that $x(t)$ has a zero mean value for all t where $\bar{x} = E[x(t)] = 0$, it follows that the linear output $y_1(t)$ also has a zero mean value for all t.

$$\bar{y}_1 = E[y_1(t)] = \int_{-\infty}^{\infty} h_1(\tau)E[x(t-\tau)]\,d\tau = 0 \tag{3.14}$$

As stated in Eq. (1.14), a linear system is said to be *stable* if every bounded input $x(t)$ produces a bounded output $y_1(t)$. A necessary and sufficient condition for a linear system to be stable is

$$\int_{-\infty}^{\infty} |h_1(\tau)|\,d\tau < \infty \tag{3.15}$$

The function $h_1(\tau)$ gives the response at time t by weighting inputs $x(t)$ occurring at time $(t - \tau)$. Hence, if the system L_1 is *physically realizable*, where it responds only to past inputs that have already occurred, then $t \geqslant (t - \tau)$, which is the same as $\tau \geqslant 0$. Thus, for physical realizability, the weighting function must satisfy

$$h_1(\tau) = 0 \qquad \text{for } \tau < 0 \tag{3.16}$$

For these situations, Eq. (3.13) is equivalent to Eq. (1.10), namely,

$$y_1(t) = \int_0^{\infty} x(t - \tau)h_1(\tau)\,d\tau \tag{3.17}$$

Without loss of generality, Eq. (3.13) will be used henceforth to describe linear systems, whether or not Eq. (3.16) applies. When measuring real data, only finite limits will be involved. The theoretical limits from minus infinity to plus infinity will be omitted in the following integrals to simplify the notation.

The calculation of $y_1(t)$ from $x(t)$ via the convolution integral expression of Eq. (3.13) is shown in Figure 3.2.

Consider now the Fourier transform of $h_1(\tau)$ denoted by

$$H_1(f) = \mathscr{F}[h_1(\tau)] = \int h_1(\tau)e^{-j2\pi f\tau}\,d\tau \tag{3.18}$$

This quantity $H_1(f)$ is the *linear frequency response function* of the system. It is also called the *first-order frequency-domain kernel* of the system. A sufficient condition for $H_1(f)$ to exist is that the linear system be stable. Fourier transforms of both sides of Eq. (3.13) give the simple frequency domain result of Eq. (1.15),

$$Y_1(f) = \mathscr{F}[y_1(t)] = H_1(f)X(f) \tag{3.19}$$

from which one can express $y_1(t)$ as the inverse Fourier transform of $Y_1(f)$, namely,

$$y_1(t) = \mathscr{F}^{-1}[Y_1(f)] = \int Y_1(f)e^{j2\pi ft}\,df \tag{3.20}$$

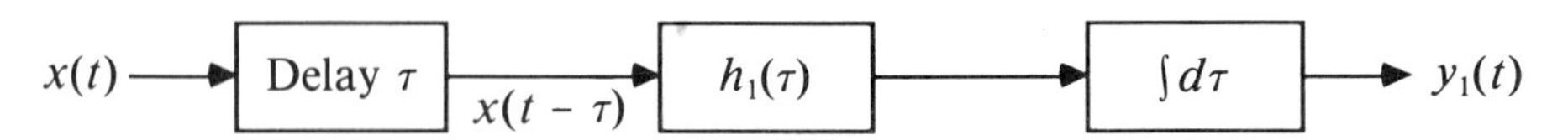

Figure 3.2 Convolution calculation of linear system output.

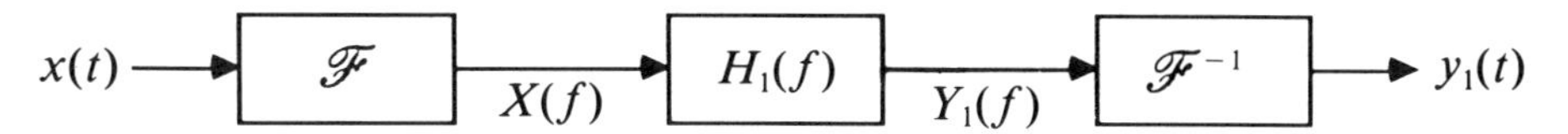

Figure 3.3 Fourier transform calculation of linear system output.

Figure 3.3 shows the calculation of $y_1(t)$ from $x(t)$ using Fourier transform procedures by Eq. (3.19), where $\mathscr{F}$ represents a suitable Fourier transform operation and $\mathscr{F}^{-1}$ denotes the corresponding inverse operation. This diagram describes the linear system shown in Figure 3.1. Note that $y_1(t) = h_1(t)$ when $x(t) = \delta(t)$. This is the same as having $Y_1(f) = H_1(f)$ when $X(f) = 1$. When the input $x(t) = \exp(j2\pi ft)$, the linear output $y_1(t) = H_1(f)x(t)$. This proves that sinusoidal frequencies are preserved under linear transformations and provides a simple way to determine $H_1(f)$ when $x(t)$ and $y_1(t)$ are input/output terms in a linear differential equation.

3.1.2 Bilinear Systems

By definition, L_2 is a *bilinear system operator* with respect to a pair of ordered inputs $\{x(t), y(t)\}$ if the operator L_2 is linear on $x(t)$ for a given $y(t)$, and conversely. In equation form, for any constants a, b, it is required that

$$\begin{aligned} L_2(x_1 + x_2, y) &= L_2(x_1, y) + L_2(x_2, y) \\ L_2(x, y_1 + y_2) &= L_2(x, y_1) + L_2(x, y_2) \end{aligned} \tag{3.21}$$

$$\begin{aligned} L_2(ax, y) &= aL_2(x, y) \\ L_2(x, by) &= bL_2(x, y) \\ L_2(ax, by) &= abL_2(x, y) \end{aligned} \tag{3.22}$$

where x represents $x(t)$ and y represents $y(t)$. Now

$$\begin{aligned} L_2(ax + by, ax + by) &= L_2(ax + by, ax) + L_2(ax + by, by) \\ &= L_2(ax, ax) + L_2(by, ax) + L_2(ax, by) + L_2(by, by) \\ &= a^2 L_2(x, x) + abL_2(y, x) + abL_2(x, y) + b^2 L_2(y, y) \end{aligned} \tag{3.23}$$

Without loss of generality, one can assume that $L_2(x, y)$ is symmetric with respect to x and y, namely,

$$L_2(x, y) = L_2(y, x) \tag{3.24}$$

Then Eq. (3.23) becomes

$$L_2(ax + by, ax + by) = a^2 L_2(x, x) + 2abL_2(x, y) + b^2 L_2(y, y) \tag{3.25}$$

To obtain Eq. (3.24) if $L_2(x, y) \neq L_2(y, x)$, one should replace $L_2(x, y)$ by $\tilde{L}_2(x, y)$ defined by

$$\tilde{L}_2(x, y) = \tfrac{1}{2}[L_2(x, y) + L_2(y, x)] \tag{3.26}$$

Now

$$\tilde{L}_2(x, y) = \tilde{L}_2(y, x) \tag{3.27}$$

so that $\tilde{L}_2(x, y)$ is symmetric with respect to x and y. The symmetric property of Eq. (3.24) will be assumed henceforth.

Consider the special case where $b = 0$. This gives

$$L_2(ax, ax) = a^2 L_2(x, x) \tag{3.28}$$

If one now defines a special bilinear system operator by

$$L_2(x) \equiv L_2(x, x) \tag{3.29}$$

then Eq. (3.28) shows that

$$L_2(ax) = a^2 L_2(x) \tag{3.30}$$

Hence, L_2 is a *nonlinear* operator on $x(t)$, since the homogeneous property of Eq. (3.9) is violated. Equation (3.30) produces an amplitude nonlinearity. Note also from Eqs. (3.24), (3.25), and (3.29) that

$$\begin{aligned} L_2(x_1 + x_2) &\equiv L_2(x_1 + x_2, x_1 + x_2) \\ &= L_2(x_1, x_1) + 2L_2(x_1, x_2) + L_2(x_2, x_2) \\ &= L_2(x_1) + 2L_2(x_1, x_2) + L_2(x_2) \end{aligned} \tag{3.31}$$

Thus, the additive property of Eq. (3.9) is violated whenever $L_2(x_1, x_2) \neq 0$.

The bilinear transformation $L_2(x_1, x_2)$ can generate sum and difference frequencies. To see this property, suppose $L_2(x_1, x_2) = x_1 x_2$ where $x_1 = \cos at$ and $x_2 = \cos bt$. Then

$$L_2(x_1, x_2) = \cos at \cos bt = (1/2)[\cos(a + b)t + \cos(a - b)t]$$

Here, the term $\cos(a - b)t$ produces an envelope nonlinearity.

From Volterra theory, let $h_2(\tau_1, \tau_2)$ be the response of the bilinear system L_2 at times τ_1, τ_2 to a pair of unit impulse function inputs $\{\delta(\tau_1), \delta(\tau_2)\}$, namely,

$$h_2(\tau_1, \tau_2) = L_2[\delta(\tau_1), \delta(\tau_2)] \tag{3.32}$$

The quantity $h_2(\tau_1, \tau_2)$ is called the *bilinear weighting function* of the system. It is also called the *second-order time-domain kernel.* From Eq. (3.24), observe that $h_2(\tau_1, \tau_2)$ is unique and satisfies

$$h_2(\tau_1, \tau_2) = h_2(\tau_2, \tau_1) \tag{3.33}$$

In words, $h_2(\tau_1, \tau_2)$ is symmetric with respect to its two variables.

As in Eq. (3.12), let $x(t)$ be approximated by

$$x(t) \approx \sum_n x(t - \tau_n)\delta(\tau_n)\Delta\tau = \sum_n a_n \delta(\tau_n)$$

where

$$a_n = x(t - \tau_n)\,\Delta\tau = x(t - n\,\Delta\tau)\Delta\tau$$

Now

$$\begin{aligned} L_2[x(t)] &\equiv L_2[x(t), x(t)] \\ &\approx L_2\left[\sum_m a_m \delta(\tau_m), \sum_n a_n \delta(\tau_n)\right] \end{aligned}$$

using different indices of summation for proper accounting. Properties of L_2 and Eq. (3.32) give

$$\begin{aligned} L_2[x(t)] &\approx \sum_m \sum_n a_m a_n L_2[\delta(\tau_m), \delta(\tau_n)] \\ &\approx \sum_m \sum_n x(t - \tau_m)x(t - \tau_n)h_2(\tau_m, \tau_m)\Delta\tau_1\,\Delta\tau_2 \end{aligned}$$

with $\tau_m = m\,\Delta\tau_1$ amd $\tau_n = n\,\Delta\tau_2$. Hence, in the limit as $\Delta\tau_1 \to 0$ and $\Delta\tau_2 \to 0$, for any input $x(t)$, one obtains the bilinear system output response $y_2(t)$ given by

$$y_2(t) = L_2[x(t)] = \iint x(t - \tau_1)x(t - \tau_2)h_2(\tau_1, \tau_2)\,d\tau_1\,d\tau_2 \tag{3.34}$$

where the possible double limits of integration are from $-\infty$ to $+\infty$. Equation (3.34) is the bilinear extension to the linear result of Eq. (3.13).

From Eq. (3.34) it is simple to verify Eqs. (3.30) and (3.31) that

$$\begin{aligned} L_2[ax(t)] &= a^2 L_2[x(t)] \\ L_2[x_1(t) + x_2(t)] &= L_2[x_1(t)] + 2L_2[x_1(t), x_2(t)] + L_2[x_2(t)] \end{aligned}$$

where

$$L_2[x_1(t), x_2(t)] = \iint x_1(t - \tau_1)x_2(t - \tau_2)h_2(\tau_1, \tau_2)\, d\tau_1\, d\tau_2 \tag{3.35}$$

Thus L_2, as defined by Eq. (3.34), is *not* a linear operator on $x(t)$. It is clear also from Eq. (3.34) that replacing $h_2(\tau_1, \tau_2)$ by $h_2(\tau_2, \tau_1)$ gives the identical result for $y_2(t)$. Finally, if the bilinear system L_2 is *physically realizable* where it responds only to past inputs, then

$$h_2(\tau_1, \tau_2) = 0 \qquad \text{for } \tau_1 < 0 \text{ and/or } \tau_2 < 0 \tag{3.36}$$

so that the lower limits in Eqs. (3.34) and (3.35) become zero instead of minus infinity. In words, the quantity $h_2(\tau_1, \tau_2)$ exists only in the first quadrant for physically realizable bilinear systems.

A bilinear system is said to be *stable* if every bounded input $x(t)$ produces a bounded output $y_2(t)$. A sufficient condition for a bilinear system to be stable is

$$\iint |h_2(\tau_1, \tau_2)|\, d\tau_1\, d\tau_2 < \infty \tag{3.37}$$

Equation (3.37) should be compared with Eq. (3.15).

For a delta function input, Eq. (3.34) gives the output

$$y_2(t) = L_2[\delta(t)] = h_2(t, t) \tag{3.38}$$

This describes the bilinear weighting function $h_2(\tau_1, \tau_2)$ only along the line $\tau_1 = \tau_2 = t$. The general description of $h_2(\tau_1, \tau_2)$ requires Eq. (3.32). To obtain this result, one can use Eq. (3.31) solved for $L_2[x_1, x_2]$.

$$L_2[x_1, x_2] = \tfrac{1}{2}(L_2[x_1 + x_2] - L_2[x_1] - L_2[x_2]) \tag{3.39}$$

As a special case of Eq. (3.39), let $x_1 = \delta(\tau_1)$ and $x_2 = \delta(\tau_2)$. Now, from Eq. (3.32),

$$h_2(\tau_1, \tau_2) = \tfrac{1}{2}(L_2[\delta(\tau_1) + \delta(\tau_2)] - L_2[\delta(\tau_1)] - L_2[\delta(\tau_2)]) \tag{3.40}$$

This calculation is pictured in Figure 3.4. Note that

$$L_2(x_1, 0) = L_2(0, x_2) = 0 \tag{3.41}$$

when either $x_2 = 0$ or $x_1 = 0$.

The calculation of $y_2(t)$ from $x(t)$ via the double convolution integral expression of Eq. (3.34) is shown in Figure 3.5.

Let $R_{xx}(\tau)$ be the autocorrelation function of $x(t)$ as defined by

$$R_{xx}(\tau) = E[x(t)x(t + \tau)] \tag{3.42}$$

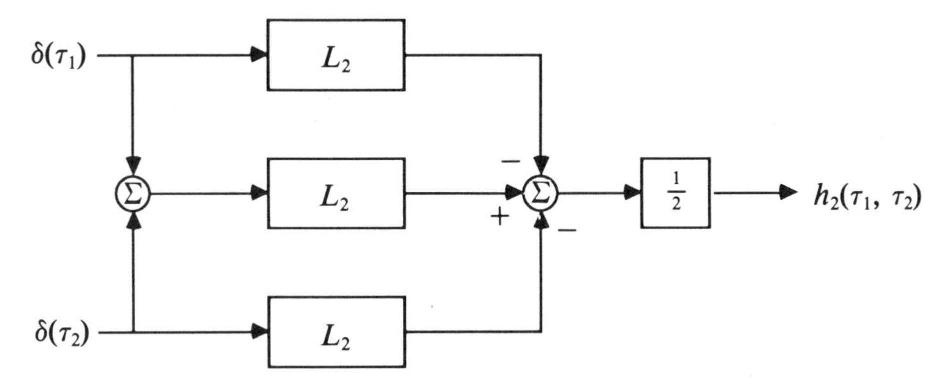

Figure 3.4 Calculation of bilinear weighting function.

and let $S_{xx}(f)$ be its associated two-sided autospectral density function of $x(t)$ satisfying

$$S_{xx}(f) = \int R_{xx}(\tau)e^{-j2\pi f\tau}\,d\tau \tag{3.43}$$

Then, from Eq. (3.34), since $R_{xx}(\tau) = R_{xx}(-\tau)$, the bilinear output mean value for all t is

$$\bar{y}_2 = E[y_2(t)] = \iint h_2(\tau_1, \tau_2)R_{xx}(\tau_2 - \tau_1)\,d\tau_1\,d\tau_2 \tag{3.44}$$

In general, $\bar{y}_2 \neq 0$.

Consider now the double Fourier transform of $h_2(\tau_1, \tau_2)$ denoted by

$$H_2(f_1, f_2) = \iint h_2(\tau_1, \tau_2)e^{-j2\pi(f_1\tau_1 + f_2\tau_2)}\,d\tau_1\,d\tau_2 \tag{3.45}$$

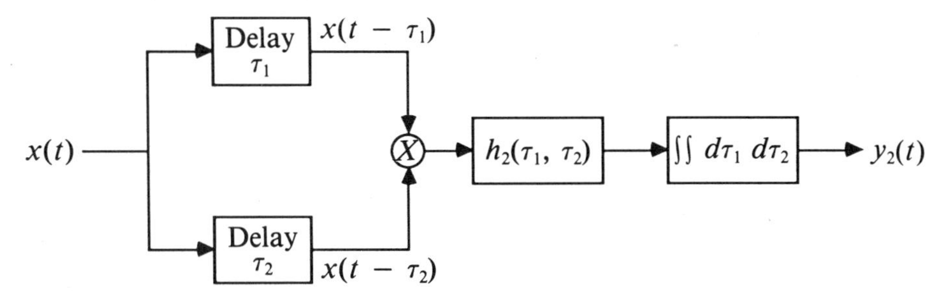

Figure 3.5 Double convolution calculation of bilinear system output.

This quantity $H_2(f_1, f_2)$ is the *bilinear frequency response function* and is called also the *second-order frequency-domain kernel.* It is symmetric with respect to its two frequency variables because of Eq. (3.33).

$$H_2(f_1, f_2) = H_2(f_2, f_1) \tag{3.46}$$

A sufficient condition for $H_2(f_1, f_2)$ to exist is that the bilinear system be stable, Eq. (3.37). From Eq. (3.45), since $h_2(\tau_1, \tau_2)$ is real-valued, it follows that the complex conjugate quantity

$$H_2^*(f_1, f_2) = H_2(-f_1, -f_2)$$

Then, using Eq. (3.46),

$$H_2^*(f, -f) = H_2(-f, f) = H_2(f, -f) \tag{3.47}$$

This proves that $H_2(f, -f)$ is a real-valued even function of f with a phase function that is zero for all f.

The inverse relation to Eq. (3.43) gives

$$R_{xx}(\tau_2 - \tau_1) = \int S_{xx}(f) e^{j2\pi f(\tau_2 - \tau_1)}\, df$$

Substitution into Eq. (3.44) shows that the bilinear output mean value is

$$\begin{aligned} \bar{y}_2 &= \int S_{xx}(f) \left[\iint h_2(\tau_1, \tau_2) e^{-j2\pi(f\tau_1 - f\tau_2)}\, d\tau_1\, d\tau_2 \right] df \\ &= \int H_2(f, -f) S_{xx}(f)\, df \end{aligned} \tag{3.48}$$

where $H_2(f, -f)$ can be found from Eq. (3.45). Thus, $\bar{y}_2$ can be determined by Eq. (3.48) in place of Eq. (3.44). This also shows that $H_2(f, -f)$ must be a real-valued even function of f.

Return to Eq. (3.34). Express $x(t - \tau_1)$ and $x(t - \tau_2)$ in terms of their inverse Fourier transforms. Then, using Eq. (3.45),

$$\begin{aligned} y_2(t) &= \iint h_2(\tau_1, \tau_2) \left[\int X(\alpha) e^{j2\pi\alpha(t - \tau_1)}\, d\alpha \right] \left[\int X(\beta) e^{j2\pi\beta(t - \tau_2)}\, d\beta \right] d\tau_1\, d\tau_2 \\ &= \iint \left[\iint h_2(\tau_1, \tau_2) e^{-j2\pi(\alpha\tau_1 + \beta\tau_2)}\, d\tau_1\, d\tau_2 \right] X(\alpha) X(\beta) e^{j2\pi(\alpha + \beta)t}\, d\alpha\, d\beta \\ &= \iint H_2(\alpha, \beta) X(\alpha) X(\beta) e^{j2\pi(\alpha + \beta)t}\, d\alpha\, d\beta \end{aligned}$$

Change variables from (α, β) to (α, f) where $f = \alpha + \beta$, $df = d\beta$. Now

$$y_2(t) = \iint H_2(\alpha, f - \alpha)X(\alpha)X(f - \alpha)e^{j2\pi ft}\, d\alpha\, df \tag{3.49}$$

But

$$y_2(t) = \int \mathscr{F}[y_2(t)]e^{j2\pi ft}\, df \tag{3.50}$$

Hence, one obtains the result

$$\mathscr{F}[y_2(t)] = \int H_2(\alpha, f - \alpha)X(\alpha)X(f - \alpha)\, d\alpha \tag{3.51}$$

Equation (3.51) has the form of a single convolution integral.

From Equation (3.6), the quantity $Y_2(f)$ is defined by

$$Y_2(f) = \mathscr{F}[y_2(t) - \bar{y}_2] = \mathscr{F}[y_2(t)] - \mathscr{F}[\bar{y}_2] \tag{3.52}$$

Note that $Y_2(f) \neq \mathscr{F}[y_2(t)]$ when $\mathscr{F}[\bar{y}_2] \neq 0$. The Fourier transform of the constant $\bar{y}_2$ is given by

$$\mathscr{F}[\bar{y}_2] = \bar{y}_2\delta(f) \approx \bar{y}_2\delta_1(f) \tag{3.53}$$

where $\delta_1(f)$ is the *finite delta function* defined by

$$\delta_1(f) = \begin{cases} T, & (-1/2T) < f < (1/2T) \\ 0, & \text{otherwise} \end{cases} \tag{3.54}$$

The quantity $\delta_1(f)$ is an approximation to the usual theoretical delta function and occurs in digital computation of Fourier transforms with subrecords of finite length T. Properties of $\delta_1(f)$ will often be used instead of $\delta(f)$ throughout this book to obtain more realistic results. From Eqs. (3.51)–(3.54), it follows that

$$Y_2(f) = \int H_2(\alpha, f - \alpha)X(\alpha)X(f - \alpha)\, d\alpha - \bar{y}_2\delta_1(f) \tag{3.55}$$

Note that $Y_2(f) = \mathscr{F}[y_2(t)]$ for all $f \neq 0$. Equation (3.55) should be compared with Eq. (3.19).

Figure 3.6 shows the calculation of $y_2(t)$ from $x(t)$ using Fourier transform procedures as required by Eqs. (3.50) and (3.51). This diagram describes the bilinear system shown in Figure 3.1. Note that $y_2(t)$ involves a multiplication of two quantities. Hence, the bilinear output $y_2(t)$ will contain mixing frequencies

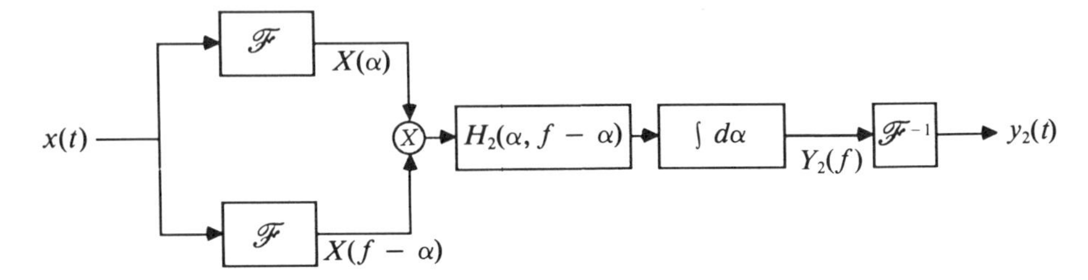

Figure 3.6 Fourier transform calculation of bilinear system output.

when the input contains frequencies f_1 and f_2. When $x(t) = \exp(j2\pi f_1 t) + \exp(j2\pi f_2 t)$, Eqs. (3.34) and (3.45) show that the bilinear output $y_2(t)$ contains the term $H_2(f_1, f_2)\exp[j2\pi(f_1 + f_2)t]$. This input/output pair can be used to determine $H_2(f_1, f_2)$ in a nonlinear differential equation.

3.1.3 Trilinear Systems

Previous ideas extend to trilinear systems (and to higher-order systems) as will now be carried out for trilinear systems. By definition, L_3 is a *trilinear system operator* with respect to a triplet of ordered inputs $\{x(t), y(t), z(t)\}$ if the operator L_3 is linear on $x(t)$ for given $y(t)$ and $z(t)$, and similarly for other combinations. In equation form, for any constants a, b, c, one has

$$\begin{aligned} L_3(a_1x_1 + a_2x_2, y, z) &= a_1L_3(x_1, y, z) + a_2L_3(x_2, y, z) \\ L_3(x, b_1y_1 + b_2y_2, z) &= b_1L_3(x, y_1, z) + b_2L_3(x, y_2, z) \\ L_3(x, y, c_1z_1 + c_2z_2) &= c_1L_3(x, y, z_1) + c_2L_3(x, y, z_2) \end{aligned} \tag{3.56}$$

This then leads without loss of generality to

$$\begin{aligned} &L_3(ax + by + cz, ax + by + cz, ax + by + cz) \\ &\quad = a^3L_3(x, x, x) + b^3L_3(y, y, y) + c^3L_3(z, z, z) \\ &\quad\quad + 3a^2bL_3(x, x, y) + 3ab^2L_3(x, y, y) + 3a^2cL_3(x, x, z) \\ &\quad\quad + 3ac^2L_3(x, z, z) + 3b^2cL_3(y, y, z) + 3bc^2L_3(y, z, z) \\ &\quad\quad + 6abcL_3(x, y, z) \end{aligned} \tag{3.57}$$

where Eq. (3.57) uses the assumption that $L_3(x, y, z)$ is symmetric with respect to interchange of variables, namely,

$$\begin{aligned} L_3(x, y, z) &= L_3(x, z, y) = L_3(y, x, z) = L_3(y, z, x) \\ &= L_3(z, x, y) = L_3(z, y, x) \end{aligned} \tag{3.58}$$

For the special case where $b = c = 0$, Eq. (3.57) becomes

$$L_3(ax, ax, ax) = a^3 L_3(x, x, x) \tag{3.59}$$

As per Eq. (3.29), define a spectral trilinear system operator by

$$L_3(x) \equiv L_3(x, x, x) \tag{3.60}$$

Then, Eq. (3.59) shows that $L_3(x)$ is not homogeneous, since

$$L_3(ax) = a^3 L_3(x) \tag{3.61}$$

Also, from Eqs. (3.60) and (3.57), $L_3(x)$ is not additive, since

$$\begin{aligned} L_3(x_1 + x_2) &\equiv L_3(x_1 + x_2, x_1 + x_2, x_1 + x_2) \\ &= L_3(x_1) + 3L_3(x_1, x_1, x_2) + 3L_3(x_1, x_2, x_2) + L_3(x_2) \end{aligned} \tag{3.62}$$

Thus, L_3 like L_2 is a *nonlinear* operator on $x(t)$.

From Volterra theory, the *trilinear weighting function* $h_3(\tau_1, \tau_2, \tau_3)$, also called the *third-order time-domain kernel*, is defined as the response of the trilinear system L_3 at times τ_1, τ_2, τ_3 to a triplet of delta function inputs $\{\delta(\tau_1), \delta(\tau_2), \delta(\tau_3)\}$, namely,

$$h_3(\tau_1, \tau_2, \tau_3) = L_3[\delta(\tau_1), \delta(\tau_2), \delta(\tau_3)] \tag{3.63}$$

From Eq. (3.58), $h_3(\tau_1, \tau_2, \tau_3)$ is unique and is symmetric with respect to its three variables. As per the derivations of Eqs. (3.13) and (3.34), one can now prove that for any input $x(t)$, the trilinear system output response $y_3(t)$ is given by

$$y_3(t) = L_3[x(t)] = \iiint x(t - \tau_1)x(t - \tau_2)x(t - \tau_3)h_3(\tau_1, \tau_2, \tau_3)\, d\tau_1\, d\tau_2\, d\tau_3 \tag{3.64}$$

where the possible triple limits of integration are from $-\infty$ to $+\infty$. Equation (3.64) is the trilinear extension to the linear result of Eq. (3.13) and to the bilinear result of Eq. (3.34).

For a delta function input, Eq. (3.64) gives the output

$$y_3(t) = L_3[\delta(t)] = h_3(t, t, t) \tag{3.65}$$

This describes the trilinear weighting function only along the line $\tau_1 = \tau_2 = \tau_3 = t$. The general description of $h_3(\tau_1, \tau_2, \tau_3)$ requires Eq. (3.63).

If the system L_3 is *physically realizable*, then

$$h_3(\tau_1, \tau_2, \tau_3) = 0$$
$$\text{for } \tau_1 < 0 \quad \text{and/or} \quad \tau_2 < 0 \quad \text{and/or} \quad \tau_3 < 0 \tag{3.66}$$

so that the lower limits in Eq. (3.64) become zero instead of minus infinity. In words, $h_3(\tau_1, \tau_2, \tau_3)$ exists only in the first octant for physically realizable trilinear systems. From the Gaussian assumption made on $x(t)$, it follows from Eq. (3.64) that the trilinear output mean value for all t is zero.

$$\bar{y}_3 = E[y_3(t)] = 0 \tag{3.67}$$

A trilinear system is said to be *stable* if every bounded input $x(t)$ produces a bounded output $y_3(t)$. A sufficient condition for a trilinear system to be stable is

$$\iiint |h_3(\tau_1, \tau_2, \tau_3)|\, d\tau_1\, d\tau_2\, d\tau_3 < \infty \tag{3.68}$$

Equation (3.68) should be compared with Eqs. (3.15) and (3.37).

The calculation of $y_3(t)$ from $x(t)$ via the triple convolution integral expression of Eq. (3.64) is shown in Figure 3.7.

Consider now the triple Fourier transform of $h_3(\tau_1, \tau_2, \tau_3)$ denoted by

$$H_3(f_1, f_2, f_3) = \iiint h_3(\tau_1, \tau_2, \tau_3) e^{-j2\pi(f_1\tau_1 + f_2\tau_2 + f_3\tau_3)}\, d\tau_1\, d\tau_2\, d\tau_3 \tag{3.69}$$

This quantity is the *trilinear frequency response function* and is also called the *third-order frequency-domain kernel.* From Eqs. (3.58) and (3.63) this kernel is symmetric with respect to its three variables. A sufficient condition for $H_3(f_1, f_2, f_3)$ to exist is that the trilinear system be stable, Eq. (3.68). A derivation similar to obtaining Eq. (3.51) proves the result

$$Y_3(f) = \mathscr{F}(y_3(t)) = \iint H_3(\alpha, \beta - \alpha, f - \beta) X(\alpha) X(\beta - \alpha) X(f - \beta)\, d\alpha\, d\beta \tag{3.70}$$

This should be compared with Eqs. (3.19) and (3.55). Note that Eq. (3.70) has the form of a double convolution integral.

Figure 3.8 shows the calculation of $y_3(t)$ from $x(t)$ using Fourier transform procedures as required by Eq. (3.70). This diagram describes the trilinear system

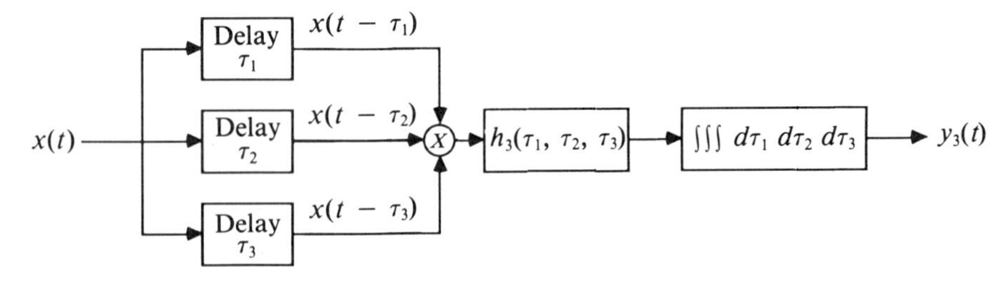

Figure 3.7 Triple convolution calculation of trilinear system output.

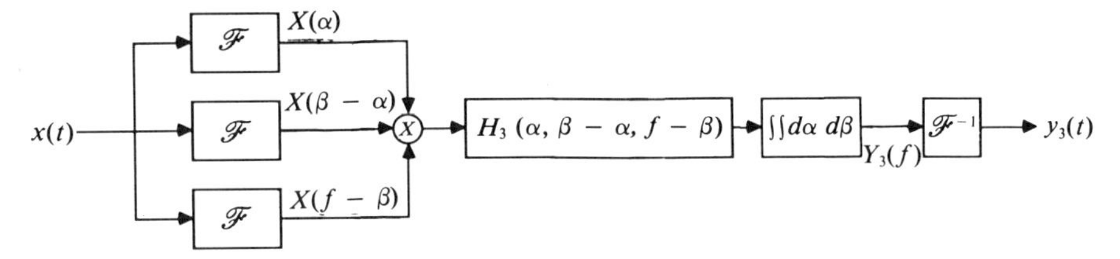

Figure 3.8 Fourier transform calculation of trilinear system output.

shown in Figure 3.1. The quantity $y_3(t)$ involves a multiplication of three quantities. Hence, the trilinear output $y_3(t)$ will contain mixing frequencies when the input contains frequencies f_1, f_2, and f_3. When $x(t) = \exp(j2\pi f_1 t) + \exp(j2\pi f_2 t) + \exp(j2\pi f_3 t)$, Eqs. (3.64) and (3.69) show that the trilinear output $y_3(t)$ contains the term $H_3(f_1, f_2, f_3)\exp[j2\pi(f_1 + f_2 + f_3)t]$. This input/output pair can be used to determine $H_3(f_1, f_2, f_3)$ in a nonlinear differential equation.

Important special cases of bilinear and trilinear systems are treated in Section 3.2, where Figures 3.5–3.8 become relatively simple. Specifically, the bilinear systems are assumed to be squaring operations that are either followed or preceded by linear systems, and the trilinear systems are assumed to be cubing operations that are either followed or preceded by linear systems. For these cases, the bilinear and trilinear frequency response functions can be replaced by linear frequency response functions that are the basis for the new nonlinear system analysis and identification procedures in Chapters 5–7.

3.1.4 Summary of Results

A brief summary follows of various basic quantities and formulas that indicate the inner mathematical structure as one proceeds from linear to bilinear to trilinear systems. All limits of integration are theoretically from minus infinity to plus infinity unless stated otherwise. The first three Volterra functionals in Eq. (3.1) are given by

$$y_1(t) = \int h_1(\tau)x(t - \tau)\,d\tau \tag{3.71}$$

$$y_2(t) = \iint h_2(\tau_1, \tau_2)x(t - \tau_1)x(t - \tau_2)\,d\tau_1\,d\tau_2 \tag{3.72}$$

$$y_3(t) = \iiint h_3(\tau_1, \tau_2, \tau_3)x(t - \tau_1)x(t - \tau_2)x(t - \tau_3)\,d\tau_1\,d\tau_2\,d\tau_3 \tag{3.73}$$

with $\bar{y}_1 = \bar{y}_3 = 0$ and $\bar{y}_2$ given by Eqs. (3.44) and (3.48). The time-domain quantities $h_1(\tau)$, $h_2(\tau_1, \tau_2)$, and $h_3(\tau_1, \tau_2, \tau_3)$ are called the *first*, *second*, and *third-order time-domain kernels*, respectively. They are also called the *linear*, *bilinear*, and *trilinear weighting functions*, respectively. Their corresponding single, double, and triple Fourier transforms are given by

$$H_1(f) = \int h_1(\tau)e^{-j2\pi f\tau}\,d\tau \tag{3.74}$$

$$H_2(f_1, f_2) = \iint h_2(\tau_1, \tau_2)e^{-j2\pi(f_1\tau_1 + f_2\tau_2)}\,d\tau_1\,d\tau_2 \tag{3.75}$$

$$H_3(f_1, f_2, f_3) = \iiint h_3(\tau_1, \tau_2, \tau_3)e^{-j2\pi(f_1\tau_1 + f_2\tau_2 + f_3\tau_3)}\,d\tau_1\,d\tau_2\,d\tau_3 \tag{3.76}$$

These frequency-domain quantities $H_1(f)$, $H_2(f_1,f_2)$, and $H_3(f_1,f_2,f_3)$ are called the *first*, *second*, and *third-order frequency-domain kernels*, respectively. They are also called the *linear*, *bilinear*, and *trilinear frequency response functions*, respectively.

Fourier transforms of Eqs. (3.71)–(3.73) with mean values satisfying Eq. (3.7) yield the formulas for the first three Fourier transform functionals in Eq. (3.4).

$$Y_1(f) = H_1(f)X(f) \tag{3.77}$$

$$Y_2(f) = \int H_2(\alpha, f - \alpha)X(\alpha)X(f - \alpha)\,d\alpha - \bar{y}_2\delta_1(f) \tag{3.78}$$

$$Y_3(f) = \iint H_3(\alpha, \beta - \alpha, f - \beta)X(\alpha)X(\beta - \alpha)X(f - \beta)\,d\alpha\,d\beta \tag{3.79}$$

Results from Eqs. (3.77)–(3.79) can return to Eqs. (3.71)–(3.73), respectively, by taking their inverse Fourier transforms. Thus, the same information is available in either the time domain or the frequency domain. In general, however, the analysis will be simpler when conducted in the frequency domain.

The above formulas describe the output terms in the general third-order nonlinear input/output model of Figure 3.1. Advantages of these formulas are that they are mathematically exact and widely applicable to many problems. Disadvantages are that the Volterra multidimensional bilinear and trilinear weighting functions and frequency response functions are usually very expensive and time-consuming to compute as well as very difficult to interpret.

3.2 EXAMPLES

A number of engineering examples follow of linear, bilinear, and trilinear systems based upon formulas in Section 3.1. The bilinear and trilinear systems considered here are finite-memory nonlinear systems represented by zero-memory square-law systems and cubic systems that are preceded or followed by linear systems. The great importance of these particular examples will be shown in Chapter 5.

3.2.1 Linear Systems

Example 3.1 Low-Pass Filter

A low-pass filter can be defined by the linear weighting function

$$h_1(\tau) = Ae^{-a\tau}, \quad A, a > 0, \tau \geqslant 0$$
$$= 0, \qquad \tau < 0 \tag{3.80}$$

The associated linear frequency response function is

$$H_1(f) = \int_0^\infty h_1(\tau)e^{-j2\pi f\tau}\,d\tau = \frac{A}{a + j2\pi f}$$
$$= \frac{A(a - j2\pi f)}{a^2 + (2\pi f)^2} = |H_1(f)|e^{-j\phi_1(f)} \tag{3.81}$$

Typical plots are shown in Figure 3.9 for the *gain factor* $|H_1(f)|$ and *phase factor* $\phi_1(f)$ given by

$$|H_1(f)| = \frac{A}{\sqrt{a^2 + (2\pi f)^2}} \tag{3.82}$$

$$\phi_1(f) = \tan^{-1}(2\pi f/a) \tag{3.83}$$

Example 3.2 Force Input/Displacement Output System (SDOF System)

References 1 and 2 derive the linear frequency response function $H_1(f)$ for a simple single-degree-of-freedom (SDOF) mechanical system consisting of a

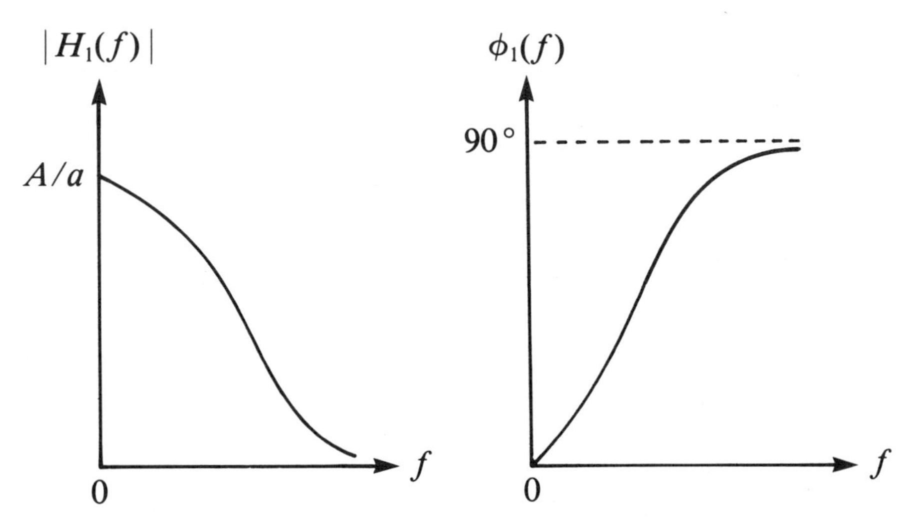

Figure 3.9 Typical gain factor and phase factor for low-pass filter.

mass, a spring with spring constant k, and a dashpot when a force is applied to the mass and the output is the resulting displacement of the mass, namely,

$$H_1(f) = \frac{1/k}{1 - (f/f_n)^2 + j2\zeta(f/f_n)} = H_1(f)|e^{-j\phi_1(f)} \tag{3.84}$$

Here, ζ is a dimensionless quantity called the *damping ratio* and f_n is the *undamped natural frequency*. Typical plots are shown in Figure 3.10 for the *gain factor* $|H_1(f)|$ and *phase factor* $\phi_1(f)$ given by

$$|H_1(f)| = \frac{1/k}{\sqrt{[1 - (f/f_n)^2]^2 + [2\zeta(f/f_n)]^2}} \tag{3.85}$$

$$\phi_1(f) = \tan^{-1}\left[\frac{2\zeta(f/f_n)}{1 - (f/f_n)^2}\right] \tag{3.86}$$

For a force input $x(t)$ with a Fourier transform $X(f)$ producing a displacement output $y_1(t)$ with Fourier transform $Y_1(f)$, one obtains

$$Y_1(f) = H_1(f)X(f) \tag{3.87}$$

using the $H_1(f)$ of Equation (3.84). This linear model is

Force Input $\longrightarrow$ $\boxed{H_1(f)}$ $\longrightarrow$ Displacement Output

Suppose the roles of $x(t)$ and $y_1(t)$ are reversed. This will give a *reverse linear model* where a displacement input $y_1(t)$ with Fourier transform $Y_1(f)$ produces a force output $x(t)$ with Fourier transform $X(f)$. Now, from Eq. (3.87),

$$X(f) = [H_1(f)]^{-1}Y_1(f) = A_1(f)Y_1(f) \tag{3.88}$$

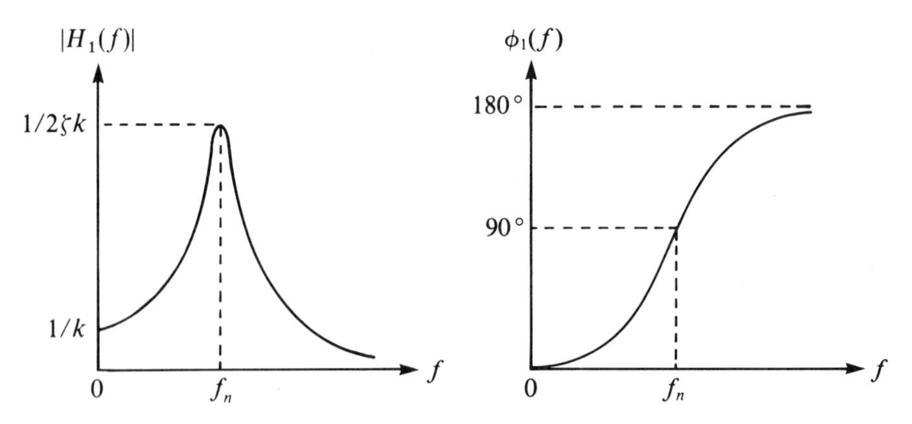

Figure 3.10 Typical gain factor and phase factor for SDOF system.

where

$$A_1(f) = [H_1(f)]^{-1} = k[1 - (f/f_n)^2 + j2\zeta(f/f_n)] \tag{3.89}$$

using the $H_1(f)$ of Eq. (3.84). Here the inverse system $A_1(f)$ exists so that a valid reverse linear model is obtained as sketched below.

Displacement Input ⟶ [$A_1(f)$] ⟶ Force Output

If the notation for $X(f)$ and $Y_1(f)$ is changed so that $X(f)$ represents the displacement input and $Y_1(f)$ represents the force output, then Eq. (3.88) takes the form of Eq. (3.87) with $H_1(f)$ replaced by $A_1(f)$, namely, $Y_1(f) = A_1(f)X(f)$. This transpose in roles preserves the usual meaning of keeping $x(t)$ and $X(f)$ for the input notation with $y_1(t)$ and $Y_1(f)$ for the output notation. Further use of this idea will be made in Section 7.6 in connection with nonlinear differential equations of motion.

3.2.2 Bilinear Systems

Example 3.3 Squarer (Square-Law) System

$x(t)$ ⟶ [Squarer] ⟶ $y_2(t) = x^2(t)$

The input $x(t)$ can be represented by

$$x(t) = \int \delta(\tau)x(t - \tau)\,d\tau \tag{3.90}$$

where $\delta(\tau)$ is the usual delta function. The squarer output $y_2(t)$ is then given by

$$\begin{aligned} y_2(t) = x^2(t) &= \left[\int \delta(\tau_1)x(t - \tau_1)\,d\tau_1\right]\left[\int \delta(\tau_2)x(t - \tau_2)\,d\tau_2\right] \\ &= \iint \delta(\tau_1)\delta(\tau_2)x(t - \tau_1)x(t - \tau_2)\,d\tau_1\,d\tau_2 \end{aligned} \tag{3.91}$$

From Eq. (3.34), the output $y_2(t)$ is also

$$y_2(t) = \iint h_2(\tau_1, \tau_2)x(t - \tau_1)x(t - \tau_2)\,d\tau_1\,d\tau_2 \tag{3.92}$$

Hence, the bilinear weighting function must be

$$h_2(\tau_1, \tau_2) = \delta(\tau_1)\delta(\tau_2) \tag{3.93}$$

Note that this function has no memory and exists only at $\tau_1 = \tau_2 = 0$. The associated bilinear frequency response function is a constant independent of frequency, namely,

$$H_2(f_1, f_2) = 1 \qquad \text{for all } f_1, f_2 \tag{3.94}$$

From Eq. (3.55), one obtains for $Y_2(f) = \mathscr{F}[y_2(t) - \bar{y}_2]$ the result

$$Y_2(f) = \int X(\alpha)X(f - \alpha)\,d\alpha - \bar{y}_2\delta_1(f) \tag{3.95}$$

with output mean value from Eq. (3.48) given by

$$\bar{y}_2 = \int S_{xx}(f)\,df = \sigma_x^2 \tag{3.96}$$

Consider a sinusoidal input at frequency f_0 where

$$x(t) = \sin(2\pi f_0 t) \tag{3.97}$$

From basic trigonometry,

$$y_2(t) = \sin^2(2\pi f_0 t) = \tfrac{1}{2} - \tfrac{1}{2}\cos 4\pi f_0 t \tag{3.98}$$

Thus, the output $y_2(t)$ contains a constant term (1/2), plus a nonlinear term at the sinusoidal $2f_0$ frequency. The output constant term shows that $y_2(t)$ will have a nonzero mean value. The output frequency $2f_0$ term shows that squaring operations always produce output frequencies that are twice the input frequencies.

Example 3.4 Squarer Preceded by Linear System

$$x(t) \longrightarrow \boxed{h(\tau)} \xrightarrow[u(t)]{} \boxed{\text{Squarer}} \longrightarrow y_2(t) = u^2(t)$$

This example involves memory operations before the squarer. The output is

$$\begin{aligned} y_2(t) = u^2(t) &= \left[\int h(\tau)x(t - \tau)\,d\tau\right]^2 \\ &= \left[\int h(\tau_1)x(t - \tau_1)\,d\tau_1\right]\left[\int h(\tau_2)x(t - \tau_2)\,d\tau_2\right] \\ &= \iint h(\tau_1)h(\tau_2)x(t - \tau_1)x(t - \tau_2)\,d\tau_1\,d\tau_2 \end{aligned} \tag{3.99}$$

From Eq. (3.34) the bilinear weighting function is here

$$h_2(\tau_1, \tau_2) = h(\tau_1)h(\tau_2) \tag{3.100}$$

Note that this bilinear weighting function is the product of two linear weighting functions.

From Eqs. (3.45) and (3.18), the associated bilinear frequency response function is

$$H_2(f_1, f_2) = H(f_1)H(f_2) \tag{3.101}$$

Note the product form of the linear frequency response functions.

As a specific case, suppose the linear system is a low-pass filter where $h(\tau)$ is a damped exponential function $Ae^{-a\tau}$. For this case, substituting Eqs. (3.80) and (3.81) into Eqs. (3.100) and (3.101) gives

$$\begin{aligned} h_2(\tau_1, \tau_2) &= A^2 e^{-a(\tau_1 + \tau_2)} \\ H_2(f_1, f_2) &= \frac{A^2}{(a + j2\pi f_1)(a + j2\pi f_2)} \end{aligned} \tag{3.102}$$

From Eq. (3.55), the Fourier transform $Y_2(f)$ is

$$Y_2(f) = \int H_2(\alpha, f - \alpha)X(a)X(f - \alpha)\,d\alpha - \bar{y}_2\delta_1(f) \tag{3.103}$$

where $H_2(\alpha, f - \alpha)$ is given here by

$$\begin{aligned} H_2(\alpha, f - \alpha) &= H(\alpha)H(f - \alpha) \\ &= \frac{A^2}{(a + j2\pi f\alpha)[a + j2\pi(f - \alpha)]} \end{aligned} \tag{3.104}$$

Also,

$$Y_2(f) = \int U(\alpha)U(f - \alpha)\,d\alpha - \bar{y}_2\delta_1(f) \tag{3.105}$$

where

$$U(\alpha) = H(\alpha)X(\alpha) = \frac{AX(\alpha)}{a + j2\pi f\alpha} \tag{3.106}$$

The output mean value $\bar{y}_2$ in this example is

$$\bar{y}_2 = E[y_2(t)] = E[u^2(t)] = \int S_{uu}(f)\,df \tag{3.107}$$

where for linear transformations defined by $H(f)$,

$$S_{uu}(f) = |H(f)|^2 S_{xx}(f) \tag{3.108}$$

Thus, for any $H(f)$ in Example 3.4,

$$\bar{y}_2 = \int |H(f)|^2 S_{xx}(f)\, df \tag{3.109}$$

From Eq. (3.48), a general formula for $\bar{y}_2$ is

$$\bar{y}_2 = \int H_2(f, -f) S_{xx}(f)\, df \tag{3.110}$$

This provides another way to derive Eq. (3.109) by substituting Eq. (3.101) into Eq. (3.110).

For the low-pass filter case of Eq. (3.102),

$$H_2(f, -f) = H(f)H(-f) = |H(f)|^2 = \frac{A^2}{a^2 + 4\pi^2 f^2} \tag{3.111}$$

Note that this $H_2(f, -f)$ satisfies the general property of being a real-valued even function of f. Here, for arbitrary $S_{xx}(f)$, Eq. (3.109) or (3.110) gives the output mean value

$$\bar{y}_2 = \int |H(f)|^2 S_{xx}(f)\, df = \int \frac{A^2 S_{xx}(f)}{a^2 + 4\pi^2 f^2}\, df \tag{3.112}$$

1. Suppose $S_{xx}(f) = 1$ for all f over $(-\infty, \infty)$. For this situation of a theoretical white noise input,

$$\bar{y}_2 = \int_{-\infty}^{\infty} \frac{A^2}{a^2 + 4\pi^2 f^2}\, df = \frac{A^2}{2a} \tag{3.113}$$

2. Suppose $S_{xx}(f) = 1$ for $-B \leqslant f \leqslant B$, and is zero outside. For this situation of a bandwidth-limited white noise input,

$$\bar{y}_2 = \int_{-B}^{B} \frac{A^2}{a^2 + 4\pi^2 f^2}\, df = \frac{A^2}{\pi a} \tan^{-1}\left(\frac{B}{a}\right) \tag{3.114}$$

As a check, Eq. (3.114) yields Eq. (3.113) as B approaches infinity.

The example of a squarer preceded by a linear system, Eq. (3.99), can be pictured as the product of the outputs from two identical linear transformations.

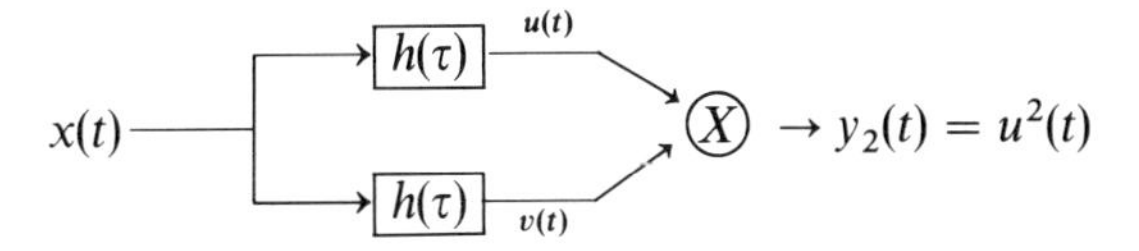

An extension of this example occurs when the two linear transformations have different weighting functions, say $h_a(\tau)$ and $h_b(\tau)$ where $h_a(\tau) \neq h_b(\tau)$. Different outputs $u_a(t)$ and $u_b(t)$ are produced where $u_a(t) \neq u_b(t)$ with $y_2(t) = u_a(t)u_b(t)$.

$x(t)$ → $h_a(\tau)$ → $u_a(t)$; $x(t)$ → $h_b(\tau)$ → $u_b(t)$; ⓧ → $y_2(t) = u_a(t)u_b(t)$

Now, instead of Eqs. (3.100) and (3.101), the results are

$$h_2(\tau_1, \tau_2) = h_a(\tau_1)h_b(\tau_2) \tag{3.115}$$

$$H_2(f_1, f_2) = H_a(f_1)H_b(f_2) \tag{3.116}$$

These bilinear functions are not symmetric. To obtain symmetric results, as shown in Eqs. (3.33) and (3.46), Eqs. (3.115) and (3.116) should be replaced by

$$h_2(\tau_1, \tau_2) = \tfrac{1}{2}[h_a(\tau_1)h_b(\tau_2) + h_a(\tau_2)h_b(\tau_1)] \tag{3.117}$$

$$H_2(f_1, f_2) = \tfrac{1}{2}[H_a(f_1)H_b(f_2) + H_a(f_2)H_b(f_1)] \tag{3.118}$$

Example 3.5 Squarer Followed by Linear System

$x(t)$ → Squarer → $v(t)$ → $h(\tau)$ → $y_2(t)$

This example involves memory operations after the squarer. The output is

$$y_2(t) = \int h(\tau_1)v(t - \tau_1)\,d\tau_1 = \int h(\tau_1)x^2(t - \tau_1)\,d\tau_1$$

where

$$x^2(t - \tau_1) = \int \delta(\tau_1 - \tau_2)x(t - \tau_1)x(t - \tau_2)\,d\tau_2$$

Hence,

$$y_2(t) = \iint h(\tau_1)\delta(\tau_1 - \tau_2)x(t - \tau_1)x(t - \tau_2)\,d\tau_1\,d\tau_2 \tag{3.119}$$

From Eq. (3.34), the bilinear weighting function is here

$$h_2(\tau_1, \tau_2) = h(\tau_1)\delta(\tau_1 - \tau_2) \tag{3.120}$$

This memory result exists only on the line $\tau_1 = \tau_2$. From Eqs. (3.45) and (3.18) the associated bilinear frequency response function is

$$H_2(f_1, f_2) = H(f_1 + f_2) \tag{3.121}$$

Note the frequency additive form of the linear frequency response function.

As a specific case, consider the linear low-pass filter system where $h(\tau) = Ae^{-a\tau}$. For this case, substituting Eqs. (3.80) and (3.81) into Eqs. (3.120) and (3.121) gives

$$\begin{aligned} h_2(\tau_1, \tau_2) &= Ae^{-a\tau_1}\delta(\tau_1 - \tau_2) \\ H_2(f_1, f_2) &= \frac{A}{a + j2\pi(f_1 + f_2)} \end{aligned} \tag{3.122}$$

Equations (3.121) and (3.122) show here that

$$H_2(\alpha, f - \alpha) = H_2(f) = \frac{A}{a + j2\pi f} \tag{3.123}$$

Hence, from Eqs. (3.55) and (3.123), the Fourier transform

$$Y_2(f) = \frac{A}{a + j2\pi f} \int X(\alpha)X(f - \alpha)\, d\alpha - \bar{y}_2\delta_1(f) \tag{3.124}$$

Also, Eqs. (3.121) and (3.122) show that

$$H_2(f, -f) = H(0) = \frac{A}{a} \tag{3.125}$$

which is real-valued as required. It follows now from Eq. (3.48) that the output mean value

$$\bar{y}_2 = H(0) \int S_{xx}(f)\, df = H(0)\sigma_x^2 \tag{3.126}$$

This result applies to any $S_{xx}(f)$.

For a bandwidth-limited white noise input where $S_{xx}(f) = 1$ exists only for $-B \leqslant f \leqslant B$, Eq. (3.126) gives the special result

$$\bar{y}_2 = H(0) \int_{-B}^{B} S_{xx}(f)\, df = 2BH(0) \tag{3.127}$$

Example 3.6 Squarer Preceded and Followed by Linear Systems

Time-Domain Relations

$$x(t) \longrightarrow \boxed{h_b(\tau)} \xrightarrow{u(t)} \boxed{\text{Squarer}} \xrightarrow{v(t)} \boxed{h_a(\tau)} \longrightarrow y_2(t)$$

$$u(t) = \int h_b(\tau)x(t-\tau)\,d\tau$$

$$v(t) = u^2(t) = \iint h_b(\tau_1)h_b(\tau_2)x(t-\tau_1)x(t-\tau_2)\,d\tau_1\,d\tau_2$$

$$y_2(t) = \int h_a(\tau)v(t-\tau)\,d\tau$$

$$v(t-\tau) = \iint h_b(\tau_1)h_b(\tau_2)x(t-\tau-\tau_1)x(t-\tau-\tau_2)\,d\tau_1\,d\tau_2$$

$$= \iint h_b(\sigma_1-\tau)h_b(\sigma_2-\tau)x(t-\sigma_1)x(t-\sigma_2)\,d\sigma_1\,d\sigma_2$$

by changing variables to $\sigma_1 = \tau + \tau_1$ and $\sigma_2 = \tau + \tau_2$. Thus,

$$y_2(t) = \iiint h_b(\sigma_1-\tau)h_b(\sigma_2-\tau)h_a(\tau)x(t-\sigma_1)x(t-\sigma_2)\,d\sigma_1\,d\sigma_2\,d\tau \tag{3.128}$$

Now, from Eq. (3.34),

$$h_2(\sigma_1, \sigma_2) = \int h_b(\sigma_1-\tau)h_b(\sigma_2-\tau)h_a(\tau)\,d\tau \tag{3.129}$$

This time-domain bilinear result is complicated to compute and interpret.

Frequency-Domain Relations

$$X(f) \longrightarrow \boxed{H_b(f)} \xrightarrow{U(f)} \boxed{\text{Squarer}} \xrightarrow{V(f)} \boxed{H_a(f)} \longrightarrow Y_2(f)$$

$$U(f) = H_a(f)X(f)$$

For $f \neq 0$,

$$V(f) = \int U(\alpha)U(f-\alpha)\,d\alpha$$

$$= \int H_b(\alpha)H_b(f-\alpha)X(\alpha)X(f-\alpha)\,d\alpha$$

Then

$$Y_2(f) = H_a(f)V(f) = H_a(f) \int H_b(\alpha)H_b(f-\alpha)X(\alpha)X(f-\alpha)\,d\alpha$$

Now, from Eq. (3.55), for $f \neq 0$,

$$H_2(\alpha, f-\alpha) = H_b(\alpha)H_b(f-\alpha)H_a(f) \tag{3.130}$$

Letting $\alpha = f_1$ and $(f-\alpha) = f_2$ yields the result

$$H_2(f_1, f_2) = H_b(f_1)H_b(f_2)H_a(f_1 + f_2) \tag{3.131}$$

This frequency-domain bilinear result is straightforward to compute and interpret compared to Eq. (3.129). Equation (3.131) is the double Fourier transform of Eq. (3.129). Special cases of Example 3.6 are Examples 3.3–3.5. More complicated examples that require intricate three-dimensional plots are discussed in references 16, 23, and 43. These and other such results are not practical to implement.

3.2.3 Trilinear Systems

Example 3.7 Cuber (Cubic) System

$$x(t) \longrightarrow \boxed{\text{Cuber}} \longrightarrow y_3(t) = x^3(t)$$

The input $x(t)$ can be represented by

$$x(t) = \int \delta(\tau)x(t-\tau)\,d\tau \tag{3.132}$$

where $\delta(\tau)$ is the usual delta function. The cuber output $y_3(t)$ is then given by

$$y_3(t) = x^3(t) = \iiint \delta(\tau_1)\delta(\tau_2)\delta(\tau_3)x(t-\tau_1)x(t-\tau_2)x(t-\tau_3)\,d\tau_1\,d\tau_2\,d\tau_3 \tag{3.133}$$

Hence, from Eq. (3.64), the trilinear weighting function is

$$h_3(\tau_1, \tau_2, \tau_3) = \delta(\tau_1)\delta(\tau_2)\delta(\tau_3) \tag{3.134}$$

Note that this function has no memory and exists only at $\tau_1 = \tau_2 = \tau_3 = 0$. The associated trilinear frequency response function is a constant independent of frequency, namely,

$$H_3(f_1, f_2, f_3) = 1 \qquad \text{for all } f_1, f_2, f_3 \tag{3.135}$$

From Eq. (3.70), the Fourier transform $Y_3(f)$ is

$$Y_3(f) = \iint X(\alpha)X(\beta - \alpha)X(f - \beta)\, d\alpha\, d\beta \tag{3.136}$$

Consider a sinusoidal input at frequency f_0 where

$$x(t) = \sin(2\pi f_0 t) \tag{3.137}$$

From basic trigonometry,

$$y_3(t) = \sin^3(2\pi f_0 t) = \tfrac{3}{4} \sin 2\pi f_0 t - \tfrac{1}{4} \sin 6\pi f_0 t \tag{3.138}$$

Here, the output $y_3(t)$ contains a linear term at the same sinusoidal input frequency f_0 plus a nonlinear term at the sinusoidal $3f_0$ frequency. The output linear term shows that $y_3(t)$ will be correlated with $x(t)$ as well as with any other linear operations on $x(t)$. The output frequency $3f_0$ term shows that cubing operations always produce output frequencies that are triple the input frequencies.

Example 3.8 Cuber Preceded by Linear System

$$x(t) \longrightarrow \boxed{h(\tau)} \xrightarrow[u(t)]{} \boxed{\text{Cuber}} \longrightarrow y_3(t)$$

This example involves memory operations before the cuber. The output is

$$y_3(t) = u^3(t) = \left[\int h(\tau)x(t - \tau)\, d\tau\right]^3$$

$$= \iiint h(\tau_1)h(\tau_2)h(\tau_3)x(t - \tau_1)x(t - \tau_2)x(t - \tau_3)\, d\tau_1\, d\tau_2\, d\tau_3 \tag{3.139}$$

From Eqs. (3.64) and (3.69), the trilinear weighting function and frequency response function are

$$h_3(\tau_1, \tau_2, \tau_3) = h(\tau_1)h(\tau_2)h(\tau_3) \tag{3.140}$$

$$H_3(f_1, f_2, f_3) = H(f_1)H(f_2)H(f_3) \tag{3.141}$$

Note the product form of the linear results as in Example 3.4. This gives

$$H_3(\alpha, \beta - \alpha, f - \beta) = H(\alpha)H(\beta - \alpha)H(f - \beta) \tag{3.142}$$

so that Eq. (3.70) becomes

$$Y_3(f) = \iint H(\alpha)H(\beta - \alpha)H(f - \beta)X(\alpha)X(\beta - \alpha)X(f - \beta)\, d\alpha\, d\beta \tag{3.143}$$

Example 3.9 Cuber Followed by Linear System

$$x(t) \longrightarrow \boxed{\text{Cuber}} \longrightarrow \boxed{h(\tau)} \longrightarrow y_3(t)$$

This example involves memory operations after the cuber. The output is

$$y_3(t) = \int h(\tau_1)v(t - \tau_1)\,d\tau_1 = \int h(\tau_1)x^3(t - \tau_1)\,d\tau_1$$

$$x^3(t - \tau_1) = \iint \delta(\tau_1 - \tau_2)\delta(\tau_1 - \tau_3)x(t - \tau_1)x(t - \tau_2)x(t - \tau_3)\,d\tau_2\,d\tau_3$$

Thus,

$$y_3(t) = \iiint h(\tau_1)\delta(\tau_1 - \tau_2)\delta(\tau_1 - \tau_3)x(t - \tau_1)x(t - \tau_2)x(t - \tau_3)\,d\tau_1\,d\tau_2\,d\tau_3 \tag{3.144}$$

From Eqs. (3.64) and (3.69), one obtains here the trilinear weighting function

$$h_3(\tau_1, \tau_2, \tau_3) = h(\tau_1)\delta(\tau_1 - \tau_2)\delta(\tau_1 - \tau_3) \tag{3.145}$$

This memory result exists only on the line $\tau_1 = \tau_2 = \tau_3$. The associated trilinear frequency response function is

$$H_3(f_1, f_2, f_3) = H(f_1 + f_2 + f_3) \tag{3.146}$$

Note the frequency additive form of the linear result as in Example 3.5. This gives

$$H_3(\alpha, \beta - \alpha, f - \beta) = H(f) \tag{3.147}$$

so that Eq. (3.70) becomes

$$x(t) = H(f) \iint X(\alpha)X(\beta - \alpha)X(f - \beta)\,d\alpha\,d\beta \tag{3.148}$$

Example 3.10 Cuber Preceded and Followed by Linear Systems

Time-Domain Relations

$$x(t) \longrightarrow \boxed{h_b(\tau)} \xrightarrow[u(t)]{} \boxed{\text{Cuber}} \xrightarrow[v(t)]{} \boxed{h_a(\tau)} \longrightarrow y_3(t)$$

$$u(t) = \int h_b(\tau)x(t - \tau)\,d\tau$$

$$v(t) = u^3(t)$$

$$= \iiint h_b(\tau_1)h_b(\tau_2)h_b(\tau_3)x(t-\tau_1)x(t-\tau_2)x(t-\tau_3)\,d\tau_1\,d\tau_2\,d\tau_3$$

$$y_3(t) = \int h_a(\tau)v(t-\tau)\,d\tau$$

One can now verify by substitutions and changes of variables that

$$y_3(t) = \iiiint h_b(\tau_1-\tau)h_b(\tau_2-\tau)h_b(\tau_3-\tau)h_a(\tau) \times x(t-\tau_1)x(t-\tau_2)x(t-\tau_3)\,d\tau_1\,d\tau_2\,d\tau_3\,d\tau \tag{3.149}$$

Equation (3.64) then gives

$$h_3(\tau_1, \tau_2, \tau_3) = \int h_b(\tau_1-\tau)h_b(\tau_2-\tau)h_b(\tau_3-\tau)h_a(\tau)\,d\tau \tag{3.150}$$

This time-domain trilinear result is complicated to compute and interpret.

Frequency-Domain Relations

$$X(f) \longrightarrow \boxed{H_b(f)} \xrightarrow{U(f)} \boxed{\text{Cuber}} \xrightarrow{V(f)} \boxed{H_a(f)} \longrightarrow Y_3(f)$$

$$U(f) = H_b(f)X(f)$$

$$V(f) = \int U(\alpha)U(\beta-\alpha)U(f-\beta)\,d\alpha\,d\beta$$

$$= \int H_b(\alpha)H_b(\beta-\alpha)H_b(f-\beta)X(\alpha)X(\beta-\alpha)X(f-\beta)\,d\alpha\,d\beta$$

$$Y_3(f) = H_a(f)V(f)$$

Substitutions then prove

$$H_3(\alpha, \beta-\alpha, f-\beta) = H_b(\alpha)H_b(\beta-\alpha)H_b(f-\beta)H_a(f) \tag{3.151}$$

Letting $\alpha = \mathrm{f}_1$, $(\beta-\alpha) = \mathrm{f}_2$, and $(f-\beta) = f_3$ yields the result

$$H_3(f_1, f_2, f_3) = H_b(f_1)H_b(f_2)H_b(f_2)H_a(f_1+f_2+f_3) \tag{3.152}$$

This frequency-domain trilinear result is straightforward to compute and interpret. Equation (3.152) is the triple Fourier transform of Eq. (3.150). Special cases of Example 3.10 are Examples 3.7–3.9. More complicated examples can require four-dimensional plots that are not practical to implement.

3.3 SYNTHESIS OF TIME-VARYING LINEAR SYSTEMS

As derived in Eq. (3.13), a *constant-parameter linear system* can be described by its weighting function $h(\tau)$, which gives the response of the system at any time to a unit impulse function applied τ time units earlier. For any input $x(t)$, the output $y(t)$ is

$$y(t) = \int h(\tau)x(t - \tau)\, d\tau \tag{3.153}$$

where the theoretical limits of integration can be from minus infinite to plus infinity. For physically realizable systems, $h(\tau) = 0$ for $\tau < 0$. A *time-varying linear system* can be described by means of a time-varying weighting function $h(\tau, t)$, which gives the response at time t to a unit impulse function applied at time $(t - \tau)$. Not only is the time interval τ important, but also the particular time t at which the system is being observed. Such physical situations are governed by linear differential-integral equations with time-varying coefficients that may be more realistic than constant-parameter coefficients. The appropriate formula for the output $y(t)$ to any input $x(t)$ follows immediately from the constant-parameter case by merely replacing $h(\tau)$ by the more general $h(\tau, t)$. Thus, the output $y(t)$ from any input $x(t)$ is now

$$y(t) = \int h(\tau, t)x(t - \tau)\, d\tau \tag{3.154}$$

For a physically realizable system, $h(\tau, t) = 0$ for $\tau < 0$.

The following theorem provides new insight on the basic nature of time-varying linear systems to constant-parameter bilinear systems.

Theorem (Schetzen) Any time-varying linear system can be synthesized by a constant-parameter bilinear system where the bilinear system will be physically realizable if and only if the time-varying linear system is physically realizable.

Proof Assume that $y(t)$ is the output of a time-varying linear system as per Eq. (3.154), where the quantity $h(\tau, t)$ is known. For the same input $x(t)$ used in Eq. (3.154), a bilinear weighting function $h_2(\tau, \beta)$ will give the output of Eq. (3.34), namely,

$$y_2(t) = \iint h_2(\tau, \beta)x(t - \beta)x(t - \tau)\, d\beta\, d\tau \tag{3.155}$$

This $y_2(t)$ will be the same as $y(t)$ of Eq. (3.154) if

$$h(\tau, t) = \int h_2(\tau, \beta)x(t - \beta)\, d\beta \tag{3.156}$$

Equation (3.156) shows that $h_2(\tau, \beta) = 0$ for $\tau < 0$ if and only if $h(\tau, t) = 0$ for $\tau < 0$ so that the bilinear system will be physically realizable if and only if the time-varying linear system is physically realizable. It will now be shown how to obtain $h_2(\tau, \beta)$ from knowledge of $h(\tau, t)$ and $x(t)$.

Let the quantity $H(\tau, g)$ be defined as the Fourier transform of $h(\tau, t)$ with respect to t holding τ constant. Then, from Eq. (3.156),

$$\begin{aligned} H(\tau, g) &= \int h(\tau, t)e^{-j2\pi gt}\, dt \\ &= \iint h_2(\tau, \beta)x(t-\beta)e^{-j2\pi gt}\, d\beta\, dt \\ &= \int\left[\int x(t-\beta)e^{-j2\pi g(t-\beta)}\, dt\right]h_2(\tau, \beta)e^{-j2\pi g\beta}\, d\beta \end{aligned} \tag{3.157}$$

For the inside integral, let $u = t - \beta$, $du = dt$ to obtain

$$H(\tau, g) = X(g)\int h_2(\tau, \beta)e^{-j2\pi g\beta}\, d\beta \tag{3.158}$$

where $X(g)$ is the Fourier transform

$$X(g) = \int x(u)e^{-j2\pi gu}\, du \tag{3.159}$$

Now define $J(f, g)$ to be the Fourier transform of $H(\tau, g)$ with respect to τ holding g constant. Then

$$\begin{aligned} J(f, g) &= \int H(\tau, g)e^{-j2\pi f\tau}\, d\tau \\ &= X(g)\iint h_2(\tau, \beta)e^{-j2\pi(f\tau+g\beta)}\, d\tau\, d\beta \\ &= X(g)H_2(f, g) \end{aligned} \tag{3.160}$$

where, from Eq. (3.45), $H_2(f, g)$ is the bilinear frequency response function of the bilinear system. Observe that $J(f, g)$ is the double Fourier transform of the given $h(\tau, t)$ for the time-varying linear system, namely,

$$J(f, g) = \iint h(\tau, t)e^{-j2\pi(f\tau+gt)}\, d\tau\, dt \tag{3.161}$$

Thus, from calculation of $J(f, g)$ and $X(g)$ one can obtain $H_2(f, g)$ by

$$H_2(f, g) = \frac{J(f, g)}{X(g)} \tag{3.162}$$

This completes the proof and defines the appropriate constant-parameter bilinear system to synthesize the given time-varying linear system from knowledge of $h(\tau, t)$ and $x(t)$. Relations derived here are special cases of results proved by Schetzen in reference 40.

3.4 HIGHER-ORDER CORRELATION AND SPECTRA

Appropriate formulas will now be stated for higher-order cross-correlation and cross-spectral density functions where results in Section 1.3 are first-order cases. These formulas will be used in Chapter 4 to help analyze and identify the dynamic input/output properties of stationary random data passing through bilinear and trilinear systems.

3.4.1 First-Order Functions

Consider any pair of real-valued zero mean value stationary random data $x(t)$ and $y(t)$. The *(first-order) cross-correlation function* between $x(t)$ and $y(t)$ is defined by

$$R_{xy}(\tau) = E[x(t)y(t + \tau)] = E[x(t - \tau)y(t)] \tag{3.163}$$

where $E[\ ]$ denotes an expected value operation. The quantities $R_{xx}(\tau)$ and $R_{yy}(\tau)$ are called *(first-order) autocorrelation functions.*

Assume that $x(t)$ and $y(t)$ are each of length T_{total} and that their Fourier transforms $X(f)$ and $Y(f)$ are obtained by dividing T_{total} into n_d disjoint parts, each of which is of length T. Now, since $x(t)$ is real-valued, the inverse Fourier transform gives

$$x(t) = \int X(f)e^{j2\pi ft}\,df = \int X^*(f)e^{-j2\pi ft}\,df$$

where $X^*(f)$ is the complex conjugate of $X(f)$. Also,

$$y(t + \tau) = \int Y(g)e^{j2\pi g(t+\tau)}\,dg$$

Thus, Eq. (3.163) is the same as

$$\begin{aligned} R_{xy}(\tau) &= E\left[\iint X^*(f)Y(g)e^{-j2\pi ft}e^{j2\pi g(t+\tau)}\,df\,dg\right] \\ &= \iint E[X^*(f)Y(g)]e^{-j2\pi(f-g)t}e^{j2\pi g\tau}\,dg\,df \\ &= \int S_{xy}(f)e^{j2\pi f\tau}\,df \end{aligned} \tag{3.164}$$

where the double integral is a function of t and the final single integral is independent of t. This equation shows that $R_{xy}(\tau)$ is the inverse Fourier transform of $S_{xy}(f)$.

For the last equation to eliminate t, $X(f)$ and $Y(g)$ should be measured at frequencies f and g that are spaced $\Delta f = (1/T)$ apart such that

$$E[X^*(f)Y(g)] = S_{xy}(f)\delta_1(f-g) \tag{3.165}$$

where, in agreement with Eq. (3.54), the *finite delta function*

$$\delta_1(f-g) = \begin{cases} T, & (-1/2T) < (f-g) < (1/2T) \\ 0, & \text{otherwise} \end{cases} \tag{3.166}$$

Hence,

$$E[X^*(f)Y(g)] = \begin{cases} TS_{xy}(f), & f = g \\ 0, & f \neq g \end{cases} \tag{3.167}$$

This proves that $S_{xy}(f)$ exists only when $f = g$ and gives the well-known formula of Eq. (1.27), namely,

$$S_{xy}(f) = \frac{1}{T} E[X^*(f)Y(f)] \tag{3.168}$$

Equation (3.168) is a direct Fourier transform way to compute $S_{xy}(f)$ without first computing $R_{xy}(\tau)$. The quantity $S_{xy}(f)$ is the *(first-order) cross-spectral density function* between $x(t)$ and $y(t)$, and is also called the *cross-spectrum.* This single-valued function is simple to compute and interpret. Equation (3.164) also shows that $S_{xy}(f)$ is the Fourier transform of $R_{xy}(\tau)$ defined by

$$S_{xy}(f) = \int R_{xy}(\tau)e^{-j2\pi f\tau}\,d\tau \tag{3.169}$$

As special cases of Eq. (3.168), *(first-order) autospectral density functions*, called also *autospectra*, are given by the formulas

$$\begin{aligned} S_{xx}(f) &= \frac{1}{T} E[X^*(f)X(f)] \\ S_{yy}(f) &= \frac{1}{T} E[Y^*(f)Y(f)] \end{aligned} \tag{3.170}$$

From Eqs. (3.163) and (3.164) the first-order autocorrelation function

$$R_{xx}(\tau) = E[x(t)x(t+\tau)] = \int S_{xx}(f)e^{j2\pi f\tau}\,df \tag{3.171}$$

At the value $\tau = 0$, one obtains the mean square value

$$R_{xx}(0) = E[x^2(t)] = \int S_{xx}(f)\,df \tag{3.172}$$

Thus, physically, the autospectrum $S_{xx}(f)$ gives a decomposition of the mean square value $E[x^2(t)]$ as a function of a single frequency variable. Specifically, $S_{xx}(f)$ represents the rate of change of the mean square value $E[x^2(t)]$ as a function of frequency. Similarly, the autospectrum $S_{yy}(f)$ gives a decomposition of the mean square value $E[y^2(t)]$ as a function of frequency. Many applications of these first-order results are in references 1 and 2.

3.4.2 Second-Order Functions

The *second-order cross-correlation function* between $x(t)$ and $y(t)$ is defined by the expected value operation

$$R_{xxy}(\tau_1, \tau_2) = E[x(t-\tau_1)x(t-\tau_2)y(t)] \tag{3.173}$$

The time-domain quantities can be represented by

$$x(t-\tau_1) = \int X^*(f)e^{-j2\pi f(t-\tau_1)}\,df$$

$$x(t-\tau_2) = \int X^*(g)e^{-j2\pi g(t-\tau_2)}\,dg$$

$$y(t) = \int Y(h)e^{j2\pi ht}\,dh$$

Thus, Eq. (3.173) is the same as

$$\begin{aligned} R_{xxy}(\tau_1, \tau_2) &= \iiint E[X^*(f)X^*(g)Y(h)]e^{-j2\pi(f+g-h)t}e^{j2\pi(f\tau_1+g\tau_2)}\,dh\,df\,dg \\ &= \iint S_{xxy}(f, g)e^{j2\pi(f\tau_1+g\tau_2)}\,df\,dg \end{aligned} \tag{3.174}$$

where the triple integral is a function of t and the final double integral is independent of t. This equation shows that $R_{xxy}(\tau_1, \tau_2)$ is the double inverse Fourier transform of $S_{xxy}(f, g)$.

For the last equation to eliminate t, $X(f)$, $X(g)$, and $Y(h)$ should be measured at frequencies f, g, and h such that

$$E[X^*(f)X^*(g)Y(h)] = S_{xxy}(f, g)\delta_1(f+g-h) \tag{3.175}$$

with $\delta_1(f)$ defined by Eq. (3.54). Hence, $S_{xxy}(f, g)$ exists only when $h = (f + g)$ and is given by the formula

$$S_{xxy}(f, g) = \frac{1}{T} E[X^*(f)X^*(g)Y(f + g)] \tag{3.176}$$

Equation (3.176) is a direct Fourier transform way to compute $S_{xxy}(f, g)$. The quantity $S_{xxy}(f, g)$ is the *second-order cross-spectral* (*bispectral*) *density function* between $x(t)$ and $y(t)$. This double-valued function is tedious to compute and interpret. Equation (3.174) proves that $S_{xxy}(f, g)$ is the double Fourier transform of $R_{xxy}(\tau_1, \tau_2)$ defined by

$$S_{xxy}(f, g) = \iint R_{xxy}(\tau_1, \tau_2)e^{-j2\pi(f\tau_1 + g\tau_2)}\, d\tau_1\, d\tau_2 \tag{3.177}$$

If $x(t)$ is a zero mean value Gaussian input to a linear system with output $y_1(t)$, then Eqs. (3.77) and (3.176) show that $S_{xxy_1}(f, g)$ will be identically zero for all f and g, since the expected value in Eq. (3.176) will involve a third-order moment of the Gaussian input data.

Special cases of Eq. (3.177) give the *second-order autospectral density functions*, also called *bispectra*, by the formulas

$$S_{xxx}(f, g) = \frac{1}{T} E[X^*(f)X^*(g)X(f + g)] \tag{3.178}$$

$$S_{yyy}(f, g) = \frac{1}{T} E[Y^*(f)Y^*(g)Y(f + g)] \tag{3.179}$$

From Eqs. (3.173) and (3.174), the *second-order autocorrelation functions* at $\tau_1 = \tau_2 = 0$ are the *skewness* quantities

$$R_{xxx}(0, 0) = E[x^3(t)] = \iint S_{xxx}(f, g)\, df\, dg \tag{3.180}$$

$$R_{yyy}(0, 0) = E[y^3(t)] = \iint S_{yyy}(f, g)\, df\, dg \tag{3.181}$$

This shows that the bispectrum $S_{xxx}(f, g)$ gives a decomposition of the skewness $E[x^3(t)]$ as a function of two frequency variables. Similarly, the bispectrum $S_{yyy}(f, g)$ gives a decomposition of the skewness $E[y^3(t)]$ as a function of two frequency variables.

For zero mean value Gaussian random data $x(t)$, the bispectrum $S_{xxx}(f, g) = 0$ for all f and g since it represents a third-order moment and all odd-order moments of such Gaussian data are zero. If an output $y(t)$ occurs by passing real-valued Gaussian random input data through a zero-memory nonlinear system, then $y(t)$ will be non-Gaussian and the output bispectrum $S_{yyy}(f, g)$ will not be zero for arbitrary f and g. Moreover, this output bispectrum $S_{yyy}(f, g)$ will be real-valued for all f and g because the nonlinear transformation has no memory. This means that if a measured output bispectrum is complex-valued, then $y(t)$ could not have come from Gaussian random data passing through a zero-memory nonlinear system. A possible model for this result would be the passage of the Gaussian random data through a finite-memory nonlinear system. Phase estimation using the bispectrum and nonlinear system identification by observing only the output are discussed in references 27 and 37. Reference 30 reviews bispectrum matters.

3.4.3 Third-Order Functions

The *third-order cross-correlation function* between $x(t)$ and $y(t)$ is defined by the expected value operation

$$R_{xxxy}(\tau_1, \tau_2, \tau_3) = E[x(t - \tau_1)x(t - \tau_2)x(t - \tau_3)y(t)] \tag{3.182}$$

As per the derivation of Eqs. (3.174) and (3.176), one can prove after a number of steps that

$$R_{xxxy}(\tau_1, \tau_2, \tau_3) = \iiint S_{xxxy}(f, g, h)e^{j2\pi(f\tau_1 + g\tau_2 + h\tau_3)}\, df\, dg\, dh \tag{3.183}$$

and also that

$$S_{xxxy}(f, g, h) = \frac{1}{T} E[X^*(f)X^*(g)X^*(h)Y(f + g + h)] \tag{3.184}$$

Equation (3.184) is a direct Fourier transform way to compute $S_{xxxy}(f, g, h)$. Equation (3.183) shows that $S_{xxxy}(f, g, h)$ is the triple Fourier transform of $R_{xxxy}(\tau_1, \tau_2, \tau_3)$ defined by

$$S_{xxxy}(f, g, h) = \iiint R_{xxxy}(\tau_1, \tau_2, \tau_3)e^{-j2\pi(f\tau_1 + g\tau_2 + h\tau_3)}\, d\tau_1\, d\tau_2\, d\tau_3 \tag{3.185}$$

The quantity $S_{xxxy}(f, g, h)$ is the *third-order cross-spectral* (*trispectral*) *density function* between $x(t)$ and $y(t)$. This triple-valued function is difficult to compute and interpret.

If $x(t)$ is a zero mean value Gaussian input to a linear system with output $y_1(t)$, then Eqs. (3.77) and (3.184) show that, in general, $S_{xxxy_1}(f, g, h) \neq 0$, since the expected value in Eq. (3.184) will involve a fourth-order moment of the Gaussian data. On the other hand, for this same input to a bilinear system with output $y_2(t)$, Eqs. (3.78) and (3.181) show that $S_{xxxy_2}(f, g, h)$ will be identically zero for all f, g, and h, since this calculation will involve odd-order moments of the Gaussian input data.

From Eq. (3.185), the *third-order autospectral density functions*, also called *trispectra*, are

$$S_{xxxx}(f, g, h) = \frac{1}{T} E[X^*(f)X^*(g)X^*(h)X(f + g + h)] \tag{3.186}$$

$$S_{yyyy}(f, g, h) = \frac{1}{T} E[Y^*(f)Y^*(g)Y^*(h)Y(f + g + h)] \tag{3.187}$$

From Eqs. (3.182) and (3.183), the *third-order autocorrelation functions* at $\tau_1 = \tau_2 = \tau_3 = 0$ are the *kurtosis* quantities

$$R_{xxxx}(0, 0, 0) = E[x^4(t)] = \iiint S_{xxxx}(f, g, h)\, df\, dg\, dh \tag{3.188}$$

$$R_{yyyy}(0, 0, 0) = E[y^4(t)] = \iiint S_{yyyy}(f, g, h)\, df\, dg\, dh \tag{3.189}$$

This shows that the trispectrum $S_{xxxx}(f, g, h)$ gives a decomposition of the kurtosis $E[x^4(t)]$ as a function of three frequency variables. Similarly, the trispectrum $S_{yyyy}(f, g, h)$ gives a decomposition of the kurtosis $E[y^4(t)]$ as a function of three frequency variables.

For zero mean value Gaussian random data $x(t)$, the trispectrum $S_{xxxx}(f, g, h)$, being a fourth-order moment of Gaussian data, reduces to products of second-order moments (autospectra). Thus the trispectrum $S_{xxxx}(f, g, h)$ is real-valued and nonnegative for all f, g, and h. Also, if an output $y(t)$ occurs by passing real-valued Gaussian random input data through a zero-memory nonlinear system, then the output trispectrum $S_{yyyy}(f, g, h)$ will be real-valued and nonnegative for all f, g, and h because the nonlinear transformation has no memory.

3.4.4 Special Bispectral and Trispectral Density Functions

Consider cases of Eqs. (3.176) and (3.184) when $f = g$ and when $f = g = h$. For these situations, one obtains special bispectral and trispectral density functions defined for stationary random data by

$$S_{xxy}(f) = S_{xxy}(f, f) = \frac{1}{T} E[X^*(f)X^*(f)Y(2f)] \tag{3.190}$$

$$S_{xxxy}(f) = S_{xxxy}(f, f, f) = \frac{1}{T} E[X^*(f)X^*(f)X^*(f)Y(3f)] \tag{3.191}$$

These two results are straightforward extensions of the usual cross-spectral density function $S_{xy}(f)$ of Eq. (3.168). They are functions of only one variable and are much easier to compute and interpret than the more generalized functions $S_{xxy}(f, g)$ of two variables and $S_{xxxy}(f, g, h)$ of three variables. The physical importance of these special results will be developed later in Chapter 5 as part of the new, simpler, practical standard ways to analyze and identify nonlinear systems.

3.4.5 Transient Random Data

For transient random data, the stationary random data results $S_{xy}(f)$, $S_{xxy}(f, g)$ and $S_{xxxy}(f, g, h)$ should be replaced by $\mathcal{S}_{xy}(f)$, $\mathcal{S}_{xxy}(f, g)$ and $\mathcal{S}_{xxxy}(f, g, h)$, respectively, where

$$\mathcal{S}_{xy}(f) = T S_{xy}(f) = E[X^*(f)Y(f)] \tag{3.192}$$

$$\mathcal{S}_{xxy}(f, g) = T^2 S_{xxy}(f, g) = T\, E[X^*(f)X^*(g)Y(f+g)] \tag{3.193}$$

$$\mathcal{S}_{xxxy}(f, g, h) = T^3 S_{xxxy}(f, g, h) = T^2 E[X^*(f)X^*(g)X^*(h)Y(f+g+h)] \tag{3.194}$$

Formulas for $\mathcal{S}_{yy}(f)$, $\mathcal{S}_{yyy}(f, g)$ and $\mathcal{S}_{yyyy}(f, g, h)$ follow when $x(t) = y(t)$. Other special formulas occur when $f = g$ and when $f = g = h$.

3.4.6 Summary of Results

The main equations for stationary random data will now be compared for their inner mathematical structure as one proceeds from first- to second- to third-order functions. By time-domain quantities, the successive cross-correlation functions are defined by the expected values

$$R_{xy}(\tau) = E[x(t-\tau)y(t)] \tag{3.195}$$

$$R_{xxy}(\tau_1, \tau_2) = E[x(t-\tau_1)x(t-\tau_2)y(t)] \tag{3.196}$$

$$R_{xxxy}(\tau_1, \tau_2, \tau_3) = E[x(t-\tau_1)x(t-\tau_2)x(t-\tau_3)y(t)] \tag{3.197}$$

Appropriate Fourier transforms of these cross-correlation functions yield the successive cross-spectral density functions

$$S_{xy}(f) = \int R_{xy}(\tau)e^{-j2\pi f\tau}\,d\tau \tag{3.198}$$

$$S_{xxy}(f_1, f_2) = \iint R_{xxy}(\tau_1, \tau_2)e^{-j2\pi(f_1\tau_1 + f_2\tau_2)}\,d\tau_1\,d\tau_2 \tag{3.199}$$

$$S_{xxxy}(f_1, f_2, f_3) = \iiint R_{xxxy}(\tau_1, \tau_2, \tau_3) e^{-j2\pi(f_1\tau_1 + f_2\tau_2 + f_3\tau_3)} d\tau_1\, d\tau_2\, d\tau_3 \quad (3.200)$$

Successive cross-spectral density functions are defined directly by frequency-domain quantities and expected values

$$S_{xy}(f) = \frac{1}{T} E[X^*(f)Y(f)] \quad (3.201)$$

$$S_{xxy}(f_1, f_2) = \frac{1}{T} E[X^*(f_1)X^*(f_2)Y(f_1 + f_2)] \quad (3.202)$$

$$S_{xxxy}(f_1, f_2, f_3) = \frac{1}{T} E[X^*(f_1)X^*(f_2)X^*(f_3)Y(f_1 + f_2 + f_3)] \quad (3.203)$$

These equations show that desired spectral density functions can be computed by two distinct methods:

Method 1. By first computing the various first-, second-, and third-order correlation functions in the time domain, and then taking their single, double, and triple Fourier transforms to move into the frequency domain.

Method 2. By computing finite Fourier transforms of the original measured data, and then taking appropriate first-, second-, and third-order expected values of these quantities. Here, all quantities are in the frequency domain without computing any associated correlation functions in the time domain. This direct method is both simpler and more efficient.

This concludes Chapter 3.

4

NONLINEAR SYSTEM INPUT/OUTPUT RELATIONSHIPS

From the basic material discussed in Chapter 3, this chapter develops input/output correlation and spectral relationships for stationary random data passing through bilinear and trilinear systems. Well-known linear system input/output relationships appear as special cases. The chapter defines linear and nonlinear coherence functions for general third-order nonlinear models consisting of linear systems in parallel with bilinear and trilinear systems. Also shown is how to identify the optimum nature of parallel linear, bilinear, and trilinear systems that minimize the output noise spectrum, based upon measurements only of input data and total output data. To obtain closed-form answers, input data are assumed to be a zero mean value Gaussian stationary random process with arbitrary spectral properties. The complexity of these techniques shows clearly the need and importance of the new simpler practical techniques in the following chapters.

4.1 LINEAR, BILINEAR, AND TRILINEAR SYSTEMS

A number of theoretical input/output correlation and spectral relations will now be developed for stationary random data passing through linear, bilinear, and trilinear systems, as shown in the three parts of Figure 4.1. The separate results are discussed in Sections 4.1.1–4.1.4 and summarized in Section 4.1.5. Implementation of these relations to analyze and identify the properties of general third-order nonlinear models is detailed in Sections 4.2 and 4.3. This implementation is important, useful material of great benefit to real problems when techniques in the following chapters are not appropriate.

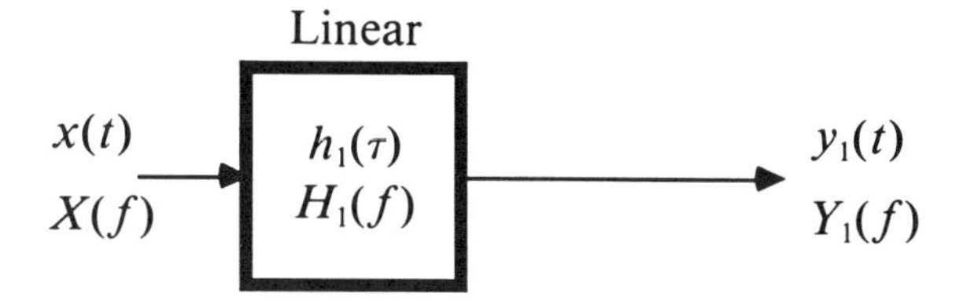

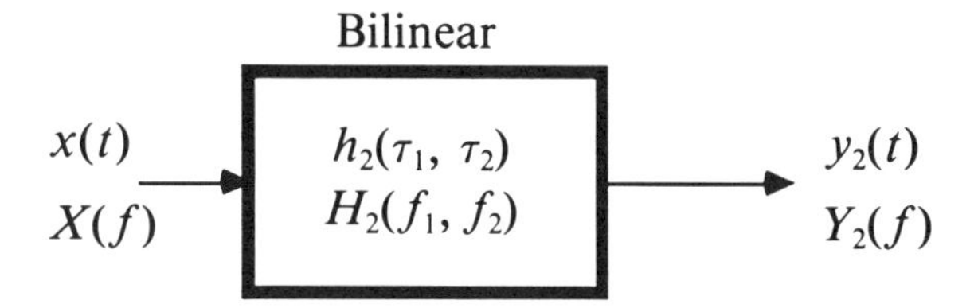

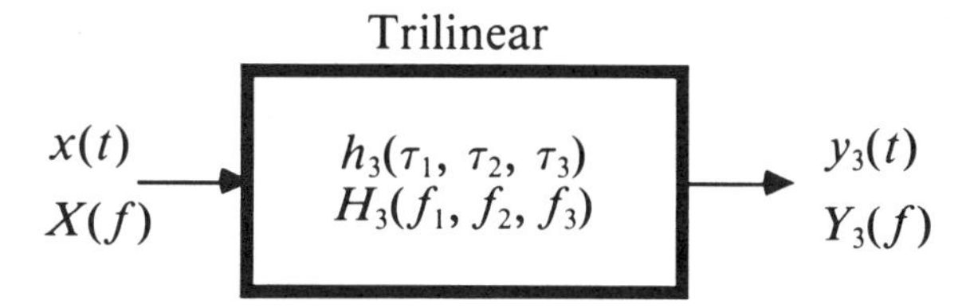

Figure 4.1 Linear, bilinear, and trilinear systems.

4.1.1 Linear Systems

For linear systems alone, from Eqs. (3.13) and (3.19), it is known that

$$y_1(t) = \int h_1(\alpha)x(t - \alpha)\, d\alpha \tag{4.1}$$

$$Y_1(f) = H_1(f)X(f) \tag{4.2}$$

where $X(f)$, $Y_1(f)$, and $H_1(f)$ are Fourier transforms of $x(t)$, $y_1(t)$, and $h_1(t)$, respectively. Hence, the input/output cross-correlation function for stationary random data is

$$\begin{aligned} R_{xy_1}(\tau) &= E[x(t)y_1(t + \tau)] = E[x(t - \tau)y_1(t)] \\ &= \int h_1(\alpha)E[x(t - \tau)x(t - \alpha)]\, d\alpha = \int h_1(\alpha)R_{xx}(\tau - \alpha)\, d\alpha \end{aligned} \tag{4.3}$$

Fourier transforms of both sides yield the input/output cross-spectrum result

$$S_{xy_1}(f) = \int R_{xy_1}(\tau)e^{-j2\pi f\tau}\,d\tau = H_1(f)S_{xx}(f) \tag{4.4}$$

This simple product relation is equivalent to the convolution integral of Eq. (4.3). From Eqs. (4.2), (1.25), and (1.27), without computing correlation functions, one can prove Eq. (4.4) more simply and directly by taking expected values of Fourier transform quantities as follows:

$$\begin{aligned} S_{xy_1}(f) &= \frac{1}{T}\,E[X^*(f)Y_1(f)] = \frac{1}{T}\,E[X^*(f)H_1(f)X(f)] \\ &= H_1(f)\left(\frac{1}{T}\,E[X^*(f)X(f)]\right) = H_1(f)S_{xx}(f) \end{aligned} \tag{4.5}$$

Thus, $H_1(f)$ is obtained from dividing the input/output cross-spectrum $S_{xy_1}(f)$ by the input autospectrum $S_{xx}(f)$, namely,

$$H_1(f) = \frac{S_{xy_1}(f)}{S_{xx}(f)} \tag{4.6}$$

The output autocorrelation function $R_{y_1y_1}(\tau)$ is

$$\begin{aligned} R_{y_1y_1}(\tau) &= E[y_1(t)y_1(t+\tau)] \\ &= E\left[\int h_1(\alpha)x(t-\alpha)\,d\alpha \int h_1(\beta)x(t+\tau-\beta)\,d\beta\right] \\ &= \iint h_1(\alpha)h_1(\beta)R_{xx}(\tau+\alpha-\beta)\,d\alpha\,d\beta \end{aligned} \tag{4.7}$$

This is a double convolution integral. Fourier transforms of both sides shows after several steps that the output autospectrum is

$$S_{y_1y_1}(f) = \int R_{y_1y_1}(\tau)e^{-j2\pi f\tau}\,d\tau = |H_1(f)|^2 S_{xx}(f) \tag{4.8}$$

Equation (4.8) is equivalent to Eq. (4.7). From Eq. (4.2), without computing correlation functions, one can prove more simply and directly that

$$S_{y_1y_1}(f) = \frac{1}{T} E[Y_1^*(f)Y_1(f)] = \frac{1}{T} E[H_1^*(f)X^*(f)H_1(f)X(f)]$$

$$= |H_1(f)|^2 \left(\frac{1}{T} E[X^*(f)X(f)]\right) = |H_1(f)|^2 S_{xx}(f) \tag{4.9}$$

Extensions of these input/output results will be carried out for bilinear and trilinear systems.

4.1.2 Definitions and Assumptions

Three preliminary matters are needed to derive the desired input/output relations for bilinear and trilinear systems. These involve proper understanding and application of the following items:

1. finite delta functions,
2. extended spectral density functions, and
3. higher-order moments for Gaussian data.

(1) Finite Delta Functions

Definitions for the finite delta functions are stated in Eqs. (3.54) and (3.166). The finite delta function has the properties

$$\delta_1(-f) = \delta_1(f) = \begin{cases} T, & (-1/2T) < f < (1/2T) \\ 0, & \text{otherwise} \end{cases}$$

$$\int_{-\infty}^{\infty} \delta_1(f)\,df = \int_{(-1/2T)}^{(1/2T)} \delta_1(f)\,df = 1 \tag{4.10}$$

Also,

$$S_{xx}(g)\delta_1(f-g) = S_{xx}(f)\delta_1(f-g) \tag{4.11}$$

when $S_{xx}(g) = S_{xx}(f)$ for $(-1/2T) < (f-g) < (1/2T)$. This is the same as measuring $S_{xx}(f)$ only at the discrete frequencies $f_k = k\Delta f = (k/T)$ apart.

(2) Extended Spectral Density Functions

Extended definitions of spectral density functions are stated in Eqs. (3.165) and (3.167). From these definitions, one has the properties

$$E[X^*(f)Y(g)] = S_{xy}(f)\delta_1(f-g) = \begin{cases} TS_{xy}(f), & f = g \\ 0, & f \neq g \end{cases} \tag{4.12}$$

$$E[X^*(f)X(g)] = S_{xx}(f)\delta_1(f-g) = \begin{cases} TS_{xx}(f), & f = g \\ 0, & f \neq g \end{cases} \tag{4.13}$$

assuming $X(f)$ measured only at frequencies $f_k = k\Delta f = (k/T)$ apart. From $X(f) = X^*(-f)$, $S_{xx}(-f) = S_{xx}(f)$, $S_{xy}(-f) = S_{yx}(f)$, and $\delta_1(-f) = \delta_1(f)$, further properties are

$$\begin{aligned} E[X(f)Y(g)] &= E[X^*(-f)Y(g)] = S_{yx}(f)\delta_1(f+g) \\ E[X(f)X(g)] &= E[X^*(-f)X(g)] = S_{xx}(f)\delta_1(f+g) \end{aligned} \tag{4.14}$$

$$\begin{aligned} E[X^*(f)Y^*(g)] &= E[X^*(f)Y(-g)] = S_{xy}(f)\delta_1(f+g) \\ E[X^*(f)X^*(g)] &= E[X^*(f)X(-g)] = S_{xx}(f)\delta_1(f+g) \end{aligned} \tag{4.15}$$

Note that for data where $X(0) = 0$, one will obtain $S_{xx}(0) = 0$.

(3) Higher-Order Moments for Gaussian Data

To obtain closed-form answers, unless stated otherwise, input data $x(t)$ henceforth will be assumed to be a zero mean value Gaussian stationary random process with arbitrary spectral properties.

From the assumed Gaussian input properties, all odd-ordered moments of $x(t)$ are zero and all even-ordered moments reduce to products of different combinations of second-order moments. Specifically, from reference 1, if $x_i \equiv x(t_i)$ and if $R_{ij} = R_{xx}(t_i - t_j) = R_{xx}(t_j - t_i)$ for the stationary random input data, then

$$E[x_1] = E[x_1x_2x_3] = E[x_1x_2x_3x_4x_5] = 0 \tag{4.16}$$

$$E[x_1x_2x_3x_4] = R_{12}R_{34} + R_{13}R_{24} + R_{14}R_{23} \tag{4.17}$$

$$\begin{aligned} E[x_1x_2x_3x_4x_5x_6] = {} & R_{12}[R_{34}R_{56} + R_{35}R_{46} + R_{36}R_{45}] \\ & + R_{13}[R_{24}R_{56} + R_{25}R_{46} + R_{26}R_{45}] \\ & + R_{14}[R_{23}R_{56} + R_{25}R_{36} + R_{26}R_{35}] \\ & + R_{15}[R_{23}R_{46} + R_{24}R_{36} + R_{26}R_{34}] \\ & + R_{16}[R_{23}R_{45} + R_{24}R_{35} + R_{25}R_{34}] \end{aligned} \tag{4.18}$$

and so on. These results apply not only to the original $x(t)$ but to any linear transformations of these $x(t)$, since linear transformations preserve the Gaussian nature of the data.

In particular, Fourier transforms are linear operations so that these higher-order moments apply to $X(f)$ in addition to $x(t)$ where

$$X(f) = \mathscr{F}[x(t)] = \int x(t)e^{-j2\pi ft}\,dt$$

They apply also to $X^*(f)$ where $X^*(f) = X(-f)$ when $x(t)$ is real-valued. Thus, if $X_i \equiv X(f_i)$ are Fourier transforms from a zero mean value Gaussian stationary random process, then

$$E[X_1] = E[X_1X_2X_3] = E[X_1X_2X_3X_4X_5] = 0 \tag{4.19}$$

$$E[X_1X_2X_3X_4] = E[X_1X_2]E[X_3X_4] + E[X_1X_3]E[X_2X_4] + E[X_1X_4]E[X_2X_3] \tag{4.20}$$

$$\begin{aligned} E[X_1X_2X_3X_4X_5X_6] = {} & E[X_1X_2]E[X_3X_4X_5X_6] + E[X_1X_3]E[X_2X_4X_5X_6] \\ & + E[X_1X_4]E[X_2X_3X_5X_6] \\ & + E[X_1X_5]E[X_2X_3X_4X_6] \\ & + E[X_1X_6]E[X_2X_3X_4X_5] \end{aligned} \tag{4.21}$$

where each fourth-order moment should then be replaced by appropriate pairs of second-order moments, and so on. All of the obtained different second-order moments then reduce to various autospectral and cross-spectral density functions as per the rules in Eqs. (4.12)–(4.15).

4.1.3 Bilinear Systems

For bilinear systems alone, from Eqs. (3.34), (3.44), and (3.48), one has the bilinear system output

$$y_2(t) = \iint h_2(\alpha, \beta)x(t-\alpha)x(t-\beta)\,d\alpha\,d\beta \tag{4.22}$$

with $\bar{y}_2 = E[y_2(t)]$ given by

$$\bar{y}_2 = \iint h_2(\alpha, \beta)R_{xx}(\alpha - \beta)\,d\alpha\,d\beta = \int H_2(f, -f)S_{xx}(f)\,df \tag{4.23}$$

In general, $\bar{y}_2 \neq 0$. Also, from Eq. (3.55),

$$Y_2(f) = \mathscr{F}[y_2(t) - \bar{y}_2] = \int H_2(u, f-u)X(u)X(f-u)\,du - \bar{y}_2\delta_1(f) \tag{4.24}$$

Observe that $Y_2(f) = \mathscr{F}[y_2(t)]$ of Eq. (3.51) for all f except $f = 0$.

Define the zero mean value bilinear system output $y_b(t)$ from $y_2(t)$ and $\bar{y}_2$ by

$$y_b(t) = y_2(t) - \bar{y}_2 \tag{4.25}$$

Then $\bar{y}_b = E[y_b(t)] = 0$ and

$$y_b(t) = \iint h_2(\alpha, \beta)[x(t-\alpha)x(t-\beta) - R_{xx}(\alpha - \beta)]\, d\alpha\, d\beta \tag{4.26}$$

with

$$Y_b(f) = \mathscr{F}[y_b(t)] = \mathscr{F}[y_2(t) - \bar{y}_2] = Y_2(f) \tag{4.27}$$

for the $Y_2(f)$ of Eq. (4.24). Also, $E[Y_b(f)] = E[Y_2(f)] = 0$. Further analysis should be conducted with $y_b(t)$ instead of $y_2(t)$. The cross-correlation functions between the input $x(t)$ and outputs $y_b(t)$ or $y_2(t)$ involve third-order moments of $x(t)$ so that from Eq. (4.16),

$$\begin{aligned} R_{xy_b}(\tau) &= R_{xy_2}(\tau) = 0 \\ S_{xy_b}(f) &= S_{xy_2}(f) = 0 \end{aligned} \tag{4.28}$$

In words, $x(t)$ is uncorrelated with both $y_b(t)$ and $y_2(t)$.

Bilinear System Output Autocorrelation and Autospectrum

The bilinear system output autocorrelation function and output autospectral density function from Eq. (4.25) satisfy

$$\begin{aligned} R_{y_by_b}(\tau) &= R_{y_2y_2}(\tau) - (\bar{y}_2)^2 \\ S_{y_by_b}(f) &= S_{y_2y_2}(f) - (\bar{y}_2)^2\delta_1(f) \end{aligned} \tag{4.29}$$

Starting from Eq. (4.26), after many steps one can prove in the time domain that

$$R_{y_by_b}(\tau) = 2\iiiint h_2(\alpha, \beta)h_2(u, v)R_{xx}(\tau + \alpha - u)R_{xx}(\tau + \beta - v)\, d\alpha\, d\beta\, du\, dv \tag{4.30}$$

This complicated four-fold integral is practically impossible to compute and apply.

In the frequency domain, as per the derivation of Eq. (3.51), by expressing $R_{xx}(\tau + \alpha - u)$ and $R_{xx}(\tau + \beta - v)$ in Eq. (4.30) in terms of their Fourier transforms, one can prove after many steps by this indirect method that

$$S_{y_by_b}(f) = 2\int |H_2(u, f-u)|^2 S_{xx}(u)S_{xx}(f-u)\, du \tag{4.31}$$

Equation (4.31) is simpler than Eq. (4.30) but still difficult to compute. Equation (4.31) can also be derived directly from Fourier transform data without using or knowing the correlation result of Eq. (4.30).

Mathematical details will now be described to derive Eqs. (4.30) and (4.31). Similar analysis procedures can be used to derive other formulas in this chapter. Readers not concerned with details of these proofs should go to Eq. (4.32).

Derivation of the Output Autocorrelation Function

From Eq. (4.26), with different variables of integration,

$$y_b(t+\tau) = \iint h_2(u, v)[x(t+\tau-u)x(t+\tau-v) - R_{xx}(u-v)]\,du\,dv$$

Multiplication of $y_b(t)$ by $y_b(t+\tau)$ and taking expected values shows that $R_{y_b y_b}(\tau)$ involves a four-fold integral where the integrand contains the expected value

$$\begin{aligned}
&E\{[x(t-\alpha)x(t-\beta)-R_{xx}(\alpha-\beta)][x(t+\tau-u)x(t+\tau-v)-R_{xx}(u-v)]\}\\
&\quad = E[x(t-\alpha)x(t-\beta)x(t+\tau-u)x(t+\tau-v)]\\
&\qquad - R_{xx}(u-v)E[x(t-\alpha)x(t-\beta)] - R_{xx}(\alpha-\beta)E[x(t+\tau-u)x(t+\tau-v)]\\
&\qquad + R_{xx}(\alpha-\beta)R_{xx}(u-v)
\end{aligned}$$

The second-order moments are

$$\begin{aligned}
E[x(t-\alpha)x(t-\beta)] &= R_{xx}(\alpha-\beta)\\
E[x(t+\tau-u)x(t+\tau-v)] &= R_{xx}(u-v)
\end{aligned}$$

From Eq. (4.17), the fourth-order moment

$$\begin{aligned}
&E[x(t-\alpha)x(t-\beta)x(t+\tau-u)x(t+\tau-v)]\\
&\quad = R_{xx}(\alpha-\beta)R_{xx}(u-v) + R_{xx}(\tau+\alpha-u)R_{xx}(\tau+\beta-v)\\
&\qquad + R_{xx}(\tau+\alpha-v)R_{xx}(\tau+\beta-u)
\end{aligned}$$

Substitution shows that only the last two terms remain in the integrand and that these two terms give the same four-fold integral, since $h_2(u, v) = h_2(v, u)$. Hence,

$$\begin{aligned}
R_{y_b y_b}(\tau) &= E[y_b(t)y_b(t+\tau)]\\
&= 2\iiiint h_2(\alpha, \beta)h_2(u, v)R_{xx}(\tau+\alpha-u)R_{xx}(\tau+\beta-v)\,d\alpha\,d\beta\,du\,dv
\end{aligned}$$

This completes the proof for the output autocorrelation function of Eq. (4.30).

Indirect Derivation of the Output Autospectral Density Function

In terms of their inverse Fourier transforms, the autocorrelation functions

$$R_{xx}(\tau + \alpha - u) = \int S_{xx}(\zeta) e^{j2\pi(\tau + \alpha - u)\zeta} \, d\zeta$$

$$R_{xx}(\tau + \beta - v) = \int S_{xx}(\eta) e^{j2\pi(\tau + \beta - v)\eta} \, d\eta$$

The exponential product

$$e^{j2\pi(\tau + \alpha - u)\zeta} e^{j2\pi(\tau + \beta - v)\eta} = e^{j2\pi(\alpha\zeta + \beta\eta)} e^{-j2\pi(u\zeta + v\eta)} e^{j2\pi(\zeta + \eta)\tau}$$

Hence, Eq. (4.30) is equivalent to

$$R_{y_b y_b}(\tau) = 2 \iint H_2(\zeta, \eta) H_2^*(\zeta, \eta) S_{xx}(\eta) e^{j2\pi(\zeta + \eta)\tau} \, d\zeta \, d\eta$$

where

$$H_2(\zeta, \eta) = \iint h_2(\alpha, \beta) e^{j2\pi(\alpha\zeta + \beta\eta)} \, d\alpha \, d\beta$$

$$H_2^*(\zeta, \eta) = \iint h_2(u, v) e^{-j2\pi(u\zeta + v\eta)} \, du \, dv$$

Thus,

$$R_{y_b y_b}(\tau) = 2 \iint |H_2(\zeta, \eta)|^2 S_{xx}(\zeta) S_{xx}(\eta) e^{j2\pi(\zeta + \eta)\tau} \, d\zeta \, d\eta$$

Change variables from (ζ, η) to (ζ, f) where $f = \zeta + \eta$, $df = d\eta$. Now

$$R_{y_b y_b}(\tau) = 2 \iint |H_2(\zeta, f - \zeta)|^2 S_{xx}(\zeta) S_{xx}(f - \zeta) e^{j2\pi f\tau} \, d\zeta \, df$$

But

$$R_{y_2 y_2}(\tau) = \int S_{y_b y_b}(f) e^{j2\pi f\tau} \, df$$

Hence, one obtains the result

$$S_{y_b y_b}(f) = 2 \int |H_2(\zeta, f - \zeta)|^2 S_{xx}(\zeta) S_{xx}(f - \zeta) \, d\zeta$$

Replacing the variable ζ by u with $d\zeta = du$ completes the indirect proof for the output autospectral density function of Eq. (4.31).

Direct Derivation of Output Autospectral Density Function

Start with Eqs. (4.24) and (4.27) where

$$Y_b(f) = Y_2(f) = \int H_2(u, f-u)X(u)X(f-u)\,du - \bar{y}_2\delta_1(f)$$

Define

$$Z(f) = \int H_2(u, f-u)X(u)X(f-u)\,du$$

This quantity $Z(f)$ is the same as $\mathscr{F}[y_2(t)]$ in Eq. (3.51). Then

$$Y_b(f) = Z(f) - \bar{y}_2\delta_1(f)$$

Expected values of both sides using $E[Y_b(f)] = 0$ gives

$$E[Z(f)] = \bar{y}_2\delta_1(f)$$

Note also that

$$E[Z^*(f)] = E[Z(f)]$$

The direct derivation of Eq. (4.31) comes from computing

$$\begin{aligned} S_{y_by_b}(f) &= \frac{1}{T}\, E[Y_b^*(f)Y_b(f)] \\ &= \frac{1}{T}\, E\{[Z^*(f) - \bar{y}_2\delta_1(f)][Z(f) - \bar{y}_2\delta_1(f)]\} \\ &= S_{zz}(f) - (\bar{y}_2)^2\delta_1(f) \end{aligned}$$

where

$$S_{zz}(f) = \frac{1}{T}\, E[Z^*(f)Z(f)]$$

and from properties of $\delta_1(f)$,

$$\frac{1}{T}\, E[(\bar{y}_2)^2\delta_1^2(f)] = (\bar{y}_2)^2\delta_1(f)$$

Use variables of integration α and β to write

$$Z(f) = \int H_2(\alpha, f-\alpha)X(\alpha)X(f-\alpha)\,d\alpha$$

$$Z^*(f) = \int H_2^*(\beta, f-\beta)X^*(\beta)X^*(f-\beta)\,d\beta$$

Now

$$S_{zz}(f) = \frac{1}{T}\iint H_2(\alpha, f-\alpha)H_2^*(\beta, f-\beta)E[X(\alpha)X(f-\alpha)X^*(\beta)X^*(f-\beta)]\,d\alpha\,d\beta$$

From Eq. (4.20) plus Eqs. (4.12)–(4.15), the fourth-order moment

$$\begin{aligned}
&E[X(\alpha)X(f-\alpha)X^*(\beta)X^*(f-\beta)] \\
&\quad = E[X(\alpha)X(f-\alpha)]E[X^*(\beta)X^*(f-\beta)] \\
&\qquad + E[X(\alpha)X^*(\beta)]E[X(f-\alpha)X^*(f-\beta)] \\
&\qquad + E[X(\alpha)X^*(f-\beta)]E[X(f-\alpha)X^*(\beta)] \\
&\quad = [S_{xx}(\alpha)\delta_1(f)][S_{xx}(\beta)\delta_1(f)] \\
&\qquad + [S_{xx}(\alpha)\delta_1(f-\alpha-\beta)][S_{xx}(f-\alpha)\delta_1(f-\alpha-\beta)] \\
&\qquad + [S_{xx}(\alpha)\delta_1(f-\alpha-\beta)][S_{xx}(f-\alpha)\delta_1(f-\alpha-\beta)]
\end{aligned}$$

Substituting into $S_{zz}(f)$ and integrating term by term gives, for the first pair of second-order moments,

$$\begin{aligned}
&\frac{1}{T}\iint H_2(\alpha, f-\alpha)H_2^*(\beta, f-\beta)S_{xx}(\alpha)S_{xx}(\beta)\delta_1^2(f)\,d\alpha\,d\beta \\
&\quad = \frac{\delta_1^2(f)}{T}\left[\int H_2(\alpha, f-\alpha)S_{xx}(\alpha)\,d\alpha\right]\left[\int H_2^*(\beta, f-\beta)S_{xx}(\beta)\,d\beta\right] \\
&\quad = (\bar{y}_2)^2\delta_1(f)
\end{aligned}$$

The second pair of second-order moments gives

$$\begin{aligned}
&\frac{1}{T}\iint H_2(\alpha, f-\alpha)H_2^*(\beta, f-\beta)S_{xx}(\alpha)S_{xx}(f-\alpha)\delta_1^2(\alpha-\beta)\,d\alpha\,d\beta \\
&\quad = \frac{\delta_1(0)}{T}\int H_2(\alpha, f-\alpha)H_2^*(\alpha, f-\alpha)S_{xx}(\alpha)S_{xx}(f-\alpha)\,d\alpha \\
&\quad = \int |H_2(\alpha, f-\alpha)|^2 S_{xx}(\alpha)S_{xx}(f-\alpha)\,d\alpha
\end{aligned}$$

The third pair of second-order moments gives

$$\frac{1}{T}\iint H_2(\alpha, f-\alpha)H_2^*(\beta, f-\beta)S_{xx}(\alpha)S_{xx}(f-\alpha)\delta_1^2(f-\alpha-\beta)\,d\alpha\,d\beta$$

$$=\frac{\delta_1(0)}{T}\int H_2(\alpha, f-\alpha)H_2^*(f-\alpha, \alpha)S_{xx}(\alpha)S_{xx}(f-\alpha)\,d\alpha$$

$$=\int |H_2(\alpha, f-\alpha)|^2 S_{xx}(\alpha)S_{xx}(f-\alpha)\,d\alpha$$

since $H_2^*(f-\alpha, \alpha) = H_2^*(\alpha, f-\alpha)$. Thus,

$$S_{zz}(f) = (\bar{y}_2)^2\delta_1(f) + 2\int |H_2(\alpha, f-\alpha)|^2 S_{xx}(\alpha)S_{xx}(f-\alpha)\,d\alpha$$

Finally,

$$S_{y_by_b}(f) = S_{zz}(f) - (\bar{y}_2)^2\delta_1(f) = 2\int |H_2(\alpha, f-\alpha)|^2 S_{xx}(\alpha)S_{xx}(f-\alpha)\,d\alpha$$

Replacing α with u gives the result of Eq. (4.31). This completes the direct proof for the output autospectral density function.

Second-Order Cross-Correlation and Cross-Spectrum

Consider now the second-order cross-correlation function $R_{xxy_b}(\tau_1, \tau_2)$. From Eqs. (3.170), (4.26), and (4.17),

$$\begin{aligned}
R_{xxy_b}(\tau_1, \tau_2) &= E[x(t-\tau_1)x(t-\tau_2)y_b(t)] \\
&= \iint h_2(\alpha, \beta)E\left[x(t-\tau_1)x(t-\tau_2)\left\{x(t-\alpha)x(t-\beta)\right.\right. \\
&\quad \left.\left. -R_{xx}(\alpha-\beta)\right\}\right]d\alpha\,d\beta \\
&= \iint h_2(\alpha, \beta)[R_{xx}(\tau_1-\alpha)R_{xx}(\tau_2-\beta) \\
&\quad + R_{xx}(\tau_1-\beta)R_{xx}(\tau_2-\alpha)]\,d\alpha\,d\beta \\
&= 2\iint h_2(\alpha, \beta)R_{xx}(\tau_1-\alpha)R_{xx}(\tau_2-\beta)\,d\alpha\,d\beta
\end{aligned} \tag{4.32}$$

The last step uses the symmetric property that $h_2(\alpha, \beta) = h_2(\beta, \alpha)$. Equation (4.32) is difficult to compute and interpret.

From Eq. (3.174), the second-order cross-spectrum (bispectrum) is defined by

$$S_{xxy_b}(f, g) = \iint R_{xxy_b}(\tau_1, \tau_2)e^{-j2\pi(f\tau_1 + g\tau_2)}\, d\tau_1\, d\tau_2 \tag{4.33}$$

Substitution of Eq. (4.32) into Eq. (4.33) after many steps gives the important bispectrum result in the frequency domain that is equivalent to Eq. (4.32), namely,

$$S_{xxy_b}(f, g) = 2H_2(f, g)S_{xx}(f)S_{xx}(g) \tag{4.34}$$

Here, $H_2(f, g)$ is the double Fourier transform of $h_2(\alpha, \beta)$ as per Eq. (3.45). Equation (4.34) can also be derived directly without computing correlation functions. Note the simplicity of Eq. (4.34) compared with Eq. (4.32).

First-order correlation and spectral density functions are identically zero between the output $y_1(t)$ from the linear system and the outputs $y_b(t)$ or $y_2(t)$ from the bilinear system. This follows from Eq. (4.28), since $y_1(t)$ is linearly related to $x(t)$. Thus,

$$\begin{aligned} R_{y_1y_b}(\tau) &= R_{y_1y_2}(\tau) = 0 \\ S_{y_1y_b}(f) &= S_{y_1y_2}(f) = 0 \end{aligned} \tag{4.35}$$

Finally, it should be noted here that the second-order correlation and spectral density functions between the input $x(t)$ and the linear output $y_1(t)$ are also identically zero,

$$\begin{aligned} R_{xxy_1}(\tau_1, \tau_2) &= 0 \\ S_{xxy_1}(f_1, f_2) &= 0 \end{aligned} \tag{4.36}$$

since they involve third-order moments of $x(t)$ that are zero from assumptions on $x(t)$.

The direct derivation of Eq. (4.34) will now be detailed. Readers not concerned with this proof should go to Section 4.1.4.

Direct Derivation of the Second-Order Cross-Spectrum

From Eqs. (3.176), (4.24), and (4.27),

$$S_{xxy_b}(f, g) = \frac{1}{T} E[X^*(f)X^*(g)Y_b(f + g)]$$

where

$$Y_b(f + g) = \int H_2(u, f + g - u)X(u)X(f + g - u)\, du - \bar{y}_2\delta_1(f + g)$$

with

$$\bar{y}_2 = \int H_2(u, -u) S_{xx}(u)\, du$$

Now,

$$S_{xxy_b}(f, g) = \frac{1}{T} \int H_2(u, f+g-u) E[X^*(f) X^*(g) X(u) X(f+g-u)]\, du$$
$$- \bar{y}_2 \delta_1(f+g) \left[\frac{1}{T} E[X^*(f) X^*(g)] \right]$$

The second-order moment

$$E[X^*(f)X^*(g)] = S_{xx}(f)\delta_1(f+g) = \begin{cases} T S_{xx}(f), & f+g=0 \\ 0, & \text{otherwise} \end{cases}$$

Hence, using Eq. (4.23), the last term in $S_{xxy_b}(f, g)$ is

$$-S_{xx}(f)\delta_1(f+g) \int H_2(u, -u) S_{xx}(u)\, du$$

The fourth-order moment in the first term of $S_{xxy_b}(f, g)$ reduces to

$$\begin{aligned} &E[X^*(f)X^*(g)X(u)X(f+g-u)] \\ &\quad = E[X^*(f)X^*(g)]E[X(u)X(f+g-u)] \\ &\qquad + E[X^*(f)X(u)]E[X^*(g)X(f+g-u)] \\ &\qquad + E[X^*(f)X(f+g-u)]E[X^*(g)X(u)] \\ &\quad = [S_{xx}(f)\delta_1(f+g)][S_{xx}(u)\delta_1(f+g)] + [S_{xx}(f)\delta_1(f-u)][S_{xx}(g)\delta_1(f-u)] \\ &\qquad + [S_{xx}(f)\delta_1(g-u)][S_{xx}(g)\delta_1(g-u)] \end{aligned}$$

Substituting into the first term in $S_{xxy_b}(f, g)$ and integrating term by term gives, for the first pair of second-order moments,

$$\frac{1}{T} \int H_2(u, f+g-u) S_{xx}(f) S_{xx}(u) \delta_1^2(f+g)\, du$$

$$= S_{xx}(f)\delta_1(f+g) \int H_2(u, -u) S_{xx}(u)\, du$$

This cancels out with the last term in $S_{xxy_b}(f, g)$. The second pair of second-order moments gives

$$\frac{1}{T}\int H_2(u, f+g-u)S_{xx}(f)S_{xx}(g)\delta_1^2(f-u)\,du = H_2(f, g)S_{xx}(f)S_{xx}(g)$$

using the fact that

$$\frac{1}{T}\iint \delta_1^2(f-u)\,du = \int \delta_1(f-u)\,du = 1$$

The third pair of second-order moments gives

$$\frac{1}{T}\int H_2(u, f+g-u)S_{xx}(f)S_{xx}(g)\delta_1^2(g-u)\,du = H_2(g, f)S_{xx}(f)S_{xx}(g)$$

where $H_2(g,f) = H_2(f, g)$ by the symmetry property. Hence,

$$S_{xxy_b}(f, g) = 2H_2(f, g)S_{xx}(f)S_{xx}(g)$$

This completes the direct proof for the important bispectrum result of Eq. (4.34).

4.1.4 Trilinear Systems

For trilinear systems alone, from Eqs. (3.64) and (3.70), the trilinear system outputs are

$$y_3(t) = \iiint h_3(\alpha, \beta, \gamma)x(t-\alpha)x(t-\beta)x(t-\gamma)\,d\alpha\,d\beta\,d\gamma \tag{4.37}$$

$$Y_3(f) = \iint H_3(u, v-u, f-v)X(u)X(v-u)X(f-v)\,du\,dv \tag{4.38}$$

From the assumed properties of the zero mean value Gaussian input $x(t)$, and the symmetric nature of $h_3(\alpha, \beta, \gamma)$ and $H_3(f_1, f_2, f_3)$, one can prove that $\bar{y}_3 = E[y_3(t)] = 0$, $E[Y_3(f)] = 0$, and that the input/output cross-correlation function is

$$R_{xy_3}(\tau) = \int c(\alpha)R_{xx}(\tau-\alpha)\,d\alpha \tag{4.39}$$

where

$$c(\alpha) = 3\iint h_3(\alpha, \beta, \gamma)R_{xx}(\beta-\gamma)\,d\beta\,d\gamma \tag{4.40}$$

Fourier transforms of both sides of Eq. (4.39) show that the corresponding input/output cross-spectral density function is

$$S_{xy_3}(f) = C(f)S_{xx}(f) \tag{4.41}$$

with $C(f)$, the Fourier transform of $c(\alpha)$, given by

$$C(f) = [S_{xy_3}(f)/S_{xx}(f)] = 3\int H_3(u, -u, f)S_{xx}(u)\,du \tag{4.42}$$

This result for $C(f)$ varies as a function of $S_{xx}(f)$. One can also show, using Eq. (4.1), that the cross-correlation function is

$$R_{y_1y_3}(\tau) = \int h_1(\beta)R_{xy_3}(\tau + \beta)\,d\beta = \iint h_1(\beta)c(\alpha)R_{xx}(\tau + \beta - \alpha)\,d\alpha\,d\beta \tag{4.43}$$

with the associated cross-spectral density function

$$S_{y_1y_3}(f) = H_1^*(f)S_{xy_3}(f) = H_1^*(f)C(f)S_{xx}(f) \tag{4.44}$$

Note that $y_3(t)$ is correlated with both $x(t)$ and $y_1(t)$. Detailed derivations for the above equations are straightforward exercises along the lines of similar derivations for bilinear systems in Section 4.1.3. All frequency-domain results can be derived directly without going through correlation functions.

To obtain uncorrelated trilinear results with $x(t)$ and $y_1(t)$, one can define the trilinear system output $y_c(t)$ from $y_3(t)$ by

$$y_c(t) = y_3(t) - \int c(\alpha)x(t - \alpha)\,d\alpha \tag{4.45}$$

Then $\bar{y}_c = E[y_c(t)] = 0$ with

$$Y_c(f) = Y_3(f) - C(f)X(f) \tag{4.46}$$

In place of Eqs. (4.39)–(4.44), one obtains

$$\begin{aligned} R_{xy_c}(\tau) &= R_{y_1y_c}(\tau) = 0 \\ S_{xy_c}(f) &= S_{y_1y_c}(f) = 0 \end{aligned} \tag{4.47}$$

Thus $y_c(t)$, unlike $y_3(t)$, will be uncorrelated with $x(t)$ and $y_1(t)$. Further analysis should be conducted with $y_c(t)$ instead of $y_3(t)$.

This quantity $y_c(t)$ is uncorrelated with the $y_b(t)$ of Eq. (4.26), since odd-order moments occur.

$$\begin{aligned} R_{y_by_c}(\tau) &= R_{y_2y_3}(\tau) = 0 \\ S_{y_by_c}(f) &= S_{y_2y_3}(f) = 0 \end{aligned} \tag{4.48}$$

One can also verify that the following second-order correlation and spectral density functions are identically zero.

$$
\begin{aligned}
R_{xxy_c}(\tau_1, \tau_2) &= R_{xxy_3}(\tau_1, \tau_2) = 0 \\
S_{xxy_c}(f, g) &= S_{xxy_3}(f, g) = 0
\end{aligned}
\tag{4.49}
$$

Trilinear System Output Autocorrelation Function

The trilinear system output autocorrelation function $R_{y_cy_c}(\tau)$ is given from Eq. (4.45) by

$$
\begin{aligned}
R_{y_cy_c}(\tau) &= E[y_c(t)y_c(t+\tau)] \\
&= R_{y_3y_3}(\tau) - \int c(\alpha)R_{xy_3}(\tau+\alpha)\,d\alpha - \int c(\beta)R_{xy_3}(\beta-\tau)\,d\beta \\
&\quad + \iint c(\alpha)c(\beta)R_{xx}(\tau+\alpha-\beta)\,d\alpha\,d\beta
\end{aligned}
$$

From Eq. (4.39),

$$
R_{xy_3}(\tau+\alpha) = \int c(\beta)R_{xx}(\tau+\alpha-\beta)\,d\beta
$$

$$
R_{xy_3}(\beta-\tau) = \int c(\alpha)R_{xx}(\beta-\tau-\alpha)\,d\alpha = \int c(\alpha)R_{xx}(\tau+\alpha-\beta)\,d\alpha
$$

Hence,

$$
R_{y_cy_c}(\tau) = \mathrm{R}_{y_3y_3}(\tau) - \iint c(\alpha)c(\beta)R_{xx}(\tau+\alpha-\beta)\,d\alpha\,d\beta \tag{4.50}
$$

An expression will now be derived for $R_{y_3y_3}(\tau)$. From Eq. (4.37),

$$
R_{y_3y_3}(\tau) = \iiiint\!\!\iint h_3(\alpha, \beta, \gamma)h_3(u, v, w)J(\alpha, \beta, \gamma, u, v, w, \tau)\,d\alpha\,d\beta\,d\gamma\,du\,dv\,dw \tag{4.51}
$$

where the integrand quantity J is defined by

$$
\begin{aligned}
&J(\alpha, \beta, \gamma, u, v, w, \tau) \\
&\quad = E[x(t-\alpha)x(t-\beta)x(t-\gamma)x(t+\tau-u)x(t+\tau-v)x(t+\tau-w)]
\end{aligned}
\tag{4.52}
$$

This breaks down into 15 terms from which symmetry properties of $h_3(\alpha, \beta, \gamma)$ and $h_3(u, v, w)$ show, after many detailed steps, that

$$J = 6R_{xx}(\tau + \alpha - u)R_{xx}(\tau + \beta - v)R_{xx}(\tau + \gamma - w)$$
$$+ 9R_{xx}(\alpha - \beta)R_{xx}(u - v)R_{xx}(\tau + \gamma - w)$$

The second term above, when substituted into Eq. (4.51), is the same from Eq. (4.40) as the second term in Eq. (4.50) except for sign. Hence, one obtains

$$R_{y_3y_3}(\tau) = R_{y_cy_c}(\tau) + \iint c(\alpha)c(\beta)R_{xx}(\tau + \alpha - \beta)\, d\alpha\, d\beta \tag{4.53}$$

where

$$R_{y_cy_c}(\tau) = 6 \iiiint\!\!\iint h_3(\alpha, \beta, \gamma)h_3(u, v, w)R_{xx}(\tau + \alpha - u)$$
$$\times R_{xx}(\tau + \beta - v)R_{xx}(\tau + \gamma - w)\, da\, d\beta\, d\gamma\, du\, dv\, dw \tag{4.54}$$

This complicated six-fold integral is practically impossible to compute and apply.

Trilinear System Output Autospectral Density Function

As per the indirect derivation of Eq. (4.31) from Eq. (4.30), one can prove after many steps that the corresponding trilinear system output autospectral density function

$$S_{y_cy_c}(f) = 6 \iint |H_3(u, v - u, f - v)|^2 S_{xx}(u)S_{xx}(v - u)S_{xx}(f - v)\, du\, dv \tag{4.55}$$

Equation (4.55) is simpler than Eq. (4.54) but still very difficult to compute.

Equation (4.55) can also be derived directly from Eqs. (3.169) and (4.46) without computing through correlation functions. From Eq. (4.53) one obtains, by taking Fourier transforms of both sides,

$$S_{y_3y_3}(f) = S_{y_cy_c}(f) + |C(f)|^2 S_{xx}(f) \tag{4.56}$$

where $C(f)$ is given by Eq. (4.42) and $S_{y_cy_c}(f)$ by Eq. (4.55).

Third-Order Cross-Correlation Function

Consider now the third-order cross-correlation function $R_{xxxy_c}(\tau_1, \tau_2, \tau_3)$. From Eqs. (3.176) and (4.45),

$$R_{xxxy_c}(\tau_1, \tau_2, \tau_3)$$
$$= E[x(t-\tau_1)x(t-\tau_2)x(t-\tau_3)y_c(t)]$$
$$= R_{xxxy_3}(\tau_1, \tau_2, \tau_3) - \int c(\alpha)E[x(t-\tau_1)x(t-\tau_2)x(t-\tau_3)x(t-\alpha)]\, d\alpha \tag{4.57}$$

where the fourth-order moment

$$
\begin{aligned}
E[x(t-\tau_1)x(t-\tau_2)x(t-\tau_3)x(t-\alpha)] \\
&= R_{xx}(\tau_1-\tau_2)R_{xx}(\tau_3-\alpha) + R_{xx}(\tau_1-\tau_3)R_{xx}(\tau_2-\alpha) \\
&\quad + R_{xx}(\tau_2-\tau_3)R_{xx}(\tau_1-\alpha)
\end{aligned}
$$

Hence, from Eq. (4.39), it follows that

$$
\begin{aligned}
R_{xxxy_c}(\tau_1, \tau_2, \tau_3) &= R_{xxxy_3}(\tau_1, \tau_2, \tau_3) - R_{xx}(\tau_1-\tau_2)R_{xy_3}(\tau_3) \\
&\quad - R_{xx}(\tau_1-\tau_3)R_{xy_3}(\tau_2) - R_{xx}(\tau_2-\tau_3)R_{xy_3}(\tau_1)
\end{aligned} \tag{4.58}
$$

Separate calculation can also be made of $R_{xxxy_3}(\tau_1, \tau_2, \tau_3)$ from

$$
\begin{aligned}
R_{xxxy_3}&(\tau_1, \tau_2, \tau_3) \\
&= E[x(t-\tau_1)x(t-\tau_2)x(t-\tau_3)y_3(t)] \\
&= \iiint h_3(\alpha, \beta, \gamma)E[x(t-\tau_1)x(t-\tau_2)x(t-\tau_3)x(t-\alpha)x(t-\beta)x(t-\gamma)]\,d\alpha\,d\beta\,d\gamma
\end{aligned} \tag{4.59}
$$

The sixth-order moment here breaks down into 15 terms from which symmetry properties of $h_3(\alpha, \beta, \gamma)$ show that

$$
\begin{aligned}
R_{xxxy_3}(\tau_1, \tau_2, \tau_3) &= 6\iiint h_3(\alpha, \beta, \gamma)R_{xx}(\tau_1-\alpha)R_{xx}(\tau_2-\beta)R_{xx}(\tau_3-\gamma)\,d\alpha\,d\beta\,d\gamma \\
&\quad + R_{xx}(\tau_1-\tau_2)R_{xy_3}(\tau_3) + R_{xx}(\tau_1-\tau_3)R_{xy_3}(\tau_2) \\
&\quad + R_{xx}(\tau_2-\tau_3)R_{xy_3}(\tau_1)
\end{aligned} \tag{4.60}
$$

This proves that the third-order cross-correlation function is

$$
\begin{aligned}
R_{xxxy_c}&(\tau_1, \tau_2, \tau_3) \\
&= 6\iiint h_3(\alpha, \beta, \gamma)R_{xx}(\tau_1-\alpha)R_{xx}(\tau_2-\beta)R_{xx}(\tau_3-\gamma)\,d\alpha\,d\beta\,d\gamma
\end{aligned} \tag{4.61}
$$

Third-Order Cross-Spectrum

From Eq. (3.182), the third-order cross-spectrum (trispectrum) is defined by

$$
S_{xxxy_c}(f, g, h) = \iiint R_{xxxy_c}(\tau_1, \tau_2, \tau_3)e^{-j2\pi(f\tau_1+g\tau_2+h\tau_3)}\,d\tau_1\,d\tau_2\,d\tau_3 \tag{4.62}
$$

Substitution of Eq. (4.61) into Eq. (4.62) after many steps gives the important trispectrum result in the frequency domain that is equivalent to Eq. (4.61), namely,

$$S_{xxxy_c}(f, g, h) = 6H_3(f, g, h)S_{xx}(f)S_{xx}(g)S_{xx}(h) \tag{4.63}$$

Here, $H_3(f, g, h)$ is the triple Fourier transform of $h_3(\tau_1, \tau_2, \tau_3)$ as per Eq. (3.69). Equation (4.63) can also be derived directly from Eq. (3.184) without computing correlation functions. Note the simplicity of Eq. (4.63) compared with Eq. (4.61).

4.1.5 Summary of Results

As in Section 3.4.6, the main equations should be compared for their inner mathematical structure as one proceeds from linear to bilinear to trilinear systems.

Time-domain cross-correlation results are

$$R_{xy_1}(\tau) = \int h_1(\alpha)R_{xx}(\tau - \alpha)\, d\alpha \tag{4.64}$$

$$R_{xxy_b}(\tau_1, \tau_2) = 2\iint h_2(\alpha, \beta)R_{xx}(\tau_1 - \alpha)R_{xx}(\tau_2 - \beta)\, d\alpha\, d\beta \tag{4.65}$$

$$R_{xxxy_c}(\tau_1, \tau_2, \tau_3) = 6\iiint h_3(\alpha, \beta, \gamma)R_{xx}(\tau_1 - \alpha)R_{xx}(\tau_2 - \beta)R_{xx}(\tau_3 - \gamma)\, d\alpha\, d\beta\, d\gamma \tag{4.66}$$

The quantities $y_b(t)$ and $y_c(t)$ are defined by Eqs. (4.25) and (4.45), respectively. Corresponding frequency-domain cross-spectrum results are

$$S_{xy_1}(f) = H_1(f)S_{xx}(f) \tag{4.67}$$

$$S_{xxy_b}(f, g) = 2H_2(f, g)S_{xx}(f)S_{xx}(g) \tag{4.68}$$

$$S_{xxxy_c}(f, g, h) = 6H_3(f, g, h)S_{xx}(f)S_{xx}(g)S_{xx}(h) \tag{4.69}$$

The output autocorrelation functions are

$$R_{y_1y_1}(\tau) = \iint h_1(\alpha)h_1(\beta)R_{xx}(\tau + \alpha - \beta)\, d\alpha\, d\beta \tag{4.70}$$

$$R_{y_by_b}(\tau) = 2\iiiint h_2(\alpha, \beta)h_2(u, v)R_{xx}(\tau + \alpha - u)R_{xx}(\tau + \beta - v)\, d\alpha\, d\beta\, du\, dv \tag{4.71}$$

$$R_{y_c y_c}(\tau) = 6 \iiiiiint h_3(\alpha, \beta, \gamma) h_3(u, v, w) R_{xx}(\tau + \alpha - u)$$
$$\times R_{xx}(\tau + \beta - v) R_{xx}(\tau + \gamma - w)\, d\alpha\, d\beta\, d\gamma\, du\, dv\, dw \qquad (4.72)$$

Corresponding output autospectral density functions are

$$S_{y_1 y_1}(f) = |H_1(f)|^2 S_{xx}(f) \qquad (4.73)$$

$$S_{y_b y_b}(f) = 2 \int |H_2(u, f - u)|^2 S_{xx}(u) S_{xx}(f - u)\, du \qquad (4.74)$$

$$S_{y_c y_c}(f) = 6 \iint |H_3(u, v - u, f - v)|^2 S_{xx}(u) S_{xx}(v - u) S_{xx}(f - v)\, du\, dv \qquad (4.75)$$

The benefits of frequency-domain results compared to time-domain results are clear from these equations.

4.2 GENERAL THIRD-ORDER NONLINEAR MODELS

4.2.1 Nonlinear Model with Correlated Outputs

Some of the preceding results will now be extended so as to analyze various desired nonlinear models excited by zero mean value Gaussian stationary random data. Consider the general third-order Volterra nonlinear time-domain model of Figure 4.2, where the total output is given by

$$y(t) = y_1(t) + y_2(t) + y_3(t) \qquad (4.76)$$

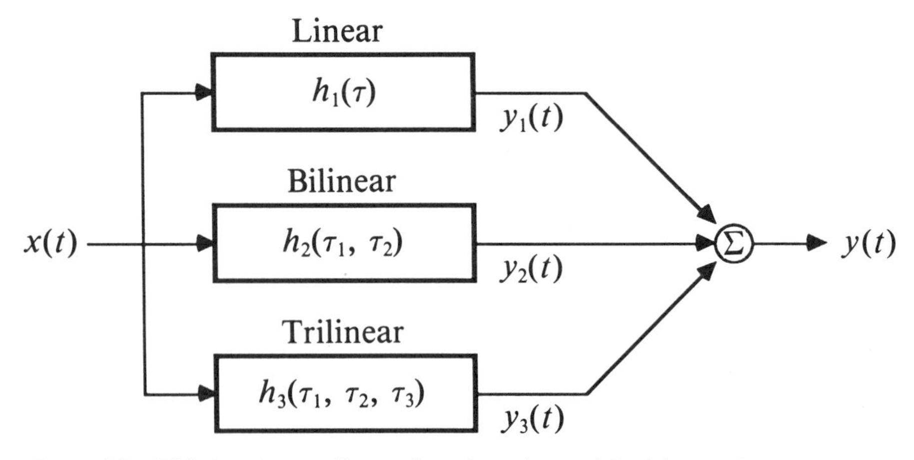

Figure 4.2 Third-order nonlinear time-domain model with correlated outputs.

Here, $y_1(t)$, $y_2(t)$, and $y_3(t)$ satisfy Eqs. (4.1), (4.22), and (4.37), respectively. The output autocorrelation function is

$$R_{yy}(\tau) = R_{y_1y_1}(\tau) + R_{y_2y_2}(\tau) + R_{y_3y_3}(\tau) + R_{y_1y_3}(\tau) + R_{y_3y_1}(\tau) \tag{4.77}$$

where $R_{y_1y_1}(\tau)$, $R_{y_2y_2}(\tau)$, $R_{y_1y_3}(\tau)$, and $R_{y_3y_3}(\tau)$ satisfy Eqs. (4.7), (4.29), (4.30), (4.43), and (4.53), respectively. These equations are complicated to apply. Also, correlation occurs between the outputs $y_1(t)$ and $y_3(t)$. One has here the mean values $\bar{y}_1 = \bar{y}_3 = 0$ with

$$\bar{y} = \bar{y}_2 = E[y(t)] = \int H_2(f, -f)S_{xx}(f)\,df \tag{4.78}$$

First-order cross-correlation functions are

$$R_{xy_1}(\tau) = \int h_1(\alpha)R_{xx}(\tau - \alpha)\,d\alpha \tag{4.79}$$

$$R_{xy_2}(\tau) = 0 \tag{4.80}$$

$$R_{xy_3}(\tau) = \int c(\alpha)R_{xx}(\tau - \alpha)\,d\alpha \tag{4.81}$$

as shown in Eqs. (4.3), (4.28), and (4.39), respectively. One can also write appropriate equations from Figure 4.2 for the second- and third-order cross-correlation functions. However, it is sufficient here to note that all of these equations are complicated and that for the assumed zero mean value Gaussian input data, the second-order functions are

$$R_{xxy_1}(\tau_1, \tau_2) = R_{xxy_3}(\tau_1, \tau_2) = 0, \qquad R_{xxy_2}(\tau_1, \tau_2) \neq 0 \tag{4.82}$$

while the third-order functions are

$$\begin{gathered} R_{xxxy_1}(\tau_1, \tau_2, \tau_3) \neq 0, \qquad R_{xxxy_3}(\tau_1, \tau_2, \tau_3) \neq 0 \\ R_{xxxy_2}(\tau_1, \tau_2, \tau_3) = 0 \end{gathered} \tag{4.83}$$

With respect to the total output $y(t)$, one has

$$R_{xy}(\tau) = R_{xy_1}(\tau) + R_{xy_3}(\tau) \tag{4.84}$$

$$R_{xxy}(\tau_1, \tau_2) = R_{xxy_2}(\tau_1, \tau_2) \tag{4.85}$$

$$R_{xxxy}(\tau_1, \tau_2, \tau_3) = R_{xxxy_1}(\tau_1, \tau_2, \tau_3) + R_{xxxy_3}(\tau_1, \tau_2, \tau_3) \tag{4.86}$$

Fourier transforms of Eq. (4.76) give the third-order Volterra nonlinear frequency-domain model of Figure 4.3 where

$$Y(f) = \mathscr{F}[y(t) - \bar{y}] = Y_1(f) + Y_2(f) + Y_3(f) \tag{4.87}$$

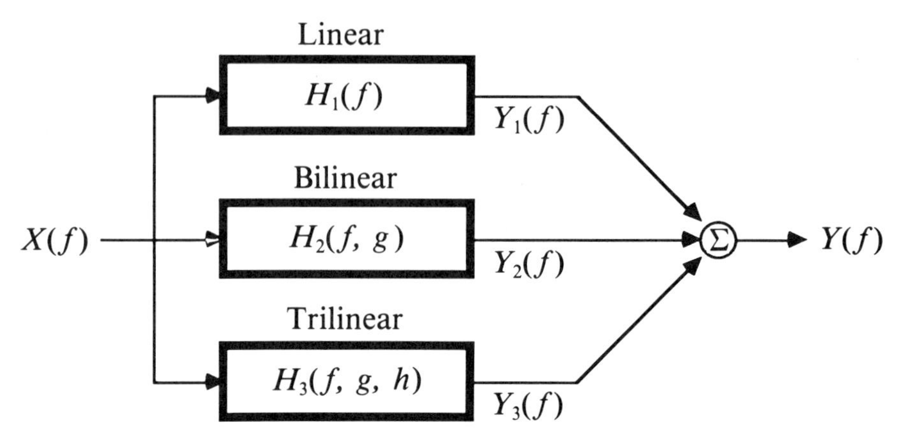

Figure 4.3 Third-order nonlinear frequency-domain model with correlated outputs.

The terms $Y_1(f)$, $Y_2(f)$, and $Y_3(f)$ satisfy Eqs. (4.2), (4.24), and (4.38), respectively.

The output autospectral density function for the general third-order nonlinear model of Figure 4.3 is as follows. From Eq. (4.77), by taking Fourier transforms,

$$S_{yy}(f) = S_{y_1y_1}(f) + S_{y_2y_2}(f) + S_{y_3y_3}(f) + S_{y_1y_3}(f) + S^*_{y_1y_3}(f) \tag{4.88}$$

where the linear system output spectrum $S_{y_1y_1}(f)$ is given by Eq. (4.9). The bilinear system output spectrum from Eq. (4.29) is

$$S_{y_2y_2}(f) = S_{y_by_b}(f) + (\bar{y}_2)^2\delta_1(f) \tag{4.89}$$

with $S_{y_by_b}(f)$ computed by Eq. (4.31) and $\bar{y}_2$ computed by Eq. (4.23). The trilinear system output spectrum from Eq. (4.56) is

$$S_{y_3y_3}(f) = S_{y_cy_c}(f) + |C(f)|^2 S_{xx}(f) \tag{4.90}$$

with $S_{y_cy_c}(f)$ computed by Eq. (4.55) and $C(f)$ computed by Eq. (4.42).

$$C(f) = 3\int H_3(u, -u, f)S_{xx}(u)\,du \tag{4.91}$$

The cross-spectrum term from Eq. (4.44) is

$$S_{y_1y_3}(f) = H_1^*(f)C(f)S_{xx}(f) \tag{4.92}$$

These correlation and spectral density relations can be replaced by simpler expressions using revised nonlinear models equivalent to Figures 4.2 and 4.3 where the outputs are uncorrelated.

4.2.2 Nonlinear Model with Uncorrelated Outputs

It is desired to decompose the zero mean value output $[y(t) - \bar{y}]$ into parts that will have zero mean values and be mutually uncorrelated. Hence, one should replace $y_1(t)$, $y_2(t)$, and $y_3(t)$ in Eq. (4.76) by $y_a(t)$, $y_b(t)$, and $y_c(t)$, respectively, defined by

$$y_a(t) = \int [h_1(\alpha) + c(\alpha)]x(t - \alpha)\, d\alpha \tag{4.93}$$

$$y_b(t) = y_2(t) - \bar{y}_2 \tag{4.94}$$

$$y_c(t) = y_3(t) - \int c(\alpha)x(t - \alpha)\, d\alpha \tag{4.95}$$

Now, Eq. (4.76) becomes, with $\bar{y} = \bar{y}_2$,

$$y(t) = \bar{y} + y_a(t) + y_b(t) + y_c(t) \tag{4.96}$$

Figure 4.2 can thus be replaced by the equivalent Figure 4.4. It will be shown later that the linear output term $y_a(t)$ is the result of passing $x(t)$ through the optimum linear system $[h_1(\alpha) + c(\alpha)]$ to predict $y(t)$ from $x(t)$.

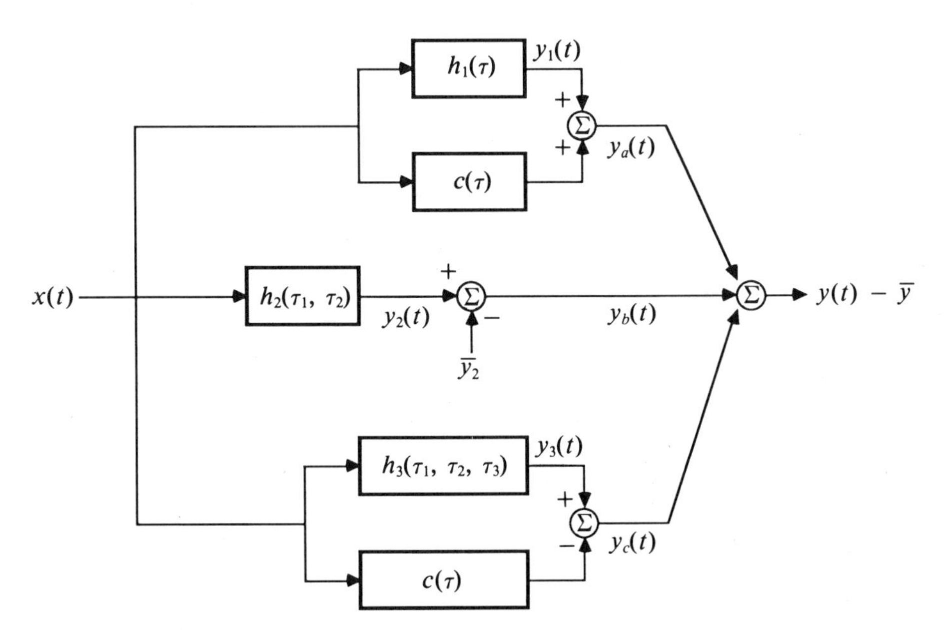

Figure 4.4 Revised third-order nonlinear time-domain model with uncorrelated outputs.

In the revised nonlinear model of Figure 4.4, the output autocorrelation function

$$R_{yy}(\tau) = (\bar{y})^2 + R_{y_a y_a}(\tau) + R_{y_b y_b}(\tau) + R_{y_c y_c}(\tau) \tag{4.97}$$

where $R_{y_a y_a}(\tau)$ is given by

$$R_{y_a y_a}(\tau) = \iint [h_1(\alpha) + c(\alpha)][h_1(\beta) + c(\beta)]R_{xx}(\tau + \alpha - \beta)\, d\alpha\, d\beta \tag{4.98}$$

and $R_{y_b y_b}(\tau)$ and $R_{y_c y_c}(\tau)$ are given by Eqs. (4.30) and (4.54), respectively. The various output cross-correlation functions and mean values are now

$$R_{y_a y_b}(\tau) = R_{y_a y_c}(\tau) = R_{y_b y_c}(\tau) = 0 \tag{4.99}$$

$$\bar{y}_a = \bar{y}_b = \bar{y}_c = 0 \tag{4.100}$$

The input/output cross-correlation functions become

$$R_{xy}(\tau) = R_{xy_a}(\tau) = \int [h_1(\alpha) + c(\alpha)]R_{xx}(\tau - \alpha)\, d\alpha \tag{4.101}$$

$$R_{xy_b}(\tau) = R_{xy_c}(\tau) = 0 \tag{4.102}$$

In words, $y_a(t)$, $y_b(t)$, and $y_c(t)$ are mutually uncorrelated, and $x(t)$ is uncorrelated with both $y_b(t)$ and $y_c(t)$.

Results for $y_a(t)$, $y_b(t)$, and $y_c(t)$ will now be expressed in the frequency domain using Fourier transforms and spectral density functions. The Fourier transform of Eq. (4.96) gives

$$Y(f) = \mathscr{F}[y(t) - \bar{y}] = Y_a(f) + Y_b(f) + Y_c(f) \tag{4.103}$$

where

$$Y_a(f) = [H_1(f) + C(f)]X(f) \tag{4.104}$$

$$Y_b(f) = Y_2(f) \tag{4.105}$$

$$Y_c(f) = Y_3(f) - C(f)X(f) \tag{4.106}$$

with the term $C(f)$ computed by Eq. (4.42), while $Y_2(f)$ and $Y_3(f)$ satisfy Eqs. (4.24) and (4.38), respectively. The result of Eq. (4.103) is displayed in the revised nonlinear model of Figure 4.5, which is equivalent to Figure 4.3.

From Eq. (4.99), the various output cross-spectral density functions will be identically zero for the mutually uncorrelated outputs, namely,

$$S_{y_a y_b}(f) = S_{y_a y_c}(f) = S_{y_b y_c}(f) = 0 \tag{4.107}$$

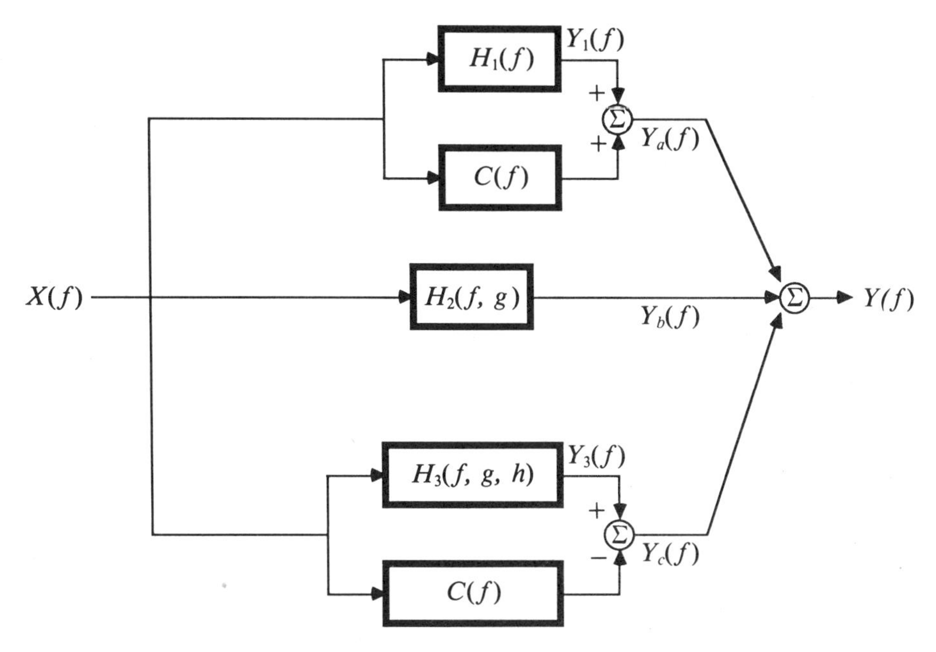

Figure 4.5 Revised third-order nonlinear frequency-domain model with uncorrelated outputs.

The total output autospectral density function in Figure 4.5 becomes

$$S_{yy}(f) = S_{y_a y_a}(f) + S_{y_b y_b}(f) + S_{y_c y_c}(f) \tag{4.108}$$

where from Eqs. (4.104), (4.31), and (4.55),

$$S_{y_a y_a}(f) = |H_1(f) + C(f)|^2 S_{xx}(f) \tag{4.109}$$

$$S_{y_b y_b}(f) = 2 \int |H_2(u, f-u)|^2 S_{xx}(u) S_{xx}(f-u)\, du \tag{4.110}$$

$$S_{y_c y_c}(f) = 6 \iint |H_3(u, v-u, f-v)|^2 S_{xx}(u) S_{xx}(v-u) S_{xx}(f-v)\, du\, dv \tag{4.111}$$

These spectral formulas are much simpler than corresponding correlation formulas. Equations (4.109)–(4.111) solve the *Output Prediction Problem* on how to predict output data properties in Figure 4.5 from knowledge of input data properties and system properties. The *Input Determination Problem* of $S_{xx}(f)$ cannot be solved here from knowledge of the system $H(f)$ and the total output data properties of $y(t)$. From Eq. (4.109), one requires knowledge of the system $[H_1(f) + C(f)]$ and the linear output term $y_a(t)$.

Fourier transforms of Eqs. (4.101) and (4.102) prove that the first-order cross-spectrum relations in Figure 4.5 are

$$S_{xy}(f) = S_{xy_a}(f) = [H_1(f) + C(f)]S_{xx}(f) \tag{4.112}$$

$$S_{xy_b}(f) = S_{xy_c}(f) = 0 \tag{4.113}$$

Second-order cross-spectrum relations from Eq. (4.34) and uncorrelated outputs are

$$S_{xxy}(f, g) = S_{xxy_b}(f, g) = 2H_2(f, g)S_{xx}(f)S_{xx}(g) \tag{4.114}$$

$$S_{xxy_a}(f, g) = S_{xxy_c}(f, g) = 0 \tag{4.115}$$

Third-order cross-spectrum relations from Eq. (4.63) and uncorrelated outputs are

$$S_{xxxy}(f, g, h) = S_{xxxy_c}(f, g, h) = 6H_3(f, g, h)S_{xx}(f)S_{xx}(g)S_{xx}(h) \tag{4.116}$$

$$S_{xxxy_a}(f, g, h) = S_{xxxy_b}(f, g, h) = 0 \tag{4.117}$$

Equations (4.42), (4.112), (4.114), and (4.116) solve the *System Identification Problem* to determine all of the systems in Figure 4.5 from measurements only of the input data $x(t)$ and the total output data $y(t)$. Furthermore, these calculations can be done using their Fourier transform quantities $X(f)$ and $Y(f)$ without ever computing any first- or higher-order correlation functions.

4.2.3 Identification of Linear, Bilinear, and Trilinear Systems

Useful procedures will now be developed to determine linear, bilinear, and trilinear system properties from measured data. Assume that only $x(t)$ and $y(t)$ are measured and that the models of Figures 4.3 and 4.5 apply except, perhaps, for extraneous zero mean value Gaussian output noise $n(t)$, which is assumed here to be uncorrelated with the mutually uncorrelated outputs $y_a(t)$, $y_b(t)$, and $y_c(t)$. This noise term represents all possible deviations from the ideal nonlinear models of Figures 4.3 and 4.5. With such output noise, they become Figures 4.6 and 4.7, where $N(f)$ is the Fourier transform of $n(t)$. In place of Eq. (4.103), one now obtains for the model of Figure 4.7 the result

$$Y(f) = Y_a(f) + Y_b(f) + Y_c(f) + N(f) \tag{4.118}$$

and in place of Eq. (4.108),

$$S_{yy}(f) = S_{y_a y_a}(f) + S_{y_b y_b}(f) + S_{y_c y_c}(f) + S_{nn}(f) \tag{4.119}$$

where $S_{nn}(f)$ is the autospectral density function of the output noise.

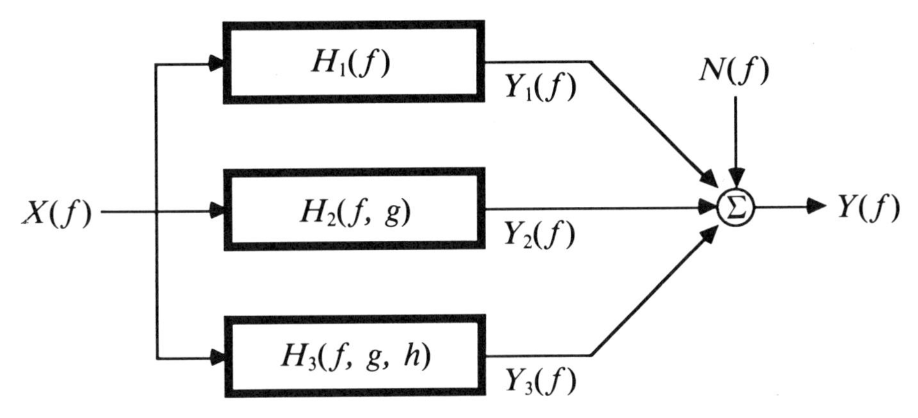

Figure 4.6 General third-order nonlinear frequency-domain model with correlated outputs.

First-, second-, and third-order spectral density functions can be computed as outlined in Section 3.4 by dividing $x(t)$ and $y(t)$ into n_d subrecords of length T each, determining $X(f)$ and $Y(f)$ for each subrecord, and performing n_d available averages so as to estimate the following functions:

$$S_{xx}(f) = \frac{1}{T} E[X^*(f)X(f)] \tag{4.120}$$

$$S_{xy}(f) = \frac{1}{T} E[X^*(f)Y(f)] \tag{4.121}$$

$$S_{xxy}(f, g) = \frac{1}{T} E[X^*(f)X^*(g)Y(f + g)] \tag{4.122}$$

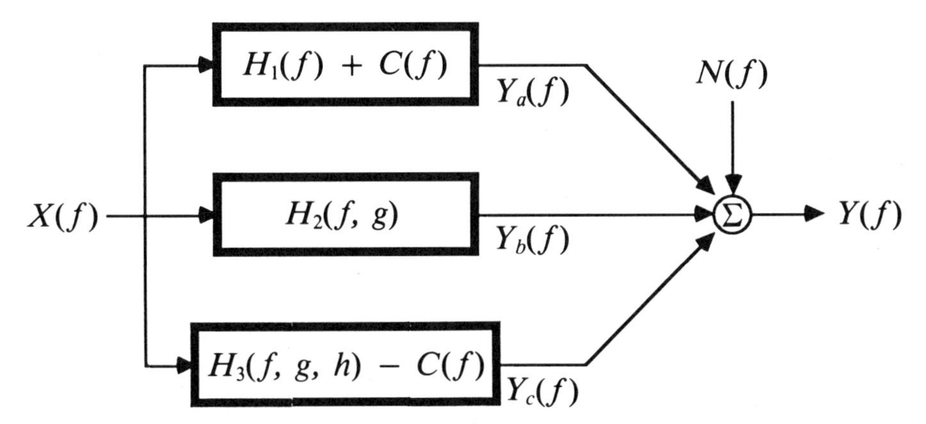

Figure 4.7 General third-order nonlinear frequency-domain model with uncorrelated outputs.

$$S_{xxxy}(f, g, h) = \frac{1}{T} E[X^*(f)X^*(g)X^*(h)Y(f + g + h)] \tag{4.123}$$

Computation of the multidimensional bispectral and trispectral density functions will generally be quite difficult.

The ratio of $S_{xy}(f)$ divided by $S_{xx}(f)$ gives the overall *optimum* linear system frequency response function $H_o(f)$ to estimate $y(t)$ from $x(t)$. Equation (4.112) then shows that

$$H_o(f) = \frac{S_{xy}(f)}{S_{xx}(f)} = H_1(f) + C(f) \tag{4.124}$$

This result for $H_o(f)$ varies as a function of $S_{xx}(f)$ when $C(f) \neq 0$. Note that $H_o(f) = H_1(f)$ if and only if $C(f) = 0$. The first-order linear system $H_1(f)$ is given by

$$H_1(f) = H_o(f) - C(f) \tag{4.125}$$

where $H_o(f)$ is known from Eq. (4.124) and $C(f)$ is still to be determined. In general, $H_1(f) \neq H_o(f)$.

The second-order bilinear system $H_2(f, g)$ is found from Eq. (4.114) by

$$H_2(f, g) = \frac{S_{xxy}(f, g)}{2S_{xx}(f)S_{xx}(g)} \tag{4.126}$$

When $f = u$ and $g = -u$, since $S_{xx}(-u) = S_{xx}(u)$, one obtains

$$H_2(u, -u) = \frac{S_{xxy}(u, -u)}{2S_{xx}^2(u)} \tag{4.127}$$

Here, from Eq. (4.122), using $X^*(-u) = X(u)$,

$$S_{xxy}(u, -u) = \frac{1}{T} E[X^*(u)X(u)Y(0)] \tag{4.128}$$

When $f = u$ and $g = f - u$,

$$H_2(u, f - u) = \frac{S_{xxy}(u, f - u)}{2S_{xx}(u)S_{xx}(f - u)} \tag{4.129}$$

where from Eq. (4.122),

$$S_{xxy}(u, f - u) = \frac{1}{T} E[X^*(u)X^*(f - u)Y(f)] \tag{4.130}$$

The third-order trilinear system $H_3(f, g, h)$ is found from Eq. (4.116) by

$$H_3(f, g, h) = \frac{S_{xxxy}(f, g, h)}{6S_{xx}(f)S_{xx}(g)S_{xx}(h)} \tag{4.131}$$

When $f = u$, $g = -u$, and $h = f$, one obtains

$$H_3(u, -u, f) = \frac{S_{xxxy}(u, -u, f)}{6S_{xx}(u)S_{xx}(u)S_{xx}(f)} \tag{4.132}$$

where from Eq. (4.123), using $X^*(-u) = X(u)$,

$$S_{xxxy}(u, -u, f) = \frac{1}{T} E[X^*(u)X(u)X^*(f)Y(f)] \tag{4.133}$$

When $f = u$, $g = v - u$, and $h = f - v$,

$$H_3(u, v - u, f - v) = \frac{S_{xxxy}(u, v - u, f - v)}{6S_{xx}(u)S_{xx}(v - u)S_{xx}(f - v)} \tag{4.134}$$

Here, from Eq. (4.123),

$$S_{xxxy}(u, v - u, f - v) = \frac{1}{T} E[X^*(u)X^*(v - u)X^*(f - v)Y(f)] \tag{4.135}$$

Substitution of Eq. (4.132) into Eq. (4.91) shows that

$$C(f) = 3\int H_3(u, -u, f)S_{xx}(u)\,du = \int \frac{S_{xxxy}(u, -u, f)}{2S_{xx}(u)S_{xx}(f)}\,du \tag{4.136}$$

The value of $C(f)$ together with $H_0(f)$ now gives $H_1(f)$ by Eq. (4.125). Thus, Eqs. (4.125), (4.126), and (4.131) solve the *System Identification Problem* on how to identify system properties from knowledge of input data properties and total output data properties.

From Eqs. (4.124), (4.128), and (4.134), it follows by substitution that the uncorrelated output autospectra terms of Eqs. (4.109)–(4.111) in Figures 4.5 and 4.7 can be computed by the formulas

$$S_{y_a y_a}(f) = \frac{|S_{xy}(f)|^2}{S_{xx}(f)} \tag{4.137}$$

$$S_{y_b y_b}(f) = \frac{1}{2}\int \frac{|S_{xxy}(u, f - u)|^2}{S_{xx}(u)S_{xx}(f - u)}\,du \tag{4.138}$$

$$S_{y_c y_c}(f) = \frac{1}{6}\iint \frac{|S_{xxxy}(u, v - u, f - v)|^2}{S_{xx}(u)S_{xx}(v - u)S_{xx}(f - v)}\,du\,dv \tag{4.139}$$

One can also calculate the correlated output autospectra terms in Figure 4.3. All of these results can be obtained solely from measurement and analysis of the input data $x(t)$ and total output data $y(t)$ that includes the extraneous output noise.

4.2.4 Linear and Nonlinear Coherence Functions

Linear and nonlinear coherence functions will now be stated for the linear and nonlinear parts of Figure 4.7. The *linear coherence function* between $x(t)$ and $y(t)$ is defined for $f \neq 0$ by

$$\gamma_{xy}^2(f) = \frac{|S_{xy}(f)|^2}{S_{xx}(f)S_{yy}(f)} = \frac{S_{y_a y_a}(f)}{S_{yy}(f)} \tag{4.140}$$

where $0 \leqslant \gamma_{xy}^2(f) \leqslant 1$ for all f. The linear output spectrum $S_{y_a y_a}(f)$ is computed by Eq. (4.109) or (4.137). This gives the proportion of $S_{yy}(f)$ due to $x(t)$ passing through the optimum linear system $H_o(f) = [H_1(f) + C(f)]$. From Eqs. (4.109), (4.124), and (4.140), the linear output spectrum

$$S_{y_a y_a}(f) = |H_o(f)|^2 S_{xx}(f) = \gamma_{xy}^2(f) S_{yy}(f) \tag{4.141}$$

The *bilinear coherence function* between $x(t)$ and $y(t)$ via the bilinear system $H_2(f, g)$ is defined for $f \neq 0$ by

$$q_{xy_b}^2(f) = \frac{S_{y_b y_b}(f)}{S_{yy}(f)} \tag{4.142}$$

where $0 \leqslant q_{xy_b}^2(f) \leqslant 1$ for all f. The bilinear output spectrum $S_{y_b y_b}(f)$ is computed by Eq. (4.110) or (4.138). This gives the proportion of $S_{yy}(f)$ due to $x(t)$ via the bilinear system $H_2(f, g)$. From Eq. (4.142), the bilinear output spectrum satisfies

$$S_{y_b y_b}(f) = q_{xy_b}^2(f) S_{yy}(f) \tag{4.143}$$

The *trilinear coherence function* between $x(t)$ and $y(t)$ via the trilinear system $[H_3(f, g, h) - C(f)]$ is defined for $f \neq 0$ by

$$q_{xy_c}^2(f) = \frac{S_{y_c y_c}(f)}{S_{yy}(f)} \tag{4.144}$$

where $0 \leqslant q_{xy_c}^2(f) \leqslant 1$ for all f. The trilinear output spectrum $S_{y_c y_c}(f)$ is computed by Eq. (4.111) or (4.139). This gives the proportion of $S_{yy}(f)$ due to $x(t)$ via the trilinear system $[H_3(f, g, h) - C(f)]$. From Eq. (4.144), the trilinear output spectrum satisfies

$$S_{y_c y_c}(f) = q_{xy_c}^2(f) S_{yy}(f) \tag{4.145}$$

Substitution of Eqs. (4.141), (4.143), and (4.145) into Eq. (4.119) now shows that the *output noise spectrum* in Figure 4.7 is given by

$$S_{nn}(f) = [1 - \gamma_{xy}^2(f) - q_{xy_b}^2(f) - q_{xy_c}^2(f)]S_{yy}(f) \tag{4.146}$$

The ratio of $S_{nn}(f)$ to $S_{yy}(f)$ gives the proportion of $S_{yy}(f)$ that is *not* due to $x(t)$ passing through any of the systems in Figure 4.7. Specifically, $S_{nn}(f)$ determines that part of the total output spectrum $S_{yy}(f)$ due to uncorrelated noise at the output, or due to $x(t)$ passing through higher-order nonlinear systems than those included in Figure 4.7.

A *goodness-of-fit* measure for the validity of this revised nonlinear model can be defined by seeing how close the sum $[\gamma_{xy}^2(f) + q_{xy_b}^2 + q_{xy_c}^2(f)]$ is to unity. This produces greatly improved results over previous work restricted to only linear system analysis where the bilinear and trilinear coherence functions are not computed and are assumed to be zero. In previous work, all of the possible bilinear and trilinear effects are included in the extraneous output noise. Thus, in general, previous work has computed an output noise spectrum that is much larger than the smaller result of Eq. (4.146). Also, previous work has computed an erroneous $H_1(f)$ when assuming incorrectly that $C(f) = 0$ in Eq. (4.125). The conclusion is that even if only linear system analysis and identification are desired, it can be made more accurate by following these procedures that allow for possible bilinear and trilinear effects. Of course, good engineering work now requires that these nonlinear effects be evaluated properly by following the procedures developed in this book, especially the simple, practical procedures in Chapters 5–7. Also, since results are functions of frequency, emphasis should be placed on those frequencies where there are peaks in $S_{yy}(f)$ as these peak values are of major physical interest where most of the energy or power occurs.

4.2.5 Summary of Results

A summary follows of some of the main results. Fourier transform relations for the third-order nonlinear frequency-domain model with correlated outputs of Figure 4.6 where $Y(f) = \mathscr{F}[y(t) - \bar{y}]$ are

$$Y(f) = Y_1(f) + Y_2(f) + Y_3(f) + N(f) \tag{4.147}$$

$$Y_1(f) = H_1(f)X(f) \tag{4.148}$$

$$Y_2(f) = \int H_2(u, f - u)X(u)X(f - u)\, du - \bar{y}_2\delta_1(f) \tag{4.149}$$

$$Y_3(f) = \iint H_3(u, v - u, f - v)X(u)X(v - u)X(f - v)\, du\, dv \tag{4.150}$$

$$\bar{y}_2 = \int H_2(f, -f)S_{xx}(f)\, df \tag{4.151}$$

These output results can all be computed from measurement of $x(t)$ and knowledge of the systems $H_1(f)$, $H_2(f, g)$, and $H_3(f, g, h)$. Note that for $f \neq 0$, the terms $\bar{y}_2$ and $\delta_1(f)$ can be omitted.

Spectral results for the third-order nonlinear frequency-domain model with correlated outputs of Figure 4.6 are

$$S_{yy}(f) = S_{y_1y_1}(f) + S_{y_2y_2}(f) + S_{y_3y_3}(f) + S_{y_1y_3}(f) + S^*_{y_1y_3}(f) + S_{nn}(f) \tag{4.152}$$

$$S_{y_1y_1}(f) = |H_1(f)|^2 S_{xx}(f) \tag{4.153}$$

$$S_{y_2y_2}(f) = S_{y_by_b}(f) + (\bar{y}_2)^2 \delta_1(f) \tag{4.154}$$

$$S_{y_by_b}(f) = 2 \int |H_2(u, f-u)|^2 S_{xx}(u) S_{xx}(f-u)\, du \tag{4.155}$$

$$S_{y_3y_3}(f) = S_{y_cy_c}(f) + |C(f)|^2 S_{xx}(f) \tag{4.156}$$

$$S_{y_cy_c}(f) = 6 \iint |H_3(u, v-u, f-v)|^2 S_{xx}(u) S_{xx}(v-u) S_{xx}(f-v)\, du\, dv \tag{4.157}$$

$$S_{y_1y_3}(f) = H_1^*(f) C(f) S_{xx}(f) \tag{4.158}$$

$$C(f) = 3 \int H_3(u, -u, f) S_{xx}(u)\, du \tag{4.159}$$

Fourier transform relations for the third-order nonlinear frequency-domain model with uncorrelated outputs of Figure 4.7 are

$$Y(f) = Y_a(f) + Y_b(f) + Y_c(f) + N(f) \tag{4.160}$$

$$Y_a(f) = [H_1(f) + C(f)] X(f) \tag{4.161}$$

$$Y_b(f) = \int H_2(u, f-u) X(u) X(f-u)\, du - \bar{y}_2 \delta_1(f) \tag{4.162}$$

$$Y_c(f) = Y_3(f) - C(f) X(f) \tag{4.163}$$

$$Y_3(f) = \iint H_3(u, v-u, f-v) X(u) X(v-u) X(f-v)\, du\, dv \tag{4.164}$$

$$C(f) = 3 \int H_3(u, -u, f) S_{xx}(u)\, du \tag{4.165}$$

$$\bar{y}_2 = \int H_2(f, -f) S_{xx}(f)\, df \tag{4.166}$$

These output results can all be computed from measurement of $x(t)$ and knowledge of the systems $H_1(f)$, $H_2(f, g)$, and $H_3(f, g, h)$. Note that for $f \neq 0$, the terms $\bar{y}_2$ and $\delta_1(f)$ can be omitted.

Spectral results for the third-order nonlinear frequency-domain model with uncorrelated outputs of Figure 4.7 are

$$S_{yy}(f) = S_{y_a y_a}(f) + S_{y_b y_b}(f) + S_{y_c y_c}(f) + S_{nn}(f) \tag{4.167}$$

$$S_{y_a y_a}(f) = |H_1(f) + C(f)|^2 S_{xx}(f) \tag{4.168}$$

$$S_{y_b y_b}(f) = 2 \int |H_2(u, f-u)|^2 S_{xx}(u) S_{xx}(f-u)\, du \tag{4.169}$$

$$S_{y_c y_c}(f) = 6 \iint |H_3(u, v-u, f-v)|^2 S_{xx}(u) S_{xx}(v-u) S_{xx}(f-v)\, du\, dv \tag{4.170}$$

Identification of the linear, bilinear, and trilinear systems from simultaneous measurements of $x(t)$ and $y(t)$ in Figures 4.5 and 4.7 can be carried out as follows. The first-order linear system $H_1(f)$ is given by

$$H_1(f) = H_o(f) - C(f) \tag{4.171}$$

where the optimum linear system is

$$H_o(f) = \frac{S_{xy}(f)}{S_{xx}(f)} \tag{4.172}$$

and where $C(f)$ from Eq. (4.136) is

$$C(f) = \int \frac{S_{xxxy}(u, -u, f)}{2S_{xx}(u) S_{xx}(f)}\, du \tag{4.173}$$

The second-order bilinear system $H_2(f, g)$ is given by

$$H_2(f, g) = \frac{S_{xxy}(f, g)}{2S_{xx}(f) S_{xx}(g)} \tag{4.174}$$

The third-order trilinear system $H_3(f, g, h)$ is given by

$$H_3(f, g, h) = \frac{S_{xxxy}(f, g, h)}{6S_{xx}(f) S_{xx}(g) S_{xx}(h)} \tag{4.175}$$

Sections 4.2.1–4.2.4 contain other useful relations.

Advantages of the above procedures are that they are mathematically exact and widely applicable to many problems. Disadvantages are that the multidimensional bispectral and trispectral density functions are usually very expensive and time-consuming to compute as well as very difficult to interpret.

4.3 OPTIMUM LINEAR AND NONLINEAR SYSTEMS

Formulas for identifying the optimum forms for the linear, bilinear, and trilinear systems $H_1(f)$, $H_2(f, g)$, and $H_3(f, g, h)$ will now be derived. These formulas minimize the output noise spectrum $S_{nn}(f)$ in the third-order nonlinear model of Figure 4.7 with respect to all possible choices of H_1, H_2, and H_3. Unlike Section 4.2.3, no assumption is made here that $n(t)$ is uncorrelated with $y_a(t)$, $y_b(t)$, and $y_c(t)$. Instead, it will be shown that this is automatically true when $S_{nn}(f)$ is minimized.

For arbitrary choices of the systems H_1, H_2, and H_3, the quantity $N(f)$ in Figure 4.7 is given by

$$N(f) = Y(f) - Y_a(f) - Y_b(f) - Y_c(f) \tag{4.176}$$

The autospectrum $S_{nn}(f)$ follows directly by multiplying $N^*(f)$ by $N(f)$, taking expected values and multiplying by the factor $(1/T)$, namely,

$$S_{nn}(f) = \frac{1}{T} E[N^*(f)N(f)] \tag{4.177}$$

From Eq. (4.176), omitting f to simplify notation,

$$\begin{aligned} E[N^*N] &= E[(Y^* - Y_a^* - Y_b^* - Y_c^*)(Y - Y_a - Y_b - Y_c)] \\ &= E[Y^*(Y - Y_a - Y_b - Y_c) - Y_a^*(Y - Y_a) - Y_b^*(Y - Y_b) - Y_c^*(Y - Y_c)] \end{aligned}$$

plus other terms whose expected values are zero because $y_a(t)$, $y_b(t)$, and $y_c(t)$ are mutually uncorrelated in Figure 4.7. Thus,

$$\begin{aligned} E[N^*N] = E[Y^*Y] - E[Y^*Y_a] - E[Y^*Y_b] - E[Y^*Y_c] - E[Y_a^*Y] \\ + E[Y_a^*Y_a] - E[Y_b^*Y] + E[Y_b^*Y_b] - E[Y_c^*Y] + E[Y_c^*Y_c] \end{aligned}$$

This proves the general result applicable to any systems $H_1(f)$, $H_2(f, g)$, and $H_3(f, g, h)$ in Figure 4.6 that

$$\begin{aligned} S_{nn}(f) = S_{yy}(f) &+ [S_{y_ay_a}(f) - S_{yy_a}(f) - S_{y_ay}(f)] \\ &+ [S_{y_by_b}(f) - S_{yy_b}(f) - S_{y_by}(f)] + [S_{y_cy_c}(f) - S_{yy_c}(f) - S_{y_cy}(f)] \end{aligned} \tag{4.178}$$

The system H_1 appears only in the first brackets, the system H_2 appears only in the second brackets, and the system H_3 appears only in the third brackets of Eq. (4.178). Hence, the minimization of $S_{nn}(f)$ with respect to H_1, H_2, and H_3 will not involve any overlapping terms and can be carried out separately.

4.3.1 Optimum Linear System

From Eq. (4.104),

$$Y_a(f) = [H_1(f) + C(f)]X(f) \tag{4.179}$$

It follows that the three spectral terms in the *first* brackets of Eq. (4.178) involving $H_1(f)$ are

$$\begin{aligned} S_{y_a y_a}(f) &= [H_1^*(f) + C^*(f)][H_1(f) + C(f)]S_{xx}(f) \\ S_{y y_a}(f) &= [H_1(f) + C(f)]S_{yx}(f) \\ S_{y_a y}(f) &= [H_1^*(f) + C^*(f)]S_{xy}(f) \end{aligned} \tag{4.180}$$

Hence, omitting f to simplify notation, the first brackets in Eq. (4.178) become

$$[S_{y_a y_a} - S_{y y_a} - S_{y_a y}] = (H_1^* + C^*)(H_1 + C)S_{xx} - (H_1 + C)S_{yx} - (H_1^* + C^*)S_{xy} \tag{4.181}$$

The optimum linear system H_1 is defined as that particular H_1 over all possible H_1 that gives a minimum value for S_{nn}. This can be determined by setting the partial derivative of S_{nn} of Eq. (4.178) with respect to H_1^* equal to zero, holding H_1 fixed. As shown in reference 1, this procedure is equivalent to the classical operations of separately setting the partial derivatives of S_{nn} with respect to the real and imaginary parts of H_1 equal to zero. The desired partial derivative of S_{nn} with respect to H_1^* is the partial derivative of Eq. (4.181) with respect to H_1^*. When set equal to zero, this gives the result

$$\frac{\partial S_{nn}}{\partial H_1^*} = (H_1 + C)S_{xx} - S_{xy} = 0 \tag{4.182}$$

Hence, the optimum linear system $H_1 = H_1(f)$ must satisfy the equation

$$H_1(f) = \frac{S_{xy}(f)}{S_{xx}(f)} - C(f) = H_o(f) - C(f) \tag{4.183}$$

This is the same as Eq. (4.124) where $H_o(f)$, the overall optimum linear system, is not the same as $H_1(f)$ when $C(f) \neq 0$.

For this optimum choice of H_1, Eq. (4.180) becomes

$$S_{y_a y_a} = (H_1^* + C^*)S_{xy} = (H_1 + C)S_{yx} \tag{4.184}$$

since $S_{y_a y_a}$ is real-valued. Thus,

$$S_{y_a y_a} = S_{y y_a} = S_{y_a y} \tag{4.185}$$

and Eq. (4.181) is the same as

$$[S_{y_a y_a} - S_{y y_a} - S_{y_a y}] = -S_{y_a y_a} \tag{4.186}$$

where $S_{y_a y_a}$ satisfies Eq. (4.141). Also, from Eq. (4.176),

$$S_{n y_a} = S_{y y_a} - S_{y_a y_a} = 0 \tag{4.187}$$

so that $n(t)$ is uncorrelated with $y_a(t)$.

4.3.2 Optimum Bilinear System

From Eqs. (4.24) and (4.27),

$$Y_b(f) = \int H_2(u, f-u) X(u) X(f-u)\, du - \bar{y}_2 \delta_1(f) \tag{4.188}$$

It follows here by a direct proof that for Gaussian input data the three spectral terms in the *second* brackets of Eq. (4.178) involving $H_2(f, g)$ are

$$S_{y_b y_b}(f) = 2 \int H_2^*(u, f-u) H_2(u, f-u) S_{xx}(u) S_{xx}(f-u)\, du \tag{4.189}$$

$$S_{y y_b}(f) = \int H_2(u, f-u) S_{xxy}^*(u, f-u)\, du$$

$$S_{y_b y}(f) = \int H_2^*(u, f-u) S_{xxy}(u, f-u)\, du$$

Now the partial derivative of S_{nn} of Eq. (4.178) with respect to H_2^* (holding H_2 fixed) when set equal to zero gives the result

$$\frac{\partial S_{nn}}{\partial H_2^*} = 2H_2(u, f-u) S_{xx}(u) S_{xx}(f-u) - S_{xxy}(u, f-u) = 0$$

Hence, the optimum bilinear system $H_2(u, f-u)$ must satisfy the equation

$$H_2(u, f-u) = \frac{S_{xxy}(u, f-u)}{2S_{xx}(u) S_{xx}(f-u)} \tag{4.190}$$

This is the same as Eqs. (4.129) and (4.126). For this choice of H_2 it follows that

$$S_{y_b y_b} = S_{y y_b} = S_{y_b y} \tag{4.191}$$

The sum of the three terms in the second brackets is now

$$[S_{y_b y_b} - S_{y y_b} - S_{y_b y}] = -S_{y_b y_b} \tag{4.192}$$

where $S_{y_b y_b}(f)$ satisfies Eq. (4.143). Also, from Eq. (4.176),

$$S_{n y_b}(f) = 0 \tag{4.193}$$

so that $n(t)$ is uncorrelated with $y_b(t)$.

4.3.3 Optimum Trilinear System

From Eqs. (4.38) and (4.46),

$$Y_c(f) = \iint H_3(u, v-u, f-v)X(u)X(v-u)X(f-v)\,du\,dv - C(f)X(f) \tag{4.194}$$

Here it follows by a direct proof that for Gaussian input data the three spectral terms in the *third* brackets of Eq. (4.178) involving $H_3(f, g, h)$ are

$$S_{y_c y_c}(f) = 6\iint |H_3(u, v-u, f-v)|^2 S_{xx}(u)S_{xx}(v-u)S_{xx}(f-v)\,du\,dv \tag{4.195}$$

$$S_{y y_c}(f) = \iint H_3(u, v-u, f-v)S^*_{xxxy}(u, v-u, f-v)\,du\,dv$$

$$S_{y_c y}(f) = \iint H^*_3(u, v-u, f-v)S_{xxxy}(u, v-u, f-v)\,du\,dv$$

Now the partial derivative of S_{nn} of Eq. (4.178) with respect to H^*_3 (holding H_3 fixed) when set equal to zero gives the result

$$\frac{\partial S_{nn}}{\partial H^*_3} = 6H_3(u, v-u, f-v)S_{xx}(u)S_{xx}(v-u)S_{xx}(f-v) - S_{xxxy}(u, v-u, f-v) = 0$$

Hence, the optimum trilinear system $H_3(u, v-u, f-v)$ must satisfy the equation

$$H_3(u, v-u, f-v) = \frac{S_{xxxy}(u, v-u, f-v)}{6S_{xx}(u)S_{xx}(v-u)S_{xx}(f-v)} \tag{4.196}$$

This is the same as Eqs. (4.134) and (4.131). For this choice of H_3, it follows that

$$S_{y_c y_c} = S_{y y_c} = S_{y_c y} \tag{4.197}$$

The sum of the three terms in the third brackets is now

$$[S_{y_c y_c} - S_{y y_c} - S_{y_c y}] = -S_{y_c y_c} \tag{4.198}$$

where $S_{y_c y_c}(f)$ satisfies Eq. (4.145). Also, from Eq. (4.176),

$$S_{n y_c} = 0 \tag{4.199}$$

so that $n(t)$ is uncorrelated with $y_c(t)$.

4.3.4 Minimum Output Noise Spectrum

Substitution of Eqs. (4.186), (4.192), and (4.198) into Eq. (4.178) proves that the minimum output noise spectrum is given by

$$\begin{aligned} S_{nn}(f) &= S_{yy}(f) - S_{y_a y_a}(f) - S_{y_b y_b}(f) - S_{y_c y_c}(f) \\ &= [1 - \gamma_{xy}^2(f) - q_{xy_b}^2(f) - q_{xy_c}^2(f)]S_{yy}(f) \end{aligned} \tag{4.200}$$

This is the same as Eq. (4.146) and provides validity for the nonlinear model of Figure 4.7, as discussed below Eq. (4.146).

In summary, Eqs. (4.183), (4.190), and (4.196) give the optimum forms for linear, bilinear, and trilinear systems, respectively, that minimize the output noise spectrum $S_{nn}(f)$ in the general third-order nonlinear model of Figure 4.7. Equations (4.187), (4.193), and (4.199) prove that the resulting output noise $n(t)$ will be automatically uncorrelated with the linear output $y_a(t)$, the bilinear output $y_b(t)$, and the trilinear output $y_c(t)$. Finally, in terms of linear and nonlinear coherence functions, Eq. (4.200) gives the minimum value for the output noise spectrum. These results are all calculated solely from measurements of only the input data $x(t)$ and the total output data $y(t)$ under the assumption that the input data represent a zero mean value Gaussian stationary random process with arbitrary spectral properties.

This concludes Chapter 4.

5

SQUARE-LAW AND CUBIC NONLINEAR SYSTEMS

This chapter studies in detail special cases of bilinear and trilinear systems that are of great physical interest. These cases consist of linear systems in parallel with nonlinear square-law systems and/or nonlinear cubic systems, where these zero-memory nonlinear systems can be followed or preceded by other linear systems. Higher-order frequency response functions for these cases become functions of only one frequency variable that are much easier to determine than the functions of two and three variables required for general bilinear and trilinear systems in Chapters 3 and 4. Associated special bispectral and trispectral density functions also become functions of only one frequency variable. The chapter shows how to decompose the output autospectral density function into uncorrelated linear and nonlinear components. Formulas state how to identify the optimum system properties in the linear and nonlinear paths from simultaneous measurements of input data and total output data. A large class of single-input/single-output nonlinear models is shown to be equivalent to three-input/single-output linear models that can be analyzed with well-developed linear theory. Simple practical procedures are outlined to analyze and identify optimum third-order polynomial least-squares models that can approximate many zero-memory and finite-memory nonlinear systems. New standard nonlinear system analysis and identification techniques are obtained that should be used to extend conventional linear system techniques.

5.1 GENERAL THIRD-ORDER NONLINEAR MODEL

Models in this chapter are special cases of the third-order nonlinear frequency-domain models of Figures 4.6 and 4.7. The total output Fourier transform $Y(f)$

in Figure 4.6 is

$$Y(f) = Y_1(f) + Y_2(f) + Y_3(f) + N(f) \tag{5.1}$$

Correlated output terms $Y_1(f)$, $Y_2(f)$, and $Y_3(f)$ in Figure 4.6 satisfy Eqs. (4.148)–(4.150). The term $N(f)$ is the Fourier transform of a zero mean value uncorrelated output noise $n(t)$. The general second-order frequency response function $H_2(f, g)$ is a function of two frequency variables; the general third-order frequency response function $H_3(f, g, h)$ is a function of three frequency variables. Formulas in Chapters 3 and 4 indicate the extensive work required to analyze and identify these general higher-order functions.

In this chapter, two broad cases of great physical interest are examined that are called Case 1 and Case 2 models. A strong mathematical basis for the importance of these two special cases is that they are finite-memory nonlinear system extensions of the optimum third-order polynomial least-squares models in Section 5.5. The Case 1 nonlinear model replaces the second-order $H_2(f, g)$ in Figure 4.6 by a zero-memory square-law system followed by a linear system $A_2(f)$, and replaces the third-order $H_3(f, g, h)$ by a zero-memory cubic system followed by a linear system $A_3(f)$. The Case 2 nonlinear model replaces the second-order $H_2(f, g)$ in Figure 4.6 by a zero-memory square-law system preceded by a linear system $B_2(f)$, and replaces the third-order $H_3(f, g, h)$ in Figure 4.6 by a zero-memory cubic system preceded by a linear system $B_3(f)$. Case 1 and Case 2 finite-memory square-law and cubic models are considered in Sections 5.2 and 5.3, respectively. Various combinations of the Case 1 and Case 2 models are discussed in Section 5.4. Earlier published work on these matters is in references 3–5.

5.2 CASE 1 SQUARE-LAW AND CUBIC MODELS

Case 1 single-input/single-output nonlinear models occur when higher-order frequency response functions can be represented by additive first-order frequency response functions such that the linear, bilinear, and trilinear systems in Figure 4.6 are

$$H_1(f) = A_1(f) \tag{5.2}$$

$$H_2(f_1, f_2) = A_2(f_1 + f_2) \tag{5.3}$$

$$H_3(f_1, f_2, f_3) = A_3(f_1 + f_2 + f_3) \tag{5.4}$$

Here,

$$H_2(f, f) = A_2(2f) \tag{5.5}$$

$$H_3(f, f, f) = A_3(3f) \tag{5.6}$$

$$H_2(f, -f) = A_2(0)$$
$$H_2(u, f-u) = A_2(f) \tag{5.7}$$

$$H_3(u, v-u, f-v) = A_3(f) \tag{5.8}$$

The $A_i(f)$ are frequency response functions of constant-parameter linear systems either known or to be determined. For this case, Eqs. (4.2), (4.24), and (4.38) become

$$Y_1(f) = A_1(f)X(f) \tag{5.9}$$

$$Y_2(f) = A_2(f)\int X(u)X(f-u)\,du - \bar{y}_2\delta_1(f) = Y_b(f) \tag{5.10}$$

$$Y_3(f) = A_3(f)\iint X(u)X(v-u)X(f-v)\,du\,dv \tag{5.11}$$

with $\bar{y}_2$ given by

$$\bar{y}_2 = A_2(0)\int S_{xx}(f)\,df = A_2(0)\sigma_x^2 \tag{5.12}$$

Equation (5.10) shows that $\bar{y}_2$ occurs only at $f = 0$.

These results have a simple physical interpretation as pictured in Figure 5.1, where zero-memory squaring and cubing operations are followed by the A for "after" linear systems. The single integral in Eq. (5.10) is the Fourier transform of $x^2(t)$, denoted by $X_2(f)$, and the double integral in Eq. (5.11) is the Fourier transform of $x^3(t)$, denoted by $X_3(f)$. Equations (5.10) and (5.11) provide the basis for direct derivations of the single-variable frequency-domain formulas in Section 5.2.1 without knowing more general formulas from Chapter 4 involving two and three variables, and without computing correlation functions first. The

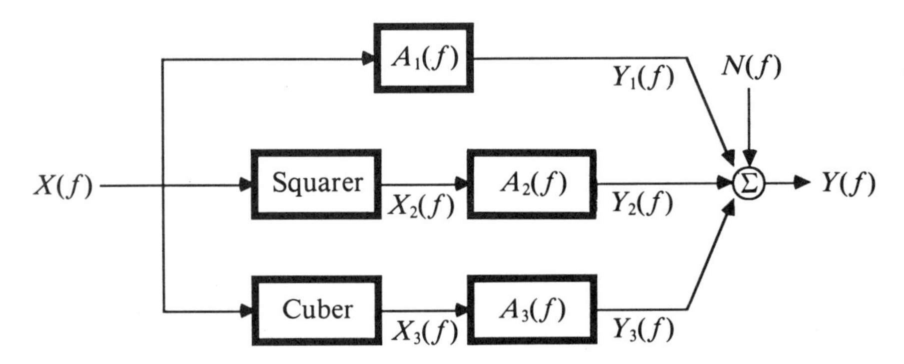

Figure 5.1 Case 1 nonlinear model with correlated outputs.

total output Fourier transform $Y(f)$ in Figure 5.1 is given by Eq. (5.1) using the terms in Eqs. (5.9)–(5.12). Figure 5.1 is a linear system in parallel with a finite-memory nonlinear system consisting of the Squarer/Cuber/$A(f)$ combination shown.

The Case 1 single-input/single-output nonlinear model of Figure 5.1 is equivalent to the three-input/single-output linear model of Figure 5.2 where the inputs $X(f)$, $X_2(f)$, and $X_3(f)$ are correlated. Recognition of this equivalence in references 3 and 4 is significant because it allows for the use of well-established linear techniques to analyze the original Case 1 nonlinear model. Multiple-input/single-output linear techniques are developed in great detail in references 1 and 2 from earlier work on these matters in references 6–9 and 18. The basis for these linear techniques is the computation of conditioned spectral density functions where the linear effects of preceding input records are removed from succeeding input records. Thus, using the notation in reference 1, the first input record $X_1(f) = X(f)$ is left alone, the second input record $X_2(f)$ is replaced by $X_{2\cdot1}(f)$, and the third input record is replaced by $X_{3\cdot2!}(f)$, where these new conditioned input records are mutually uncorrelated. There is no requirement that the input data $X(f)$ be Gaussian to identify the linear frequency response functions $A_1(f)$, $A_2(f)$, and $A_3(f)$ in Figures 5.1 and 5.2 from measurement of $X(f)$ and the total output $Y(f)$.

These general linear techniques from references 1 and 2 cannot be applied from measurement of $X(f)$ and $Y(f)$ to identify the linear frequency response functions $B_1(f)$, $B_2(f)$, and $B_3(f)$ in Case 2 nonlinear models. Instead, Section 5.3 shows how a Case 2 single-input/three-output nonlinear model with correlated outputs can be replaced by a revised Case 2 single-input/three-output nonlinear model with uncorrelated outputs. Then special bispectral and tri-

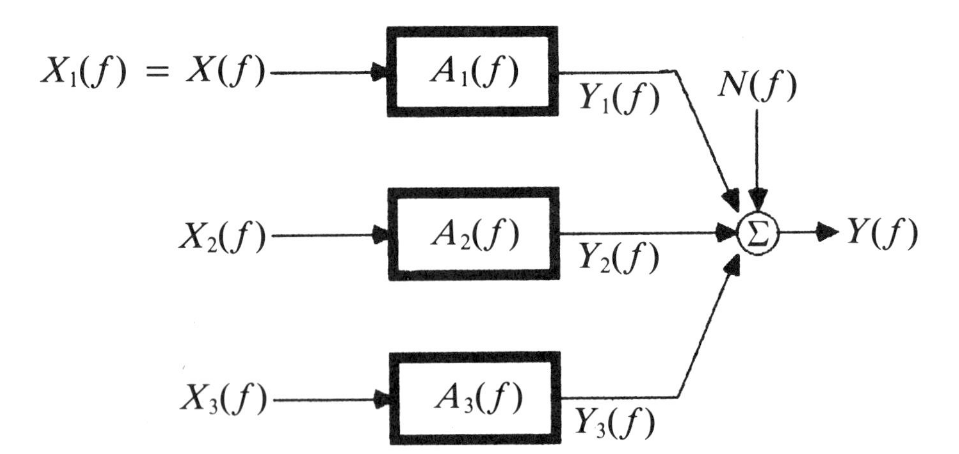

Figure 5.2 Three-input/single-output linear model with correlated inputs equivalent to Figure 5.1.

spectral density functions are used to solve the system identification problem for Case 2 nonlinear models.

Note also that it is not possible to identify the linear, bilinear, and trilinear systems in Figures 4.6 or 4.7 by creating equivalent three-input/single-output linear models. Instead, the linear, bilinear, and trilinear systems are identified by replacing the three correlated outputs $Y_1(f)$, $Y_2(f)$, and $Y_3(f)$ in Figure 4.6 with three mutually uncorrelated outputs $Y_a(f)$, $Y_b(f)$, and $Y_c(f)$ in Figure 4.7 and using the procedure in Section 4.2.3.

The case 1 nonlinear models of Figures 5.1 and 5.2 can be generalized further by letting $X_2(f)$ be the Fourier transform of any zero-memory nonlinear function $g_2[x(t)]$ and by letting $X_3(f)$ be the Fourier transform of any different zero-memory nonlinear function $g_3[x(t)]$. Figures 5.1 and 5.2 are important special cases where $g_2[x(t)] = x^2(t)$ and $g_3[x(t)] = x^3(t)$. Similarly, single-input/single-output nonlinear models with an arbitrary number of parallel paths can be analyzed when $X_2(f)$ and $X_3(f)$ extend to a sum of different $X_i(f)$ that are Fourier transforms of different $g_i[x(t)]$, $i = 2, 3, \ldots, q$, with each of these nonlinear systems then followed by different linear systems $A_i(f)$, $i = 2, 3, \ldots, q$. This extended Case 1 type nonlinear model can be replaced by a multiple-input/single-output linear model with the correlated inputs $X(f)$ amd $X_i(f)$, $i = 2, \ldots, q$, and the $A_i(f)$ can be identified by the conditioned spectral density techniques in references 1 and 2. Nonlinear system zero-memory (amplitude-domain) operations are defined by the $g_i[x(t)]$ separately from the nonlinear system frequency-domain operations that are defined by the $A_i(f)$.

New techniques discussed here in Chapter 5 represent alternative practical ways to analyze Case 1 and Case 2 nonlinear models that do not require computations of conditioned spectral density functions. Various derived formulas are strictly applicable only to problems when the input data $X(f)$ are Gaussian. For Gaussian input data, the mutually uncorrelated conditioned input records for the model in Figure 5.2 become $X_{2\cdot 1}(f) = X_2(f)$ and $X_{3\cdot 2!}(f) = X_3(f) - 3\sigma_x^2 X(f)$. These particular conditioned input records are obtained later in Figure 5.4 without using or referring to material in references 1 and 2. This material is required for non-Gaussian input data.

5.2.1 Special Bispectral and Trispectral Density Functions

From Eqs. (4.114) and (4.116), it follows for Case 1 models of Figure 5.1 that the special bispectral density function $S_{xxy}(f)$ defined in Eq. (3.190) is given by

$$S_{xxy}(f) = S_{xxy}(f, f) = 2A_2(2f)S_{xx}^2(f) \tag{5.13}$$

and the special trispectral density function $S_{xxxy}(f)$ defined in Eq. (3.191) is given by

$$S_{xxxy}(f) = S_{xxxy}(f, f, f) = 6A_3(3f)S_{xx}^3(f) \tag{5.14}$$

These special functions of only one frequency variable are much simpler to compute than the general functions $S_{xxy}(f, g)$ and $S_{xxxy}(f, g, h)$ of two and three variables. Note that Eq. (5.13) has $A_2(2f)$ instead of $A_2(f)$ because of the preceding squaring operation, and Eq. (5.14) has $A_3(3f)$ instead of $A_3(f)$ because of the preceding cubing operation.

A direct derivation of these special bispectrum and trispectrum functions will now be made by substituting $Y(2f)$ from Eq. (5.1) into Eq. (4.122), and by substituting $Y(3f)$ from Eq. (5.1) into Eq. (4.123). The resulting higher-order moments will then be reduced to products of second-order moments. Readers not concerned with details of these proofs should go to Section 5.2.2.

Direct Derivation of Special Bispectrum

Rather than merely listing the special bispectrum $S_{xxy}(f)$ of Eq. (5.13) as a particular case of the previous more general theory in Chapter 4, a direct derivation will now be carried out. One can determine $S_{xxy}(f)$ directly from Eq. (3.190) by computing

$$S_{xxy}(f) = \frac{1}{T} E[X^*(f)X^*(f)Y(2f)]$$

using $Y(2f)$ from Eq. (5.1). Thus,

$$S_{xxy}(f) = \frac{1}{T} E[X^*(f)X^*(f)\{Y_1(2f) + Y_2(2f) + Y_3(2f) + N(2f)\}]$$

Now, the expected values involving $Y_1(2f)$ and $Y_3(2f)$ will be zero because they involve odd-ordered moments of zero mean value Gaussian data, and the expected value involving $N(2f)$ will be zero because of assumptions on $n(t)$. Hence,

$$S_{xxy}(f) = \frac{1}{T} E[X^*(f)X^*(f)Y_2(2f)]$$

where from Eq. (5.10),

$$Y_2(2f) = A_2(2f) \int X(u)X(2f - u)\, du - \bar{y}_2\delta_1(2f)$$

Thus,

$$S_{xxy}(f) = \frac{A_2(2f)}{T} \int E[X^*(f)X^*(f)X(u)X(2f - u)]\, du - \bar{y}_2 E[X^*(f)X^*(f)]$$

consists of two distinct terms.

From Eq. (4.15), the second-order moment

$$E[X^*(f)X^*(f)] = S_{xx}(f)\delta_1(2f) = S_{xx}(0)\delta_1(2f)$$

Hence, the second term in $S_{xxy}(f)$ becomes

$$\bar{y}_2 E[X^*(f)X^*(f)] = \bar{y}_2 S_{xx}(0)\delta_1(2f)$$

From Eq. (4.20) followed by Eqs. (4.12)–(4.15), the fourth-order moment inside the integral of the first term in $S_{xxxy}(f)$ is

$$\begin{aligned} E[X^*(f)X^*(f)X(u)X(2f-u)] &= E[X^*(f)X^*(f)]E[X(u)X(2f-u)] \\ &\quad + E[X^*(f)X(u)]E[X^*(f)X(2f-u)] \\ &\quad + E[X^*(f)X(2f-u)]E[X^*(f)X(u)] \\ &= S_{xx}(f)S_{xx}(u)\delta_1^2(2f) + 2S_{xx}^2(f)\delta_1^2(f-u) \end{aligned}$$

The integral in $S_{xxy}(f)$ from properties of $\delta_1(f)$ now contains the two parts

$$\int S_{xx}(f)S_{xx}(u)\delta_1^2(2f)\,du = S_{xx}(f)\sigma_x^2\delta_1^2(2f)$$

$$\int 2S_{xx}^2(f)\delta_1^2(f-u)\,du = 2TS_{xx}^2(f)$$

Multiplication of these two parts by the factor $A_2(2f)/T$ gives

$$\frac{A_2(2f)}{T}\,[S_{xx}(f)\sigma_x^2\delta_1^2(2f)] = A_2(0)S_{xx}(0)\sigma_x^2\delta_1(2f) = \bar{y}_2 S_{xx}(0)\delta_1(2f)$$

$$\frac{A_2(2f)}{T}\,[2TS_{xx}^2(f)] = 2A_2(2f)S_{xx}^2(f)$$

Finally, combining the above items from the first and second terms of $S_{xxy}(f)$ proves that the special bispectrum

$$S_{xxy}(f) = 2A_2(2f)S_{xx}^2(f)$$

This is the same result as Eq. (5.13).

Alternative Derivation of Special Bispectrum

A simple alternative way to derive the special bispectrum of Eq. (5.13) is to let $x_2(t) = x^2(t)$ in Figure 5.1. Then

$$X_2(f) = \mathscr{F}[x^2(t)] = \int X(\alpha)X(f-\alpha)\,d\alpha$$

and for $f \neq 0$, one can prove that

$$S_{x_2x_2}(f) = \frac{1}{T} E[X_2^*(f)X_2(f)] = 2 \int S_{xx}(\alpha)S_{xx}(f-\alpha)\,d\alpha$$

Here, one way to compute $S_{x_2y}(f)$ is by the formula

$$S_{x_2y}(f) = \frac{1}{T} E[X_2^*(f)Y(f)] = \int S_{xxy}(\alpha, f-\alpha)\,d\alpha$$

where

$$S_{xxy}(\alpha, f-\alpha) = \frac{1}{T} E[X^*(\alpha)X^*(f-\alpha)Y(f)]$$

In Figure 5.1, the output Fourier transform $Y_2(f)$ for $f \neq 0$ is

$$Y_2(f) = A_2(f)X_2(f)$$

with

$$S_{x_2y_2}(f) = A_2(f)S_{x_2x_2}(f)$$

But

$$S_{x_2y}(f) = S_{x_2y_2}(f)$$

since for Gaussian input data

$$S_{x_2y_1}(f) = S_{x_2y_3}(f) = S_{x_2n}(f) = 0$$

Hence, a second way to compute $S_{x_2y}(f)$ is by the formula

$$S_{x_2y}(f) = A_2(f)S_{x_2x_2}(f) = 2A_2(f) \int S_{xx}(\alpha)S_{xx}(f-\alpha)\,d\alpha$$

Equating the two different ways to compute $S_{x_2y}(f)$ gives

$$\int S_{xxy}(\alpha, f-\alpha)\,d\alpha = 2A_2(f) \int S_{xx}(\alpha)S_{xx}(f-\alpha)\,d\alpha$$

Thus,

$$S_{xxy}(\alpha, f-\alpha) = 2A_2(f)S_{xx}(\alpha)S_{xx}(f-\alpha)$$

Let $\alpha = g$ and $(f - \alpha) = g$. Then $f = 2g$ and

$$S_{xxy}(g) = S_{xxy}(g, g) = 2A_2(2g)S_{xx}^2(g)$$

This is the same result as Eq. (5.13) when g is replaced by f.

Direct Derivation of Special Trispectrum

One can determine the special trispectrum $S_{xxxy}(f)$ of Eq. (5.14) directly from Eq. (3.191) by computing

$$S_{xxxy}(f) = \frac{1}{T} E[X^*(f)X^*(f)X^*(f)Y(3f)]$$

using $Y(3f)$ from Eq. (5.1). Thus,

$$S_{xxxy}(f) = \frac{1}{T} E[X^*(f)X^*(f)X^*(f)\{Y_1(3f) + Y_2(3f) + Y_3(3f) + N(3f)\}]$$

Now, the expected value involving $Y_2(3f)$ will be zero because it involves an odd-ordered moment of zero mean value Gaussian data, and the expected value involving $N(3f)$ will be zero because of assumptions on $n(t)$. There still remain expected values with $Y_1(3f)$ and $Y_3(3f)$. The expected value involving $Y_1(3f)$ is

$$\begin{aligned}
\frac{1}{T} E[X^*(f)X^*(f)X^*(f)Y_1(3f)] &= \frac{A_1(3f)}{T} E[X^*(f)X^*(f)X^*(f)X(3f)] \\
&= \frac{3A_1(3f)}{T} E[X^*(f)X^*(f)]E[X^*(f)X(3f)] \\
&= 3A_1(3f)S_{xx}(f)S_{xx}(0)\delta_1(f)
\end{aligned}$$

This term will be zero when $\bar{x} = 0$ because $S_{xx}(0)$ will be zero when $\bar{x} = 0$. Also, this term exists only at $f = 0$ when $S_{xx}(0) \neq 0$.

The expected value involving $Y_3(3f)$ is

$$\frac{A_3(3f)}{T} \iint E[X^*(f)X^*(f)X^*(f)X(u)X(v - u)X(3f - v)]\, du\, dv$$

The sixth-order moment in the integrand

$$\begin{aligned}
&E[X^*(f)X^*(f)X^*(f)X(u)X(v - u)X(3f - v)] \\
&\qquad = 6E[X^*(f)X(u)]E[X^*(f)X(v - u)]E[X^*(f)X(3f - v)]
\end{aligned}$$

plus other terms that exist only at $f = 0$. The factor of 6 comes from considering all of the ways the above three second-order moments can occur. These three second-order moments from Eq. (4.13) are

$$E[X^*(f)X(u)] = S_{xx}(f)\delta_1(f-u)$$
$$E[X^*(f)X(v-u)] = S_{xx}(f)\delta_1(f+u-v)$$
$$E[X^*(f)X(3f-v)] = S_{xx}(f)\delta_1(2f-v)$$

Thus, for $f \neq 0$, the special trispectrum

$$S_{xxxy}(f) = \frac{6A_3(3f)}{T} \iint S_{xx}^3(f)\delta_1(f-u)\delta_1(f+u-v)\delta_1(2f-v)\,du\,dv$$

where $S_{xx}^3(f)$ can be taken in front of the double integral. The double integral from properties of $\delta_1(f)$ becomes

$$\iint \delta_1(f-u)\delta_1(f+u-v)\delta_1(2f-v)\,du\,dv = \int \delta_1^2(2f-v)\,dv = T$$

Hence, the special trispectrum

$$S_{xxxy}(f) = 6A_3(3f)S_{xx}^3(f)$$

This is the same result as Eq. (5.14).

5.2.2 Output Autospectral Density Functions

From knowledge of $A_1(f)$, $A_2(f)$, and $A_3(f)$, Eqs. (5.2)–(5.8) together with Eqs. (4.152)–(4.159) show that the total output autospectral density function in Figures 5.1 and 5.2 is given by

$$S_{yy}(f) = S_{y_1y_1}(f) + S_{y_2y_2}(f) + S_{y_3y_3}(f) + S_{y_1y_3}(f) + S_{y_1y_3}^*(f) + S_{nn}(f) \tag{5.15}$$

where

$$S_{y_1y_1}(f) = |A_1(f)|^2 S_{xx}(f) \tag{5.16}$$
$$S_{y_2y_2}(f) = S_{y_by_b}(f) + (\bar{y}_2)^2\delta_1(f) \tag{5.17}$$
$$S_{y_by_b}(f) = 2|A_2(f)|^2 \int S_{xx}(u)S_{xx}(v-u)\,du \tag{5.18}$$
$$S_{y_3y_3}(f) = S_{y_cy_c}(f) + |C(f)|^2 S_{xx}(f) \tag{5.19}$$
$$S_{y_cy_c}(f) = 6|A_3(f)|^2 \iint S_{xx}(u)S_{xx}(v-u)S_{xx}(f-v)\,du\,dv \tag{5.20}$$

$$S_{y_1y_3}(f) = A_1^*(f)C(f)S_{xx}(f) = A_1^*(f)S_{xy_3}(f) \tag{5.21}$$

$$C(f) = 3A_3(f)\int S_{xx}(u)\,du = 3\sigma_x^2 A_3(f) \tag{5.22}$$

Equation (5.15) shows when $S_{nn}(f) = 0$ that the total output spectrum $S_{yy}(f)$ does not equal the sum of the terms for $S_{y_1y_1}(f)$ plus $S_{y_2y_2}(f)$ plus $S_{y_3y_3}(f)$ because the output terms $Y_1(f)$ and $Y_3(f)$ are correlated. To achieve a more desirable decomposition, the model of Figure 5.1 should be revised as shown in Figure 5.3, where the mutually uncorrelated outputs $Y_a(f)$, $Y_b(f)$, and $Y_c(f)$ replace the previous correlated outputs $Y_1(f)$, $Y_2(f)$, and $Y_3(f)$. The steps required are the same as in Chapter 4 when Figure 4.6 is replaced by Figure 4.7.

The revised Case 1 nonlinear model of Figure 5.3 is equivalent to the three-input/single-output linear model of Figure 5.4, where the three inputs $X(f)$, $X_2(f)$, and $Z(f) = X_3(f) - 3\sigma_x^2 X(f)$ are uncorrelated. Thus, it is simple here to identify the linear systems $H_o(f)$, $A_2(f)$, and $A_3(f)$ from measurement of $X(f)$ and $Y(f)$ since Figure 5.4 represents three separate single-input/single-output linear systems. Pertinent practical equations are given later in Section 5.2.4. Note in Figures 5.3 and 5.4 that $H_o(f)$ is a function of the input variance σ_x^2 and so will vary with different input spectra $S_{xx}(f)$. However, $A_1(f)$, $A_2(f)$, and $A_3(f)$ will be independent of $S_{xx}(f)$ when Figures 5.3 and 5.4 are appropriate nonlinear models.

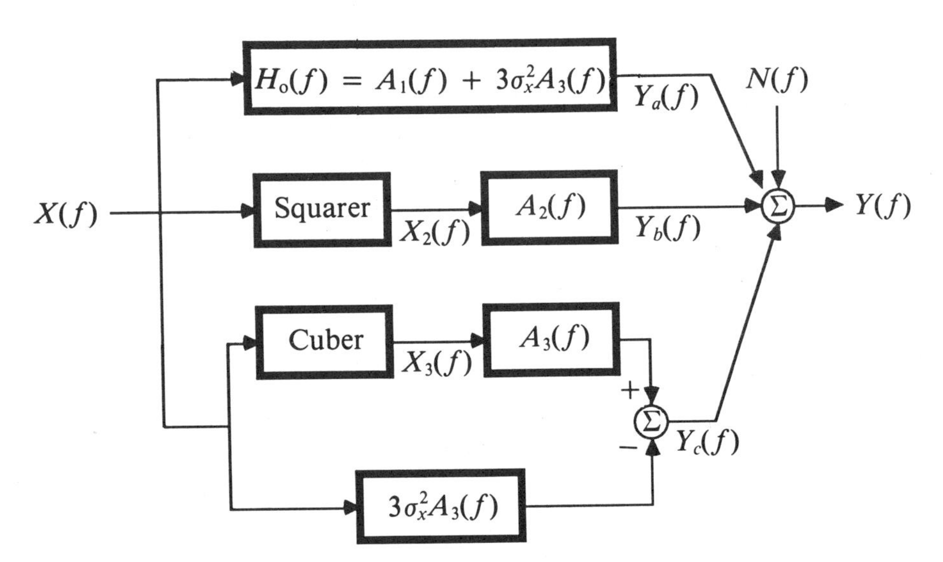

Figure 5.3 Revised Case 1 nonlinear model with uncorrelated outputs.

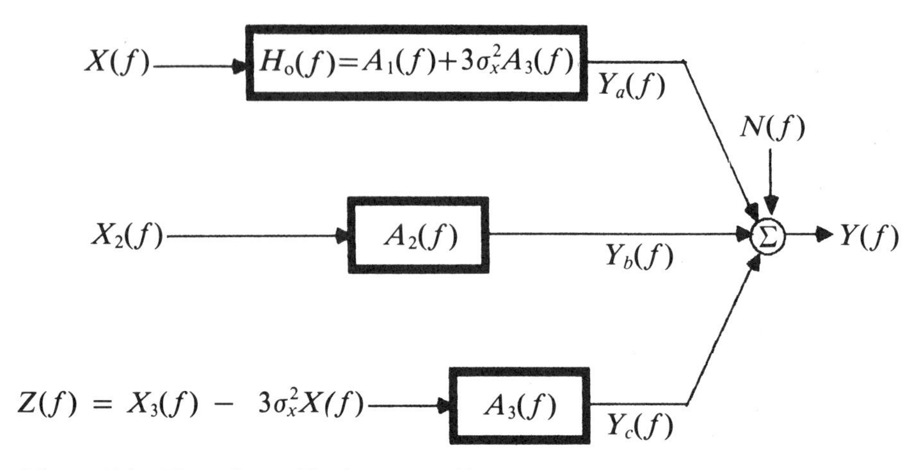

Figure 5.4 Three-input/single-output linear model with uncorrelated inputs equivalent to Figure 5.3.

For the models in Figure 5.3 and 5.4, the total output Fourier transform is given by

$$Y(f) = Y_a(f) + Y_b(f) + Y_c(f) + N(f) \tag{5.23}$$

The total output autospectral density function is

$$S_{yy}(f) = S_{y_a y_a}(f) + S_{y_b y_b}(f) + S_{y_c y_c}(f) + S_{nn}(f) \tag{5.24}$$

where from Eqs. (4.109)–(4.111), the separate spectral terms are

$$S_{y_a y_a}(f) = |H_o(f)|^2 S_{xx}(f) \tag{5.25}$$

$$S_{y_b y_b}(f) = 2|A_2(f)|^2 \int S_{xx}(u) S_{xx}(f-u)\, du \tag{5.26}$$

$$S_{y_c y_c}(f) = 6|A_3(f)|^2 \iint S_{xx}(u) S_{xx}(v-u) S_{xx}(f-v)\, du\, dv \tag{5.27}$$

From Eqs. (4.124), (5.2), and (5.22), the optimum linear system $H_o(f)$ satisfies here the relation

$$H_o(f) = \frac{S_{xy}(f)}{S_{xx}(f)} = A_1(f) + C(f) = A_1(f) + 3\sigma_x^2 A_3(f) \tag{5.28}$$

Equation (5.25) represents the output autospectral density function for the uncorrelated output term $y_a(t)$ in the optimum linear system path. Equation (5.26) represents the output autospectral density function for the uncorrelated output term $y_b(t)$ in the finite-memory nonlinear square-law path, and Eq. (5.27) represents the output autospectral density function for the uncorrelated output term $y_c(t)$ in the finite-memory nonlinear cubic path.

Equation (5.28) shows that the optimum linear system $H_o(f) \neq A_1(f)$ when $C(f) \neq 0$. Thus, conventional linear procedures that consider only single-input/single-output linear models and compute only $H_o(f)$ give erroneous results for the desired $A_1(f)$ and fail to identify the systems $A_2(f)$ and $A_3(f)$ in Figures 5.1–5.4.

Useful properties of Figures 5.3 and 5.4 can be assessed by computing the linear and nonlinear coherence functions that determine the proportion of the output spectrum that comes from the three paths. From Eqs. (4.140), (5.24) and (5.25), the linear coherence function

$$\gamma_{xy}^2(f) = \frac{S_{y_a y_a}(f)}{S_{yy}(f)} \tag{5.29}$$

where $0 \leqslant \gamma_{xy}^2(f) \leqslant 1$ for all f states the proportion via the optimum linear path. From Eqs. (4.142), (5.24), and (5.26), the nonlinear coherence function

$$q_{xy_b}^2(f) = \frac{S_{y_b y_b}(f)}{S_{yy}(f)} \tag{5.30}$$

where $0 \leqslant q_{xy_b}^2(f) \leqslant 1$ for all f states the proportion via the finite-memory square-law path. Finally, from Eqs. (4.144), (5.23), and (5.26), the nonlinear coherence function

$$q_{xy_c}^2(f) = \frac{S_{y_c y_c}(f)}{S_{yy}(f)} \tag{5.31}$$

where $0 \leqslant q_{xy_c}^2(f) \leqslant 1$ for all f states the proportion via the finite-memory cubic path. The models of Figures 5.3 and 5.4 will be valid at frequencies where the sum of these three coherence functions is close to unity. The output noise spectrum $S_{nn}(f)$ in Figures 5.3 and 5.4 is given by Eq. (4.146) using the linear and nonlinear coherence functions of Eqs. (5.29)–(5.31). Computation of $S_{nn}(f)$ provides a quantitative evaluation of Figures 5.3 and 5.4.

The linear coherence function $\gamma_{xy}^2(f)$ of Eq. (5.29) is the same as the ordinary coherence function between the input $X(f)$ and the total output $Y(f)$. The nonlinear coherence function $q_{xy_b}^2(f)$ of Eq. (5.30) is the same as the ordinary coherence function between the second mutually uncorrelated input, that is, $X_2(f)$ in Figure 5.4 and the total output $Y(f)$. The nonlinear coherence function $q_{xy_c}^2(f)$ of Eq. (5.31) is the same as the ordinary coherence function between the third mutually uncorrelated input, that is, $Z(f)$ in Figure 5.4 and the total output $Y(f)$.

Rather than merely listing the nonlinear output spectra terms from Eqs. (5.26) and (5.27) as particular cases of the previous more general theory in Chapter 4, these important formulas will now be derived directly without computing correlation functions. This will be done using Fourier transform frequency-domain procedures assuming the input to be a zero mean value Gaussian stationary random process with arbitrary spectral properties. Readers not concerned with details of these proofs should go to Section 5.2.3.

Direct Derivation of Output Spectrum from Finite-Memory Square-Law Path

A direct derivation for Eq. (5.26) will now be carried out similar to the direct derivation for Eq. (5.13). From Eq. (5.10), using different variables of integration,

$$Y_b(f) = A_2(f) \int X(\alpha)X(f-\alpha)\,d\alpha - \bar{y}_2\delta_1(f)$$

$$Y_b^*(f) = A_2^*(f) \int X^*(\beta)X^*(f-\beta)\,d\beta - \bar{y}_2\delta_1(f)$$

One can determine $S_{y_b y_b}(f)$ directly by computing

$$\begin{aligned} S_{y_b y_b}(f) &= \frac{1}{T} E[Y_2^*(f)Y_2(f)] \\ &= \frac{|A_2(f)|^2}{T} \iint E[X^*(\beta)X^*(f-\beta)X(\alpha)X(f-\alpha)]\,d\alpha\,d\beta \\ &\quad - \frac{\bar{y}_2\delta_1(f)}{T} A_2(f) \int E[X(\alpha)X(f-\alpha)]\,d\alpha \\ &\quad - \frac{\bar{y}_2\delta_1(f)}{T} A_2^*(f) \int E[X^*(\beta)X^*(f-\beta)]\,d\beta + \frac{1}{T}(\bar{y}_2)^2\delta_1^2(f) \end{aligned}$$

From the properties of $\delta_1(f)$,

$$\frac{1}{T}(\bar{y}_2)^2\delta_1^2(f) = (\bar{y}_2)^2\delta_1(f)$$

$$\frac{\bar{y}_2\delta_1(f)}{T} A_2(f) = \frac{\bar{y}_2\delta_1(f)}{T} A_2^*(f) = \bar{y}_2 A_2(0)$$

From Eqs. (4.14) and (4.15), the second-order moments

$$E[X(\alpha)X(f-\alpha)] = S_{xx}(\alpha)\delta_1(f)$$

$$E[X^*(\beta)X^*(f-\beta)] = S_{xx}(\beta)\delta_1(f)$$

The single integrals in $S_{y_b y_b}(f)$ become

$$\int E[X(\alpha)X(f-\alpha)]\,d\alpha = \int E[X^*(\beta)X^*(f-\beta)]\,d\beta = \sigma_x^2\delta_1(f)$$

Also,

$$\bar{y}_2 A_2(0)\sigma_x^2 \delta_1(f) = (\bar{y}_2)^2 \delta_1(f)$$

since $\bar{y}_2 = A_2(0)\sigma_x^2$. Thus,

$$S_{y_b y_b}(f) = \frac{|A_2(f)|^2}{T} \iint E[X^*(\beta)X^*(f-\beta)X(\alpha)X(f-\alpha)]\,d\alpha\,d\beta - (\bar{y}_2)^2\delta_1(f)$$

From Eq. (4.20) followed by Eqs. (4.12)–(4.15), the fourth-order moment

$$\begin{aligned}
&E[X^*(\beta)X^*(f-\beta)X(\alpha)X(f-\alpha)] \\
&\quad = E[X^*(\beta)X^*(f-\beta)]E[X(\alpha)X(f-\alpha)] \\
&\qquad + E[X^*(\beta)X(\alpha)]E[X^*(f-\beta)X(f-\alpha)] \\
&\qquad + E[X^*(\beta)X(f-\alpha)]E[X^*(f-\beta)X(\alpha)] \\
&\quad = S_{xx}(\beta)S_{xx}(\alpha)\delta_1^2(f) + S_{xx}(\beta)S_{xx}(f-\beta)[\delta_1^2(\alpha-\beta) + \delta_1^2(f-\alpha-\beta)]
\end{aligned}$$

From properties of $\delta_1(f)$, the double integral in $S_{y_b y_b}(f)$ now contains the three terms

$$\iint S_{xx}(\beta)S_{xx}(\alpha)\delta_1^2(f)\,d\alpha\,d\beta = \sigma_x^4\delta_1^2(f)$$

$$\iint S_{xx}(\beta)S_{xx}(f-\beta)\delta_1^2(\alpha-\beta)\,d\alpha\,d\beta = \delta_1(0)\int S_{xx}(\alpha)S_{xx}(f-\alpha)\,d\alpha$$

$$\iint S_{xx}(\beta)S_{xx}(f-\beta)\delta_1^2(f-\alpha-\beta)\,d\alpha\,d\beta = \delta_1(0)\int S_{xx}(\alpha)S_{xx}(f-\alpha)\,d\alpha$$

Multiplication by the factor $|A_2(f)|^2/T$ yields one term

$$\frac{|A_2(f)|^2}{T}\sigma_x^4\delta_1^2(f) = A_2^2(0)\sigma_x^4\delta_1(f) = (\bar{y}_2)^2\delta_1(f)$$

and two identical terms

$$\frac{|A_2(f)|^2}{T}\delta_1(0)\int S_{xx}(\alpha)S_{xx}(f-\alpha)\,d\alpha = |A_2(f)|^2\int S_{xx}(\alpha)S_{xx}(f-\alpha)\,d\alpha$$

Substitutions give the final result:

$$S_{y_b y_b}(f) = 2|A_2(f)|^2\int S_{xx}(\alpha)S_{xx}(f-\alpha)\,d\alpha$$

This completes the proof for Eq. (5.26) where α is replaced by u.

Direct Derivation of Output Spectrum from Finite-Memory Cubic Path

A direct derivation of Eq. (5.27) can be carried out as follows. From Eqs. (4.106) and (5.11),

$$Y_c(f) = Y_3(f) - C(f)X(f)$$

where

$$Y_3(f) = A_3(f)\iint X(u)X(v-u)X(f-v)\,du\,dv$$

From Eq. (5.22), the function $C(f)$ is

$$C(f) = 3\sigma_x^2 A_3(f)$$

The cross-spectrum $S_{xy_3}(f)$ occurs by computing

$$S_{xy_3}(f) = \frac{1}{T}\,E[X^*(f)Y_3(f)] = \frac{A_3(f)}{T}\iint E[X^*(f)X(u)X(v-u)X(f-v)]\,du\,dv$$

Reduction of the fourth-order moment into products of pairs of second-order moments gives the three terms

$$\begin{aligned}
E[X^*(f)X(u)]E[X(v-u)X(f-v)] &= S_{xx}(f)S_{xx}(u-v)\delta_1^2(f-u)\\
E[X^*(f)X(v-u)]E[X(u)X(f-v)] &= S_{xx}(f)S_{xx}(u)\delta_1^2(f+u-v)\\
E[X^*(f)X(f-v)]E[X(u)X(v-u)] &= S_{xx}(f)S_{xx}(u)\delta_1^2(v)
\end{aligned}$$

The double integral of each of these three terms yields the same result, $T\sigma_x^2 S_{xx}(f)$. Hence,

$$S_{xy_3}(f) = 3\sigma_x^2 A_3(f)S_{xx}(f) = C(f)S_{xx}(f)$$

A direct computation of $S_{y_cy_c}(f)$ is given by the formula

$$\begin{aligned}
S_{y_cy_c}(f) &= \frac{1}{T}\,E[Y_c^*(f)Y_c(f)]\\
&= \frac{1}{T}\,E[\{Y_3^*(f) - C^*(f)X^*(f)\}\{Y_3(f) - C(f)X(f)\}]\\
&= S_{y_3y_3}(f) - C^*(f)S_{xy_3}(f) - C(f)S_{y_3x}(f) + |C(f)|^2 S_{xx}(f)\\
&= S_{y_3y_3}(f) - |C(f)|^2 S_{xx}(f)
\end{aligned}$$

The autospectrum $S_{y_3y_3}(f)$ is computed directly by

$$S_{y_3y_3}(f) = \frac{1}{T} E[Y_3^*(f)Y_3(f)]$$
$$= \frac{|A_3(f)|^2}{T} \iiiint E[X^*(u)X^*(v-u)X^*(f-v)$$
$$\times X(\alpha)X(\beta-\alpha)X(f-\beta)]\, d\alpha\, d\beta\, du\, dv$$

Reduction of the sixth-order moment produces 15 terms where nine terms are of a similar nature and six terms are of a similar nature. Substitution of these 15 terms into the formula for $S_{y_3y_3}(f)$ followed by performing the four-fold integration shows after many steps that the nine similar terms yield

$$9\sigma_x^4|A_3(f)|^2 S_{xx}(f) = |C(f)|^2 S_{xx}(f)$$

and the six similar terms yield

$$6|A_3(f)|^2 \iint S_{xx}(u)S_{xx}(v-u)S_{xx}(f-v)\, du\, dv$$

Thus,

$$S_{y_3y_3}(f) = |C(f)|^2 S_{xx}(f) + S_{y_cy_c}(f)$$

where $S_{y_cy_c}(f)$ is given by

$$S_{y_cy_c}(f) = 6|A_3(f)|^2 \iint S_{xx}(u)S_{xx}(v-u)S_{xx}(f-v)\, du\, dv$$

This completes the proof for Eq. (5.27).

5.2.3 System Identification in Case 1 Models

To identify the A-systems in Case 1 models from measurements of $x(t)$ and $y(t)$ only, one can proceed as outlined in Section 4.2.3. Here, in place of Eqs. (4.122) and (4.123), one should compute the special bispectral and trispectral density functions of single variables

$$S_{xxy}(f) = S_{xxy}(f, f) = \frac{1}{T} E[X^*(f)X^*(f)Y(2f)] \tag{5.32}$$

$$S_{xxxy}(f) = S_{xxxy}(f, f, f) = \frac{1}{T} E[X^*(f)X^*(f)X^*(f)Y(3f)] \tag{5.33}$$

These functions defined earlier in Eqs. (3.190) and (3.191) are direct extensions of the ordinary cross-spectral density functions $S_{xy}(f)$ of Eq. (3.168) or (4.121). Both $S_{xy}(f)$ and $S_{xx}(f)$ should also be computed.

From Eq. (5.28), at frequency f,

$$A_1(f) = H_o(f) - 3\sigma_x^2 A_3(f) \tag{5.34}$$

where the optimum linear system $H_o(f)$ is known from $S_{xx}(f)$ and $S_{xy}(f)$. Note that computation of $A_1(f)$ requires knowledge of $A_3(f)$. From Eq. (5.13), at frequency $2f$,

$$A_2(2f) = \frac{S_{xxy}(f)}{2S_{xx}^2(f)} \tag{5.35}$$

To determine $A_2(f)$ instead of $A_2(2f)$, one should replace f by $(f/2)$ above. From Eq. (5.14), at frequency $3f$,

$$A_3(3f) = \frac{S_{xxxy}(f)}{6S_{xx}^3(f)} \tag{5.36}$$

To determine $A_3(f)$ instead of $A_3(3f)$, one should replace f by $(f/3)$ above. Thus, to obtain $A_2(f)$ and $A_3(f)$, one should compute the following quantities:

$$A_2(f) = \frac{S_{xxy}(f/2)}{2S_{xx}^2(f/2)} \tag{5.37}$$

$$A_3(f) = \frac{S_{xxxy}(f/3)}{6S_{xx}^3(f/3)} \tag{5.38}$$

where $S_{xxy}(f/2)$ and $S_{xxy}(f/3)$ are calculated from Eqs. (5.32) and (5.33) with f replaced as shown by $(f/2)$ and $(f/3)$, respectively. Substitution of $A_3(f)$ into Eq. (5.34) gives $A_1(f)$.

This completes the procedure for identifying the A-systems in Case 1 models based upon computing the special bispectral and trispectral functions of Eqs. (5.32) and (5.33). From the work in Section 4.3, these A-systems represent the optimum systems for minimizing the output noise spectrum in Case 1 models.

5.2.4 Linear Procedure for System Identification in Case 1 Models

Consider Figure 5.4 containing the same A-systems as Figure 5.3. Without computing the special bispectral and trispectral functions of Eqs. (5.32) and (5.33), a practical linear analysis procedure will now be outlined to identify the A-systems in Case 1 models and determine the output autospectral density functions. From input and output records of Figure 5.3, for any $f \neq 0$, begin by computing the following Fourier transforms:

$$X(f) = \mathscr{F}[x(t)], \qquad Y(f) = \mathscr{F}[y(t)] \tag{5.39}$$

$$X_2(f) = \mathscr{F}[x^2(t)], \qquad X_3(f) = \mathscr{F}[x^3(t)] \tag{5.40}$$

Then combine to obtain

$$Z(f) = X_3(f) - 3\sigma_x^2 X(f) \tag{5.41}$$

The quantities $X(f)$, $X_2(f)$, and $Z(f)$ for Gaussian $X(f)$ represent the mutually uncorrelated input data in the three-input/single-output linear model shown in Figure 5.4, that together are equivalent to the single-input/single-output nonlinear model of Figure 5.3. The total measured output quantity $Y(f)$ in Figure 5.4 is the same as the measured $Y(f)$ in Figure 5.3, where $Y(f)$ satisfies Eq. (5.23). Recognition of this equivalence between Figures 5.3 and 5.4 makes system identification in Case 1 models a straightforward linear problem using conventional procedures. One needs to compute only ordinary autospectral and cross-spectral density functions as listed in the following Eqs. (5.42)–(5.46).

By suitable expected value operations for any $f \neq 0$, compute the usual two-sided spectral density functions below where $x_2(t) = x^2(t)$ and $x_3(t) = x^3(t)$.

$$S_{xx}(f) = \frac{1}{T} E[X^*(f)X(f)], \qquad S_{xy}(f) = \frac{1}{T} E[X^*(f)Y(f)] \tag{5.42}$$

$$S_{x_2x_2}(f) = \frac{1}{T} E[X_2^*(f)X_2(f)], \qquad S_{x_2y}(f) = \frac{1}{T} E[X_2^*(f)Y(f)] \tag{5.43}$$

$$S_{x_3x_3}(f) = \frac{1}{T} E[X_3^*(f)X_3(f)], \qquad S_{x_3y}(f) = \frac{1}{T} E[X_3^*(f)Y(f)] \tag{5.44}$$

Compute also the quantities

$$S_{zz}(f) = \frac{1}{T} E[Z^*(f)Z(f)] = S_{x_3x_3}(f) - 9\sigma_x^4 S_{xx}(f) \tag{5.45}$$

$$S_{zy}(f) = \frac{1}{T} E[Z^*(f)Y(f)] = S_{x_3y}(f) - 3\sigma_x^2 S_{xy}(f) \tag{5.46}$$

In Figures 5.3 and 5.4, it now follows directly that the linear systems $H_o(f)$, $A_2(f)$, and $A_3(f)$ for any $f \neq 0$ can be identified by the single-input/single-output formulas

$$H_o(f) = \frac{S_{xy}(f)}{S_{xx}(f)} \tag{5.47}$$

$$A_2(f) = \frac{S_{x_2y}(f)}{S_{x_2x_2}(f)} \tag{5.48}$$

$$A_3(f) = \frac{S_{zy}(f)}{S_{zz}(f)} \tag{5.49}$$

The system $A_1(f)$ in Figures 5.3 and 5.4 is then computed by the formula

$$A_1(f) = H_o(f) - 3\sigma_x^2 A_3(f) \tag{5.50}$$

The linear and nonlinear output autospectral density functions in Figures 5.3 and 5.4 for any $f \neq 0$ satisfy the simple formulas

$$S_{y_ay_a}(f) = |H_o(f)|^2 S_{xx}(f) \tag{5.51}$$

$$S_{y_by_b}(f) = |A_2(f)|^2 S_{x_2x_2}(f) \tag{5.52}$$

$$S_{y_cy_c}(f) = |A_3(f)|^2 S_{zz}(f) \tag{5.53}$$

The A-systems determined by Eqs. (5.50), (5.48), and (5.49) are identical to the A-systems determined by Eqs. (5.34), (5.37), and (5.38), respectively. Hence, these results also represent the optimum systems for minimizing the output noise spectrum in Case 1 models. The linear and nonlinear output spectra results in Eqs. (5.51)–(5.53) are identical to the corresponding linear and nonlinear output spectra results in Eqs. (5.25)–(5.27), respectively. Linear and nonlinear coherence functions follow as in Eqs. (5.29) to (5.31).

This completes the practical linear procedure for identifying the A-systems in Case 1 models. This procedure should be used in preference to computing the special bispectra and trispectra of Eqs. (5.32) and (5.33) whenever appropriate. A similar linear procedure results from the conditioned spectral density techniques in references 1 and 2 when input data $X(f)$ are non-Gaussian and when $X_2(f)$ and $X_3(f)$ are given by Eq. (5.40).

5.2.5 Summary of Results

A summary follows on some of the main results for Case 1 models where linear systems follow the squaring and cubing operations. Fourier transform relations for the Case 1 nonlinear model with correlated outputs of Figures 5.1 and 5.2 are

$$Y(f) = Y_1(f) + Y_2(f) + Y_3(f) + N(f) \tag{5.54}$$

where

$$Y_1(f) = A_1(f)X(f) \tag{5.55}$$

$$Y_2(f) = A_2(f)\int X(u)X(f-u)\,du - \bar{y}_2\delta_1(f) \tag{5.56}$$

$$Y_3(f) = A_3(f) \iint X(u)X(v-u)X(f-v)\,du\,dv \tag{5.57}$$

These correlated output results can be computed from measurement of the input $x(t)$ and knowledge of the linear frequency response functions $A_1(f)$, $A_2(f)$, and $A_3(f)$.

Spectral results for the Case 1 nonlinear model with correlated outputs of Figures 5.1 and 5.2 are

$$S_{yy}(f) = S_{y_1y_1}(f) + S_{y_2y_2}(f) + S_{y_3y_3}(f) + S_{y_1y_3}(f) + S^*_{y_1y_3}(f) + S_{nn}(f) \tag{5.58}$$

where

$$S_{y_1y_1}(f) = |A_1(f)|^2 S_{xx}(f) \tag{5.59}$$

$$S_{y_2y_2}(f) = 2|A_2(f)|^2 \int S_{xx}(u)S_{xx}(f-u)\,du + (\bar{y}_2)^2\delta_1(f) \tag{5.60}$$

$$S_{y_3y_3}(f) = S_{y_cy_c}(f) + 9\sigma_x^4|A_3(f)|^2 S_{xx}(f) \tag{5.61}$$

$$S_{y_cy_c}(f) = 6|A_3(f)|^2 \int S_{xx}(u)S_{xx}(v-u)S_{xx}(f-v)\,du\,dv \tag{5.62}$$

$$S_{y_1y_3}(f) = 3\sigma_x^2 A_1^*(f)A_3(f)S_{xx}(f) \tag{5.63}$$

Fourier transform relations for the revised Case 1 nonlinear model with uncorrelated outputs of Figures 5.3 and 5.4 are

$$Y(f) = Y_a(f) + Y_b(f) + Y_c(f) + N(f) \tag{5.64}$$

where

$$Y_a(f) = [A_1(f) + 3\sigma_x^2 A_3(f)]X(f) = H_o(f)X(f) \tag{5.65}$$

$$Y_b(f) = A_2(f) \int X(u)X(f-u)\,du - \bar{y}_2\delta_1(f) \tag{5.66}$$

$$Y_c(f) = A_3(f) \iint X(u)X(v-u)X(f-v)\,du\,dv - 3\sigma_x^2 A_3(f)X(f) \tag{5.67}$$

These uncorrelated output results can be computed from measurement of $x(t)$ and knowledge of the linear frequency response functions $A_1(f)$, $A_2(f)$, and $A_3(f)$.

Spectral results for the revised Case 1 nonlinear model with uncorrelated outputs of Figures 5.3 and 5.4 are

$$S_{yy}(f) = S_{y_ay_a}(f) + S_{y_by_b}(f) + S_{y_cy_c}(f) + S_{nn}(f) \tag{5.68}$$

where

$$S_{y_a y_a}(f) = |H_o(f)|^2 S_{xx}(f) = |S_{xy}(f)|^2 / S_{xx}(f) \tag{5.69}$$

$$S_{y_b y_b}(f) = 2|A_2(f)|^2 \int S_{xx}(u) S_{xx}(f-u)\, du \tag{5.70}$$

$$S_{y_c y_c}(f) = 6|A_3(f)|^2 \iint S_{xx}(u) S_{xx}(v-u) S_{xx}(f-v)\, du\, dv \tag{5.71}$$

Equivalent ways of computing the last two results are stated in Eqs. (5.52) and (5.53).

For simultaneous measurement of the input $x(t)$ and the total output $y(t)$ the systems $A_1(f)$, $A_2(f)$, and $A_3(f)$, in Figures 5.1–5.4 can be identified by two different procedures as explained in Sections 5.2.3 and 5.2.4. The alternative practical linear procedure in Section 5.2.4 is simpler because it does not require computation of special bispectral and trispectral density functions. Also, this preferred procedure applies to either Gaussian or non-Gaussian input data.

5.3 CASE 2 SQUARE-LAW AND CUBIC MODELS

Case 2 single-input/single-output nonlinear models occur when higher-order frequency response functions can be represented by the product of first-order frequency response functions such that the linear, bilinear, and trilinear systems in Figures 4.6 and 4.7 are

$$H_1(f) = B_1(f) \tag{5.72}$$

$$H_2(f_1, f_2) = B_2(f_1) B_2(f_2) \tag{5.73}$$

$$H_3(f_1, f_2, f_3) = B_3(f_1) B_3(f_2) B_3(f_3) \tag{5.74}$$

Here,

$$H_2(f, f) = B_2^2(f) \tag{5.75}$$

$$H_3(f, f, f) = B_3^3(f) \tag{5.76}$$

$$\begin{aligned} H_2(f, -f) &= |B_2(f)|^2 \\ H_2(u, f-u) &= B_2(u) B_2(f-u) \end{aligned} \tag{5.77}$$

$$H_3(u, v-u, f-v) = B_3(u) B_3(v-u) B_3(f-v) \tag{5.78}$$

The $B_i(f)$ are frequency response functions of constant-parameter linear systems either known or to be determined. For this case, Eqs. (4.2), (4.24), and (4.38) become

$$Y_1(f) = B_1(f)X(f) \tag{5.79}$$

$$Y_2(f) = \int U_2(\alpha)U_2(f-\alpha)\,d\alpha - \bar{y}_2\delta_1(f) = Y_b(f) \tag{5.80}$$

$$Y_3(f) = \iint U_3(\alpha)U_3(\beta-\alpha)U_3(f-\beta)\,d\alpha\,d\beta \tag{5.81}$$

where

$$U_2(f) = B_2(f)X(f) \tag{5.82}$$

$$U_3(f) = B_3(f)X(f) \tag{5.83}$$

$$\bar{y}_2 = \sigma_2^2 = \int S_{u_2u_2}(f)\,df = \int |B_2(f)|^2 S_{xx}(f)\,df \tag{5.84}$$

$$\sigma_3^2 = \int S_{u_3u_3}(f)\,df = \int |B_3(f)|^2 S_{xx}(f)\,df \tag{5.85}$$

Note that $\bar{y}_2$ in Eq. (5.84) for Case 2 models is quite different from the $\bar{y}_2$ in Eq. (5.12) for Case 1 models.

These results have a simple physical interpretation as pictured in Figure 5.5, where zero-memory squaring and cubing operations are preceded by the B for "before" linear systems. The quantity $u_2(t)$ with Fourier transform $U_2(f)$ occurs by passing $x(t)$ through $B_2(f)$, and the single integral in Eq. (5.80) is the Fourier transform of $u_2^2(t)$. The quantity $u_3(t)$ with Fourier transform $U_3(f)$ occurs by passing $x(t)$ through $B_3(f)$, and the double integral in Eq. (5.81) is the Fourier transform of $u_3^3(t)$. Equations (5.80) and (5.81) provide the basis for direct

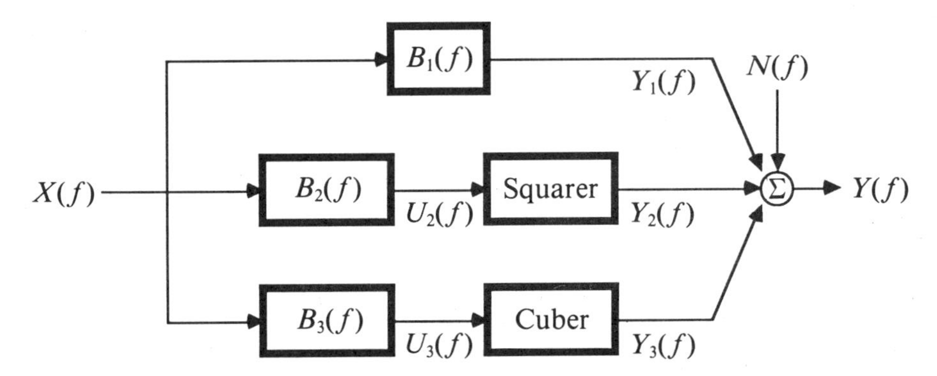

Figure 5.5 Case 2 nonlinear model with correlated outputs.

derivations of the single-variable formulas in Section 5.3.1 without knowing more general formulas from Chapter 4 involving two and three variables, and without computing correlation functions first. The total output $Y(f)$ in Figure 5.5 is given by Eq. (5.1) using the terms in Eqs. (5.79)–(5.85). Figure 5.5 is a linear system in parallel with a finite-memory nonlinear system consisting of the $B(f)$/Squarer/Cuber combination shown.

The quantities $U_2(f)$ and $U_3(f)$ cannot be determined from measurement of only $x(t)$ but also require $B_2(f)$ and $B_3(f)$, in contrast to Case 1 models where $X_2(f)$ and $X_3(f)$ can be determined from measurement of only $x(t)$ without knowing $A_2(f)$ and $A_3(f)$. Thus, in general, the Case 2 nonlinear model of Figure 5.5 cannot be replaced by an equivalent three input/single output linear model as done previously by Figure 5.2 for the Case 1 nonlinear model of Figure 5.1. Hence, the conditional spectral density techniques in references 1 and 2 are not able to solve for the systems $B_1(f)$, $B_2(f)$, and $B_3(f)$ in Case 2 models.

5.3.1 Special Bispectral and Trispectral Density Functions

From Eqs. (4.114) and (4.116), it follows for Case 2 models that the special bispectral density function $S_{xxy}(f)$ of Eq. (3.190) is

$$S_{xxy}(f) = S_{xxy}(f, f) = 2B_2^2(f)S_{xx}^2(f) \tag{5.86}$$

and the special trispectral density function $S_{xxxy}(f)$ of Eq. (3.191) is

$$S_{xxxy}(f) = S_{xxxy}(f, f, f) = 6B_3^3(f)S_{xx}^3(f) \tag{5.87}$$

These results should be compared with Eqs. (5.13) and (5.14). Note that Eq. (5.86) has $B_2^2(f)$ instead of $B_2(f)$ because of the following squaring operation, and Eq. (5.87) has $B_3^3(f)$ instead of $B_3(f)$ because of the following cubing operation. These important formulas can be derived directly from the Fourier transform relations governing Figure 5.5. Details for these derivations are not shown here.

5.3.2 Output Autospectral Density Functions

From knowledge of $B_1(f)$, $B_2(f)$, and $B_3(f)$, Eqs. (5.72)–(5.78) together with Eqs. (4.152)–(4.159) show that the total output spectral density function in Figure 5.5 is

$$S_{yy}(f) = S_{y_1y_1}(f) + S_{y_2y_2}(f) + S_{y_3y_3}(f) + S_{y_1y_3}(f) + S_{y_1y_3}^*(f) + S_{nn}(f) \tag{5.88}$$

where

$$S_{y_1y_1}(f) = |B_1(f)|^2 S_{xx}(f) \tag{5.89}$$

$$S_{y_2y_2}(f) = S_{y_by_b}(f) + (\bar{y}_2)^2\delta_1(f) \tag{5.90}$$

$$S_{y_3y_3}(f) = S_{y_cy_c}(f) + |C(f)|^2 S_{xx}(f) \tag{5.91}$$

$$S_{y_1y_3}(f) = B_1^*(f)C(f)S_{xx}(f) \tag{5.92}$$

$$C(f) = 3B_3(f)\int |B_3(u)|^2\,du = 3\sigma_3^2 B_3(f) \tag{5.93}$$

with further formulas for $S_{y_by_b}(f)$ and $S_{y_cy_c}(f)$ as stated in Eqs. (5.96) and (5.97). Note that $C(f)$ in Eq. (5.93) for Case 2 models is quite different from the $C(f)$ in Eq. (5.22) for Case 1 models.

Equation (5.92) shows that the output terms $Y_1(f)$ and $Y_3(f)$ are correlated so that it is difficult to decompose the total output spectrum of Eq. (5.88) into distinct parts. To achieve this goal, the model of Figure 5.5 should be revised as shown in Fig. 5.6, where the mutually uncorrelated outputs $Y_a(f)$, $Y_b(f)$, and $Y_c(f)$ replace the previous correlated outputs $Y_1(f)$, $Y_2(f)$, and $Y_3(f)$. The steps required are the same as when Figure 5.1 is replaced by Figure 5.3, and when Figure 4.3 is replaced by Figure 4.5.

For the model in Figure 5.6, the total output autospectral density function is

$$S_{yy}(f) = S_{y_ay_a}(f) + S_{y_by_b}(f) + S_{y_cy_c}(f) + S_{nn}(f) \tag{5.94}$$

where from Eqs. (4.109)–(4.111),

$$S_{y_ay_a}(f) = |H_o(f)|^2 S_{xx}(f) \tag{5.95}$$

$$S_{y_by_b}(f) = 2\int S_{u_2u_2}(\alpha)S_{u_2u_2}(f-\alpha)\,d\alpha \tag{5.96}$$

$$S_{y_cy_c}(f) = 6\iint S_{u_3u_3}(\alpha)S_{u_3u_3}(\beta-\alpha)S_{u_3u_3}(f-\beta)\,d\alpha\,d\beta \tag{5.97}$$

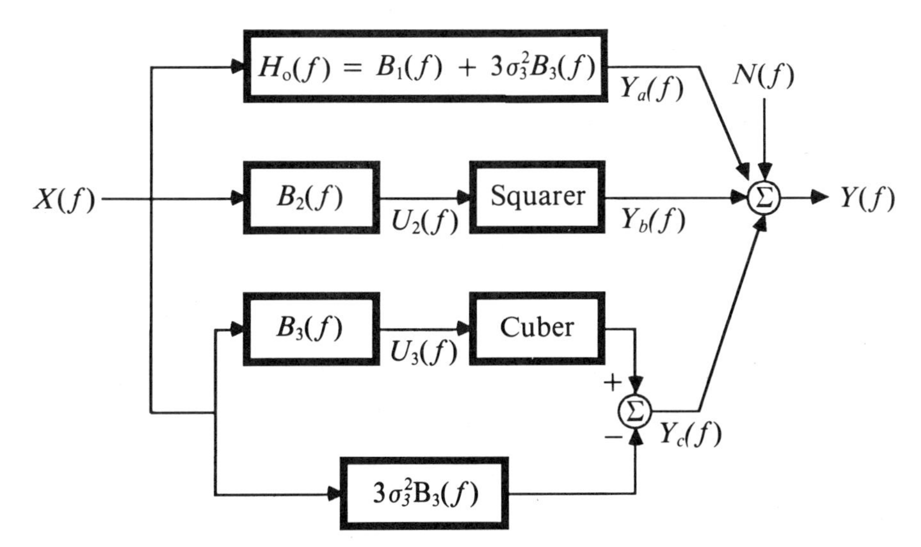

Figure 5.6 Revised Case 2 nonlinear model with uncorrelated outputs.

with

$$H_o(f) = B_1(f) + C(f) = B_1(f) + 3\sigma_3^2 B_3(f) \tag{5.98}$$

$$S_{u_2u_2}(f) = |B_2(f)|^2 S_{xx}(f) \tag{5.99}$$

$$S_{u_3u_3}(f) = |B_3(f)|^2 S_{xx}(f) \tag{5.100}$$

Linear and nonlinear coherence functions can now be computed exactly as per Eqs. (5.29)–(5.31) to assess the validity of the revised Case 2 model of Figure 5.6. These results can be compared with similar results from the revised Case 1 model of Figure 5.3 to help decide which of these nonlinear models is more appropriate to fit the measured data in physical problems.

5.3.3 System Identification in Case 2 Models

To identify the B-systems in Case 2 models from measurements of $x(t)$ and $y(t)$ only, one should follow the procedure outlined in Sections 4.2.3 and 5.2.3. As before, begin by computing the spectral quantities

$$S_{xx}(f) = \frac{1}{T} E[|X(f)|^2] \tag{5.101}$$

$$S_{xy}(f) = \frac{1}{T} E[X^*(f)Y(f)] \tag{5.102}$$

$$S_{xxy}(f) = \frac{1}{T} E[X^*(f)X^*(f)Y(2f)] \tag{5.103}$$

$$S_{xxxy}(f) = \frac{1}{T} E[X^*(f)X^*(f)X^*(f)Y(3f)] \tag{5.104}$$

From Eq. (5.98), one obtains

$$B_1(f) = H_o(f) - 3\sigma_3^2 B_3(f) \tag{5.105}$$

where $H_o(f)$ is found by the usual formula

$$H_o(f) = \frac{S_{xy}(f)}{S_{xx}(f)} \tag{5.106}$$

Note that computation of $B_1(f)$ requires knowledge of $B_3(f)$. In general, $H_o(f) \neq B_1(f)$.

From Eq. (5.86) involving the special bispectral density function,

$$B_2^2(f) = \frac{S_{xxy}(f)}{2S_{xx}^2(f)} = |B_2(f)|^2 e^{-2j\phi_2(f)} \tag{5.107}$$

This gives $B_2(f)$ by the formula

$$B_2(f) = \frac{[S_{xxy}(f)/2]^{1/2}}{S_{xx}(f)} = |B_2(f)|e^{-j\phi_2(f)} \tag{5.108}$$

From Eq. (5.87) involving the special trispectral density function,

$$B_3^3(f) = \frac{S_{xxxy}(f)}{6S_{xx}^3(f)} = |B_3(f)|^3 e^{-3j\phi_3(f)} \tag{5.109}$$

This gives $B_3(f)$ by the formula

$$B_3(f) = \frac{[S_{xxxy}(f)/6]^{1/3}}{S_{xx}(f)} = |B_3(f)|e^{-j\phi_3(f)} \tag{5.110}$$

This completes the procedure for identifying the B-systems in Case 2 models. From the work in Section 4.3, these B-systems represent the optimum systems for minimizing the output noise spectrum in Case 2 models. Note that identification of the B-systems in Case 2 models requires computation of the special bispectral and trispectral density functions of Eqs. (5.103) and (5.104). An alternative practical procedure similar to Section 5.2.4 that does not require such computations is not available for Case 2 models.

5.3.4 Summary of Results

Fourier transform relations for the Case 2 nonlinear model with correlated outputs of Figure 5.5 are

$$Y(f) = Y_1(f) + Y_2(f) + Y_3(f) + N(f) \tag{5.111}$$

where $Y_1(f)$, $Y_2(f)$, and $Y_3(f)$ satisfy Eqs. (5.79), (5.80), and (5.81), respectively. These correlated output results can be computed from measurement of the input $x(t)$ and from knowledge of the linear frequency response functions $B_1(f)$, $B_2(f)$, and $B_3(f)$.

Spectral results for the Case 2 nonlinear model with correlated outputs of Figure 5.5 are listed in Eqs. (5.88)–(5.93). All of these formulas can be derived directly from the Fourier transform relations.

Fourier transform relations for the Case 2 nonlinear model with uncorrelated outputs of Figure 5.6 are

$$Y(f) = Y_a(f) + Y_b(f) + Y_c(f) + N(f) \tag{5.112}$$

where

$$Y_a(f) = [B_1(f) + 3\sigma_3^2 B_3(f)]X(f) = H_o(f)X(f) \tag{5.113}$$

$$Y_b(f) = \int U_2(\alpha)U_2(f-\alpha)\,d\alpha - \sigma_2^2\delta_1(f) \tag{5.114}$$

$$Y_c(f) = \iint U_3(\alpha)U_3(\beta-\alpha)U_3(f-\beta)\,d\alpha\,d\beta - 3\sigma_3^2 B_3(f)X(f) \tag{5.115}$$

These uncorrelated output results can be computed from measurement of $x(t)$ and from knowledge of the linear frequency response functions $B_1(f)$, $B_2(f)$, and $B_3(f)$.

Spectral results for the Case 2 nonlinear model with uncorrelated outputs of Figure 5.6 are listed in Eqs. (5.94)–(5.100). All of these formulas can be derived directly from the Fourier transform relations.

From simultaneous measurements of the input $x(t)$ and the total output $y(t)$, the systems $B_1(f)$, $B_2(f)$, and $B_3(f)$ can be identified using the procedure explained in Section 5.3.3. These B-systems are the optimum systems that minimize the output noise spectrum in Figures 5.5 and 5.6.

5.4 OTHER SQUARE-LAW AND CUBIC MODELS

This section discusses various combinations of the Case 1 and Case 2 models. Readers interested in other matters should go to Section 5.5. In the Case 1 nonlinear model, the A_2 and A_3 systems are *after* the squaring and cubing operations, respectively. In the Case 2 nonlinear model, the B_2 and B_3 systems are *before* the squaring and cubing operations, respectively. Additional Cases 3 and 4 can be considered depending upon different orders of the A and B systems as listed below. Four other possible choices, namely, (B_1, A_2, A_3), (A_1, B_2, B_3), (B_1, B_2, A_3), and (A_1, A_2, B_3), give equivalent results to Cases 1–4. Table 5.1 lists the four cases.

Case 1 is covered by figures and equations in Section 5.2. Case 2 is covered by figures and equations in Section 5.3. Case 3 is a mixture of Cases 1 and 2

TABLE 5.1 Four Cases of Square-Law and Cubic Models

Case	Linear Path	Square-Law Path	Cubic Path
1	A_1	A_2	A_3
2	B_1	B_2	B_3
3	A_1	B_2	A_3
4	B_1	A_2	B_3

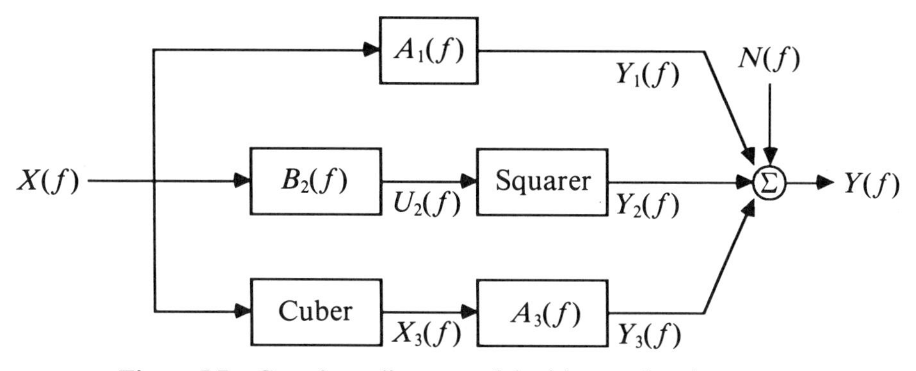

Figure 5.7 Case 3 nonlinear model with correlated outputs.

involving A_1, B_2, and A_3 as shown in Figure 5.7. This model has correlated outputs. The revised Case 3 nonlinear model with uncorrelated outputs is shown in Figure 5.8.

The following equations apply to Figures 5.7 and 5.8. The system

$$A_1(f) = H_o(f) - 3\sigma_x^2 A_3(f) \tag{5.116}$$

where $H_o(f)$ is computed by Eq. (5.28). The system

$$B_2(f) = \frac{[S_{xxy}(f)/2]^{1/2}}{S_{xx}(f)} \tag{5.117}$$

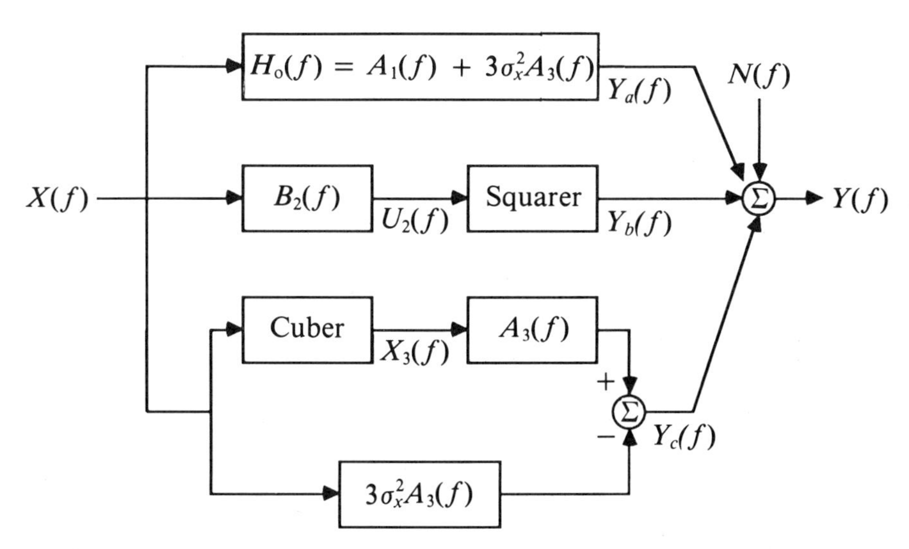

Figure 5.8 Revised Case 3 nonlinear model with uncorrelated outputs.

where the special bispectrum $S_{xxy}(f)$ is computed by Eq. (5.32). The system $A_3(f)$ can be computed by the practical system identification procedure in Section 5.2.4 or by the formula

$$A_3(f) = \frac{S_{xxxy}(f/3)}{6S_{xx}^3(f/3)} \tag{5.118}$$

where the special trispectrum $S_{xxxy}(f/3)$ is computed by Eq. (5.33). The following quantities are defined:

$$S_{y_1y_1}(f) = |A_1(f)|^2 S_{xx}(f) \tag{5.119}$$

$$S_{y_2y_2}(f) = S_{y_by_b}(f) + (\bar{y}_2)^2 \delta_1(f) \tag{5.120}$$

$$S_{y_3y_3}(f) = S_{y_cy_c}(f) + 9\sigma_x^4 |A_3(f)|^2 S_{xx}(f) \tag{5.121}$$

$$S_{u_2u_2}(f) = |B_2(f)|^2 S_{xx}(f) \tag{5.122}$$

$$\bar{y}_2 = \sigma_2^2 = \int S_{u_2u_2}(f)\, df \tag{5.123}$$

$$S_{y_ay_a}(f) = |H_o(f)|^2 S_{xx}(f) \tag{5.124}$$

$$S_{y_by_b}(f) = 2 \int S_{u_2u_2}(\alpha) S_{u_2u_2}(f-\alpha)\, d\alpha \tag{5.125}$$

$$S_{y_cy_c}(f) = 6|A_3(f)|^2 \iint S_{xx}(u) S_{xx}(v-u) S_{xx}(f-v)\, du\, dv \tag{5.126}$$

$$S_{yy}(f) = S_{y_ay_a}(f) + S_{y_by_b}(f) + S_{y_cy_c}(f) + S_{nn}(f) \tag{5.127}$$

A different mixture of Cases 1 and 2 is shown in Figure 5.9 involving B_1, A_2, and B_3, called the Case 4 nonlinear model. Like Figure 5.7, this model has correlated outputs. The revised Case 4 nonlinear model with uncorrelated outputs is shown in Figure 5.10.

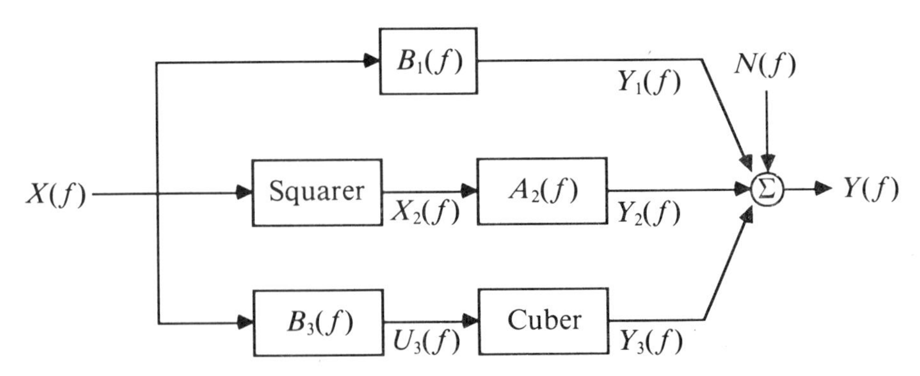

Figure 5.9 Case 4 nonlinear model with correlated outputs.

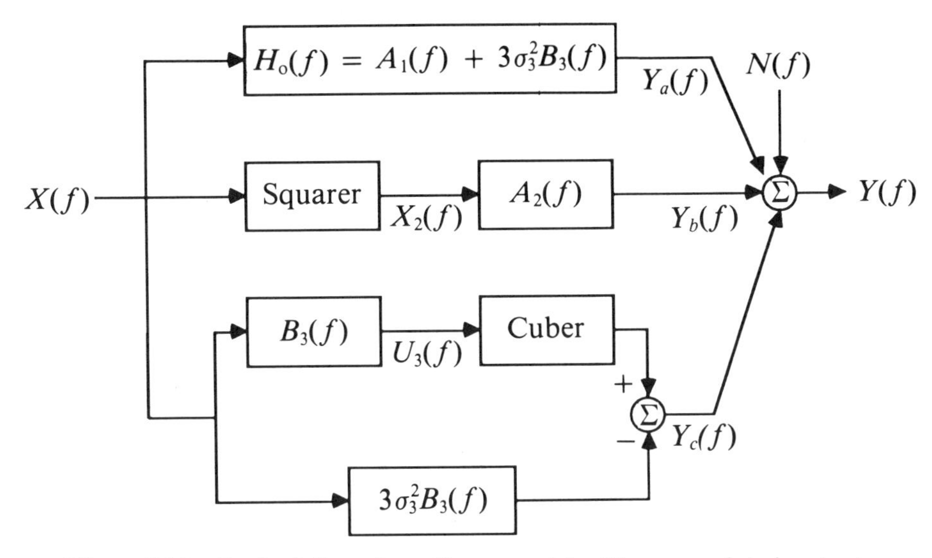

Figure 5.10 Revised Case 4 nonlinear model with uncorrelated outputs.

The following equations apply to Figures 5.9 and 5.10. The system

$$B_1(f) = H_o(f) - 3\sigma_3^2 B_3(f) \tag{5.128}$$

The system $A_2(f)$ can be computed by the practical system identification procedure in Section 5.2.4 or by the formula

$$A_2(f) = \frac{S_{xxy}(f/2)}{2S_{xx}^2(f/2)} \tag{5.129}$$

where $S_{xxy}(f/2)$ is computed by Eq. (5.32). The system

$$B_3(f) = \frac{[S_{xxxy}(f)/6]^{1/3}}{S_{xx}(f)} \tag{5.130}$$

where $S_{xxxy}(f)$ is computed by Eq. (5.33). The following quantities are defined:

$$H_o(f) = \frac{S_{xy}(f)}{S_{xx}(f)} \tag{5.131}$$

$$S_{y_1y_1}(f) = |B_1(f)|^2 S_{xx}(f) \tag{5.132}$$

$$S_{y_2y_2}(f) = S_{y_by_b}(f) + (\bar{y}_2)^2 \delta_1(f) \tag{5.133}$$

$$S_{y_3y_3}(f) = S_{y_cy_c}(f) + 9\sigma_3^4 |B_3(f)|^2 S_{xx}(f) \tag{5.134}$$

$$\bar{y}_2 = A_2(0)\sigma_x^2 \tag{5.135}$$

$$S_{u_3u_3}(f) = |B_3(f)|^2 S_{xx}(f) \tag{5.136}$$

$$\sigma_3^2 = \int S_{u_3u_3}(f)\,df \tag{5.137}$$

$$S_{y_ay_a}(f) = |H_o(f)|^2 S_{xx}(f) \tag{5.138}$$

$$S_{y_by_b}(f) = 2|A_2(f)|^2 \int S_{xx}(u) S_{xx}(f-u)\,du \tag{5.139}$$

$$S_{y_cy_c}(f) = 6 \iint S_{u_3u_3}(\alpha) S_{u_3u_3}(\beta-\alpha) S_{u_3u_3}(f-\beta)\,d\alpha\,d\beta \tag{5.140}$$

$$S_{yy}(f) = S_{y_ay_a}(f) + S_{y_by_b}(f) + S_{y_cy_c}(f) + S_{nn}(f) \tag{5.141}$$

5.4.1 Linear System in Parallel with Square-Law System

Two special cases of physical interest are the cases of a linear system in parallel *only* with a square-law system, where the squarer might be followed or preceded by a linear system as in Figures 5.11 and 5.12. For Gaussian input data, the output terms $y_1(t)$ and $y_2(t)$ will be uncorrelated in both Figures 5.11 and 5.12. The same values are obtained for the linear system frequency response functions $A_1(f)$ and $B_1(f)$. However, different results are generally obtained for $A_2(f)$ and $B_2(f)$. When this occurs, Figures 5.11 and 5.12 produce the same values for the output spectra $S_{y_1y_1}(f)$, but they produce different values for the output spectra $S_{y_2y_2}(f)$ because of the frequency effects of $A_2(f)$ and $B_2(f)$.

When the nonzero mean values are removed from $y_2(t)$ and $y(t)$, Figures 5.11 and 5.12 can be replaced by the equivalent revised Figures 5.13 and 5.14, where the output terms $y_a(t)$ and $y_b(t)$ will be uncorrelated in both Figures 5.13 and 5.14. Figures 5.13 and 5.14 produce the same values for the output spectra $S_{y_ay_a}(f)$, but they produce different values for the output spectra $S_{y_by_b}(f)$. Thus, one can distinguish results from Figure 5.13 versus Figure 5.14, similar to how one can distinguish results from Figure 5.11 versus Figure 5.12.

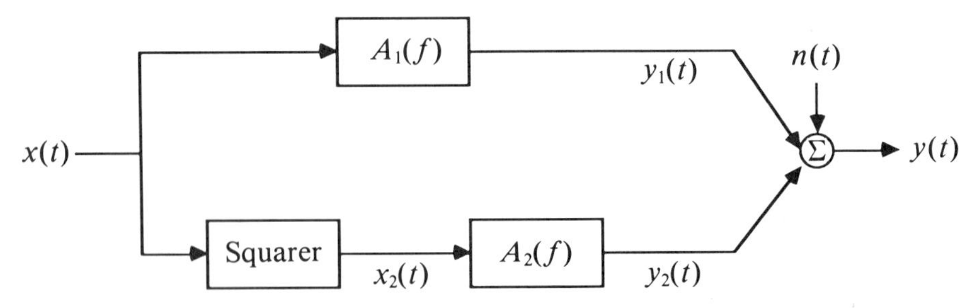

Figure 5.11 Parallel linear and square-law systems with squarer followed by linear system.

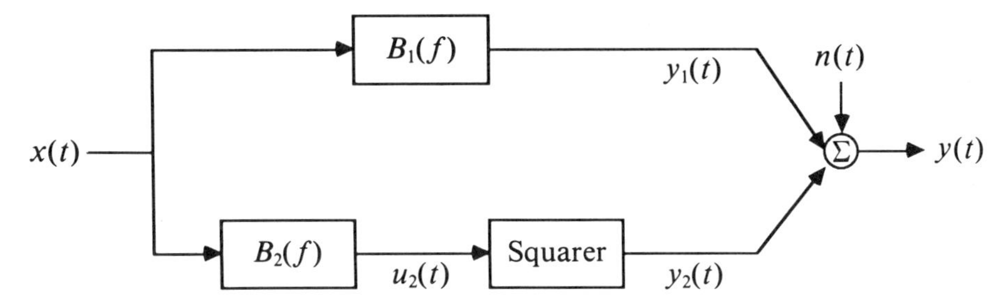

Figure 5.12 Parallel linear and square-law systems with squarer preceded by linear system.

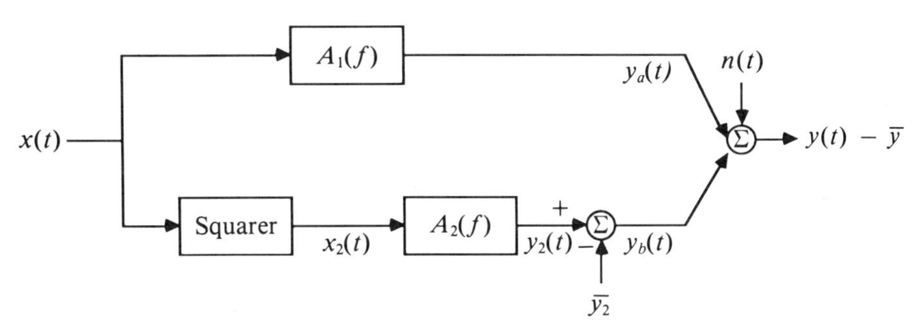

Figure 5.13 Revised Figure 5.11 with zero mean value outputs.

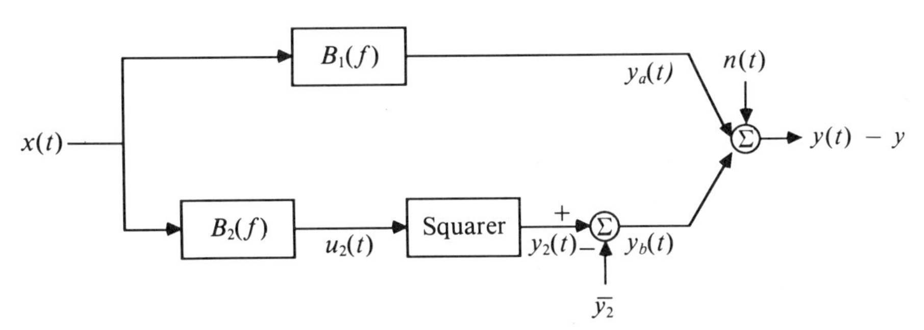

Figure 5.14 Revised Figure 5.12 with zero mean value outputs.

Figures 5.11 and 5.13 are special cases of Figures 5.9 and 5.10, respectively, where $B_3(f) = 0$. Figures 5.12 and 5.14 are special cases of Figures 5.7 and 5.8, respectively, where $A_3(f) = 0$. Figures 5.11 and 5.13 are also equivalent to cases of Figures 5.2 and 5.4, where $A_3(f) = 0$.

The following equations apply to Figures 5.11 and 5.13. The system

$$A_1(f) = H_o(f) = \frac{S_{xy}(f)}{S_{xx}(f)} \tag{5.142}$$

The system $A_2(f)$ can be computed by the practical system identification procedure in Section 5.2.4 or by the formula

$$A_2(f) = \frac{S_{xxy}(f/2)}{2S_{xx}^2(f/2)} \tag{5.143}$$

where $S_{xxy}(f/2)$ is computed by Eq. (5.32). Here, the quantities

$$S_{y_1y_1}(f) = |A_1(f)|^2 S_{xx}(f) = S_{y_ay_a}(f) \tag{5.144}$$

$$S_{y_2y_2}(f) = S_{y_by_b}(f) + (\bar{y}_2)^2 \delta_1(f) \tag{5.145}$$

$$\bar{y}_2 = A_2(0)\sigma_x^2 \tag{5.146}$$

$$S_{y_by_b}(f) = 2|A_2(f)|^2 \int S_{xx}(u) S_{xx}(f-u)\, du \tag{5.147}$$

$$S_{yy}(f) = S_{y_ay_a}(f) + S_{y_by_b}(f) + S_{nn}(f) \tag{5.148}$$

The following equations apply to Figures 5.12 and 5.14. The systems

$$B_1(f) = H_o(f) = \frac{S_{xy}(f)}{S_{xx}(f)} \tag{5.149}$$

$$B_2(f) = \frac{[S_{xxy}(f)/2]^{1/2}}{S_{xx}(f)} \tag{5.150}$$

where $S_{xxy}(f)$ is computed by Eq. (5.32). Here, the quantities

$$S_{y_1y_1}(f) = |B_1(f)|^2 S_{xx}(f) = S_{y_ay_a}(f) \tag{5.151}$$

$$S_{y_2y_2}(f) = S_{y_by_b}(f) + \sigma_2^4 \delta_1(f) \tag{5.152}$$

$$\sigma_2^2 = \int S_{u_2u_2}(f)\, df \tag{5.153}$$

$$S_{u_2u_2}(f) = |B_2(f)|^2 S_{xx}(f) \tag{5.154}$$

$$S_{y_by_b}(f) = 2 \int S_{u_2u_2}(\alpha) S_{u_2u_2}(f-\alpha)\, d\alpha \tag{5.155}$$

$$S_{yy}(f) = S_{y_ay_a}(f) + S_{y_by_b}(f) + S_{nn}(f) \tag{5.156}$$

5.4.2 Linear System in Parallel with Cubic System

Two other special cases of physical interest are cases of a linear system in parallel *only* with a cubic system, where the cuber might be followed or preceded by a linear system as in Figures 5.15 and 5.16. For Gaussian input data, the output terms $y_1(t)$ and $y_3(t)$ will be correlated in both Figures 5.15 and 5.16. Here, unlike Figures 5.11 and 5.12, different values are obtained for the linear

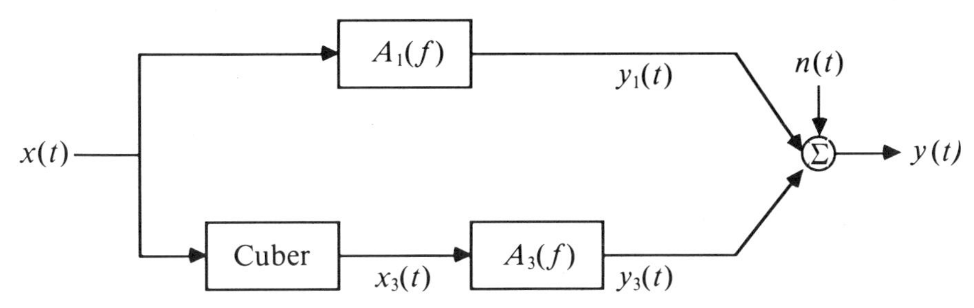

Figure 5.15 Parallel linear and cubic systems with cuber followed by linear system.

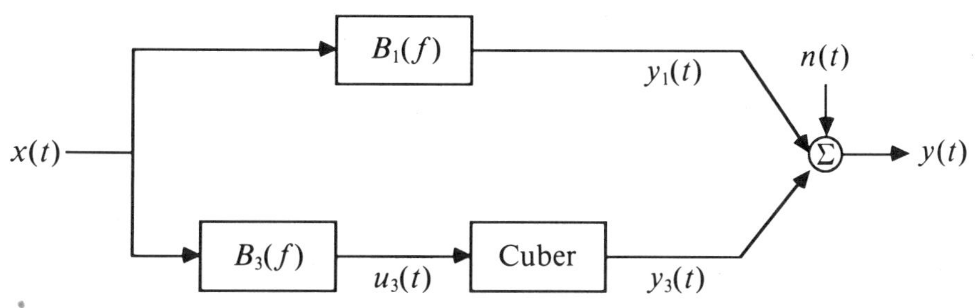

Figure 5.16 Parallel linear and cubic systems with cuber preceded by linear system.

system frequency response functions $A_1(f)$ and $B_1(f)$ because different correlation functions exist between the $y_1(t)$ and $y_3(t)$ of Figure 5.15 versus the $y_1(t)$ and $y_3(t)$ of Figure 5.16. Also, Figures 5.15 and 5.16 give different results for the terms $A_3(f)$ and $B_3(f)$, as well as different results for the output spectra $S_{y_1y_1}(f)$ and $S_{y_3y_3}(f)$.

Figures 5.15 and 5.16 can be revised as shown in Figures 5.17 and 5.18 to produce the uncorrelated outputs $y_a(t)$ and $y_c(t)$. Figures 5.17 and 5.18 produce the same values for the output spectra $S_{y_ay_a}(f)$ but different values for the output spectra $S_{y_cy_c}(f)$, so it easy to distinguish results from these two situations.

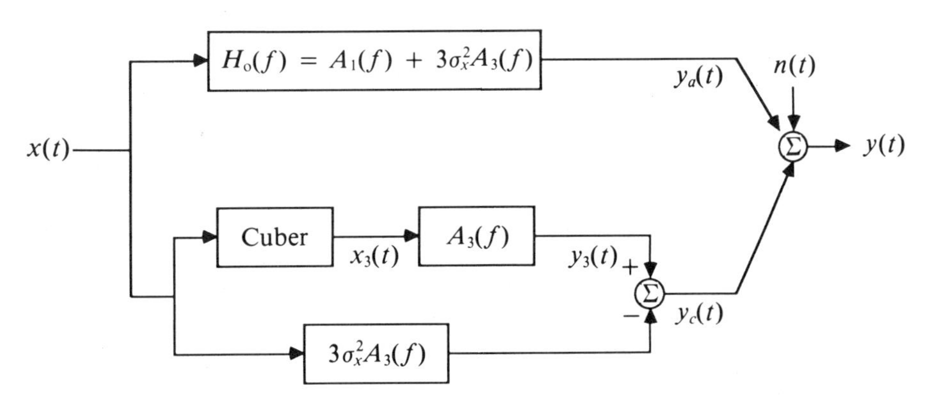

Figure 5.17 Revised Figure 5.15 with uncorrelated outputs.

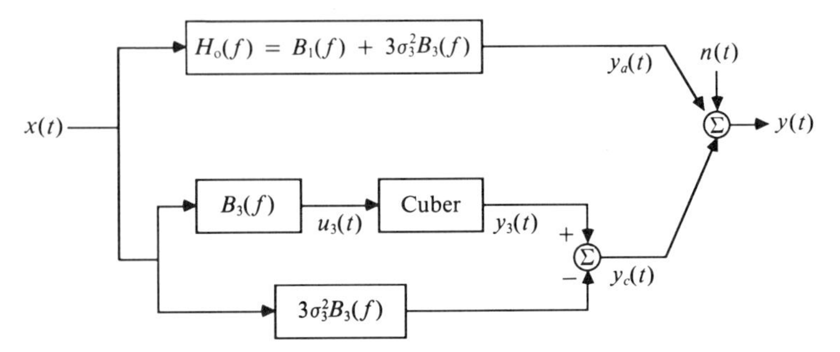

Figure 5.18 Revised Figure 5.16 with uncorrelated outputs.

Figures 5.15 and 5.17 are special cases of Figures 5.7 and 5.8, respectively, where $B_2(f) = 0$. Figures 5.16 and 5.18 are special cases of Figures 5.9 and 5.10, respectively, where $B_2(f) = 0$. Figures 5.15 and 5.17 are also equivalent to cases of Figures 5.2 and 5.4, where $A_2(f) = 0$.

The following equations apply to Figures 5.15 and 5.17. The systems

$$A_1(f) = H_o(f) - 3\sigma_x^2 A_3(f) \tag{5.157}$$

$$H_o(f) = \frac{S_{xy}(f)}{S_{xx}(f)} \tag{5.158}$$

$$A_3(f) = \frac{S_{xxxy}(f/3)}{6S_{xx}^3(f/3)} \tag{5.159}$$

where $S_{xxxy}(f/3)$ is computed by Eq. (5.33). The practical system identification procedure in Section 5.2.4 can also be used here to compute $A_3(f)$. The quantities

$$S_{y_1y_1}(f) = |A_1(f)|^2 S_{xx}(f) \tag{5.160}$$

$$S_{y_3y_3}(f) = S_{y_cy_c}(f) + 9\sigma_x^4 |A_3(f)|^2 S_{xx}(f) \tag{5.161}$$

$$S_{y_ay_a}(f) = |H_o(f)|^2 S_{xx}(f) \tag{5.162}$$

$$S_{y_cy_c}(f) = 6|A_3(f)|^2 \iint S_{xx}(u) S_{xx}(v-u) S_{xx}(f-v)\, du\, dv \tag{5.163}$$

$$S_{yy}(f) = S_{y_ay_a}(f) + S_{y_cy_c}(f) + S_{nn}(f) \tag{5.164}$$

The following equations apply to Figures 5.16 and 5.18:

$$B_1(f) = H_o(f) - 3\sigma_3^2 B_3(f) \tag{5.165}$$

$$H_o(f) = \frac{S_{xy}(f)}{S_{xx}(f)} \tag{5.166}$$

$$B_3(f) = \frac{[S_{xxxy}(f)/6]^{1/3}}{S_{xx}(f)} \tag{5.167}$$

where $S_{xxxy}(f)$ is computed by Eq. (5.33). The quantities

$$S_{y_1y_1}(f) = |B_1(f)|^2 S_{xx}(f) \tag{5.168}$$

$$S_{y_3y_3}(f) = S_{y_cy_c}(f) + 9\sigma_3^4 |B_3(f)|^2 S_{xx}(f) \tag{5.169}$$

$$S_{u_3u_3} = |B_3(f)|^2 S_{xx}(f) \tag{5.170}$$

$$\sigma_3^2 = \int S_{u_3u_3}(f)\, df \tag{5.171}$$

$$S_{y_ay_a}(f) = |H_0(f)|^2 S_{xx}(f) \tag{5.172}$$

$$S_{y_cy_c}(f) = 6 \iint S_{u_3u_3}(\alpha) S_{u_3u_3}(\beta - \alpha) S_{u_3u_3}(f - \beta)\, d\alpha\, d\beta \tag{5.173}$$

$$S_{yy}(f) = S_{y_ay_a}(f) + S_{y_cy_c}(f) + S_{nn}(f) \tag{5.174}$$

5.4.3 Square-Law System with Sign

Consider the example of a square-law system with sign (squarer with sign) where an input $x(t)$ produces an output $x(t)|x(t)|$. The physical meaning of this particular nonlinear system will be used in the nonlinear wave force model of Section 7.4. From Eq. (2.196), when $x(t)$ represents zero mean value Gaussian data with variance σ_x^2, the optimum third-order polynomial least-squares approximation for $x(t)|x(t)|$ is given by $v(t)$ where

$$v(t) = a_1 x(t) + a_3 x^3(t) \approx x(t)|x(t)| \tag{5.175}$$

As shown in Figure 2.34, $v(t)$ gives a small error compared to the true value $x(t)|x(t)|$ when the constants a_1 and a_3 have the values

$$\begin{aligned} a_1 &= \sigma_x \sqrt{(2/\pi)} = 3a_3\sigma_x^2 \\ a_3 &= \sqrt{(2/\pi)}/3\sigma_x \end{aligned} \tag{5.176}$$

Thus, the nonlinear operation $x|x|$ can be approximated closely by the zero-memory nonlinear model shown in Figure 5.19, consisting of a linear system in parallel with a cubic system. The Fourier transform relation for Eq. (5.175) and Figure 5.19 is

$$V(f) = a_1 X(f) + a_3 X_3(f) \tag{5.177}$$

where $V(f)$, $X(f)$, and $X_3(f)$ are Fourier transforms of $v(t)$, $x(t)$, and $x_3(t) = x^3(t)$, respectively. The quantities $v(t)$ and $V(f)$ can be predicted solely from measurement of $x(t)$ plus knowledge of a_1 and a_3.

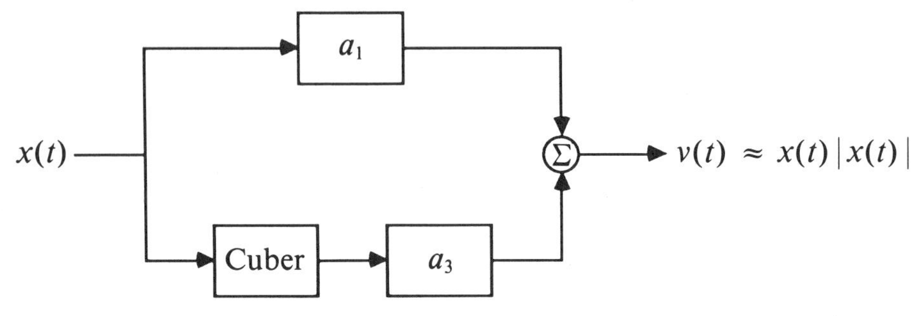

Figure 5.19 Zero-memory third-order polynomial nonlinear model for squarer with sign.

Suppose the input $x(t)$ passes through the zero-memory nonlinear model for $x(t)|x(t)|$ that yields $v(t)$ in Figure 5.19, and then $v(t)$ passes through a constant-parameter linear system with frequency response function $A(f)$. This gives the Case 1 finite-memory nonlinear model for $x(t)|x(t)|$ shown in Figure 5.20 where the output $y(t)$ has a Fourier transform $Y(f)$ satisfying

$$Y(f) = A(f)V(f) \tag{5.178}$$

using the $V(f)$ of Eq. (5.177). Figure 5.20 is a special example of Figure 5.15.

A different Case 2 finite-memory nonlinear system for $x(t)|x(t)|$ is drawn in Figure 5.21, where Figure 5.19 is preceded by a linear system with frequency response function $B(f)$. Figure 5.21 is a special example of Figure 5.16.

5.5 THIRD-ORDER POLYNOMIAL LEAST-SQUARES MODEL

The zero-memory nonlinear model in Figure 5.19 representing $x(t)|x(t)|$ and related models in Figures 5.20 and 5.21 of finite-memory nonlinear systems can be extended to apply to other nonlinear situations. A simple, practical procedure will now be outlined for analyzing general third-order polynomial least-squares

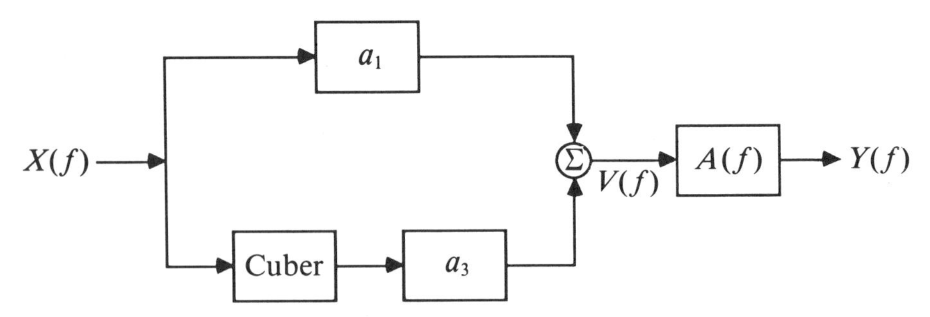

Figure 5.20 Case 1 finite-memory nonlinear model for squarer with sign.

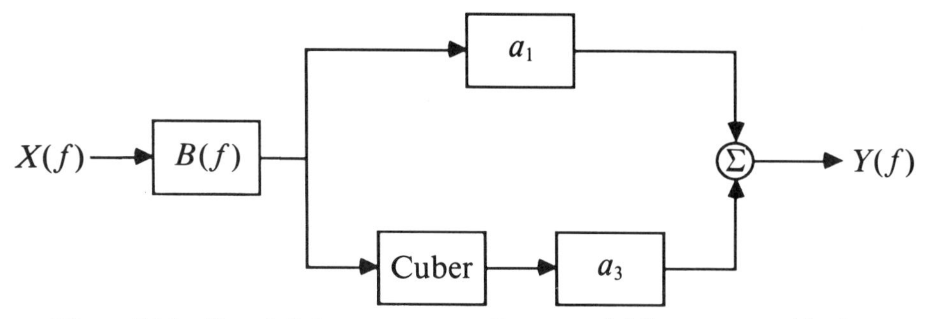

Figure 5.21 Case 2 finite-memory nonlinear model for squarer with sign.

models that can approximate many zero- and finite-memory nonlinear systems. This procedure broadens the scope of the material in Section 5.2.

Following the ideas in Section 2.9, let $y(t) = g[x(t)]$ be an arbitrary zero-memory nonlinear system, and let $v = v(t)$ be the optimum third-order polynomial least-squares approximation to $y = y(t)$ under the requirement that $x = x(t)$ represents a zero mean value Gaussian stationary random process. Then

$$v(t) = a_1 x(t) + a_2 x^2(t) + a_3 x^3(t) \tag{5.179}$$

where the constant coefficients a_1, a_2, and a_3 are determined by Eqs. (2.184)–(2.186). These equations require computation of the expected values

$$E[xy] = E[xg(x)] = \int xg(x)p(x)\,dx \tag{5.180}$$

$$E[x^2y] = E[x^2g(x)] = \int x^2g(x)p(x)\,dx \tag{5.181}$$

$$E[x^3y] = E[x^3g(x)] = \int x^3g(x)p(x)\,dx \tag{5.182}$$

where $p(x)$ is the Gaussian probability density function of Eq. (2.4).

When $v(t)$ is deemed to be a suitable approximation to $g[x(t)]$, the zero-memory nonlinear system for $g[x(t)]$ can be achieved by the third-order polynomial least-squares model of Figure 5.22, where the outputs $v_1(t)$ and $v_3(t)$ will be correlated. The Fourier transform relation for Eq. (5.179) and Figure 5.22 is the frequency-domain relation for Figure 2.33, namely,

$$V(f) = a_1 X(f) + a_2 X_2(f) + a_3 X_3(f) \tag{5.183}$$

where $V(f)$, $X(f)$, $X_2(f)$, and $X_3(f)$ are Fourier transforms of $v(t)$, $x(t)$, $x^2(t)$, and $x^3(t)$, respectively. Thus, $v(t)$ and $V(f)$ can be predicted solely from measurement of $x(t)$ plus knowledge of a_1, a_2, and a_3.

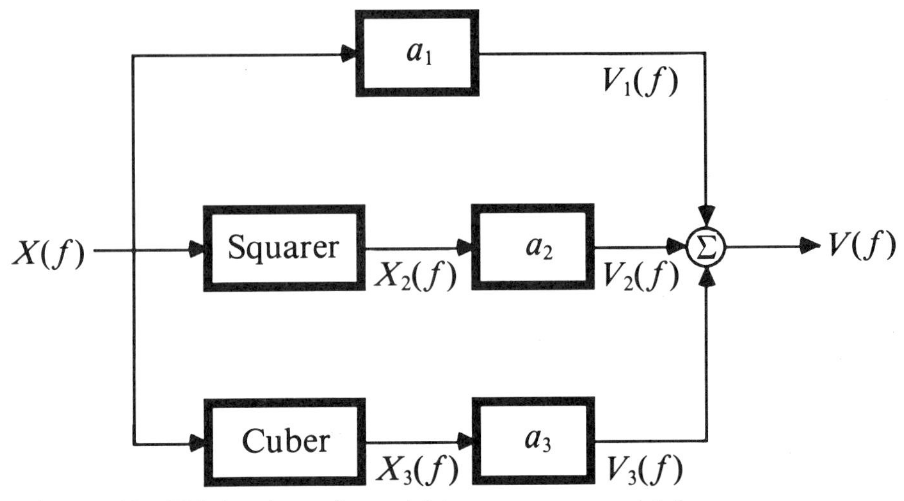

Figure 5.22 Third-order polynomial least-squares model for a zero-memory nonlinear system.

Suppose the measured input $x(t)$ passes through the zero-memory nonlinear system for $g[x(t)]$ that yields $v(t)$ in Figure 5.22, and then $v(t)$ passes through a constant-parameter linear system with frequency response function $A(f)$ to yield an output $w(t)$. Let this unmeasured output $w(t)$ be corrupted by extraneous uncorrelated noise $n(t)$ to produce the total measured output $y(t)$. This gives the third-order polynomial least-squares model for a finite-memory nonlinear system shown in Figure 5.23 where $Y(f)$, $W(f)$, and $N(f)$ are Fourier transforms of $y(t)$, $w(t)$, and $n(t)$, respectively. The Fourier transform relation for Figure 5.23 is

$$Y(f) = W(f) + N(f) = A(f)V(f) + N(f) \tag{5.184}$$

where $V(f)$ is known from Eq. (5.183).

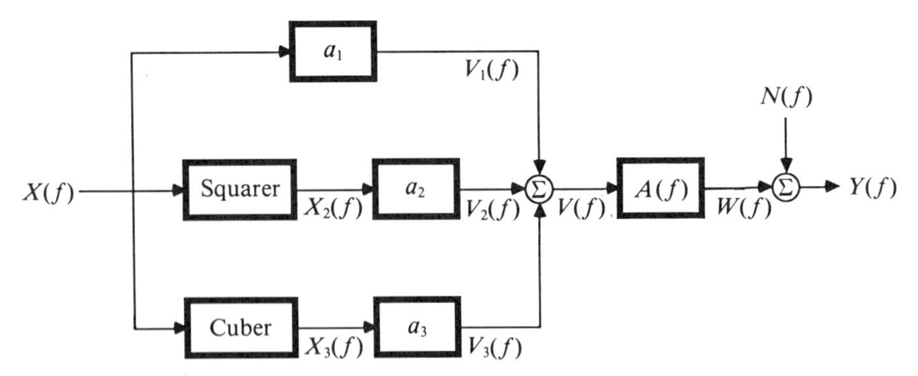

Figure 5.23 Third-order polynomial least-squares model for a finite-memory nonlinear system.

Note that Figure 5.23 is a special case of Figure 5.1 where $A_1(f) = a_1 A(f)$, $A_2(f) = a_2 A(f)$, and $A_3(f) = a_3 A(f)$. For arbitrary $A_1(f)$, $A_2(f)$, and $A_3(f)$, Figure 5.1 represents a logical expansion of Figure 5.23 that enables Figure 5.1 to be more widely applicable than Figure 5.23.

5.5.1 Linear Analysis Procedure

Similar to the procedure in Section 5.2.4, a practical linear way to analyze the finite-memory third-order polynomial of Figure 5.23 now follows. As discussed in Section 1.3, assume that n_d associated records of $x(t)$ and $y(t)$ are available, where each record is of length T. For any input record $x(t)$, one can compute $x_2(t) = x^2(t)$ and $x_3(t) = x^3(t)$, and then compute their finite Fourier transforms for any $f \neq 0$ by

$$X(f) = \mathscr{F}[x(t)] \tag{5.185}$$

$$X_2(f) = \mathscr{F}[x^2(t)] \tag{5.186}$$

$$X_3(f) = \mathscr{F}[x^3(t)] \tag{5.187}$$

These quantities represent the input data for the three-input/single-output linear model of Figure 5.2. Expected value operations of these results over the n_d available records give the spectral density functions

$$S_{xx}(f) = \frac{1}{T} E[X^*(f)X(f)] \tag{5.188}$$

$$S_{x_2x_2}(f) = \frac{1}{T} E[X_2^*(f)X_2(f)] \tag{5.189}$$

$$S_{x_3x_3}(f) = \frac{1}{T} E[X_3^*(f)X_3(f)] \tag{5.190}$$

$$S_{xx_3}(f) = \frac{1}{T} E[X^*(f)X_3(f)] \tag{5.191}$$

The quantity $S_{xx_2}(f)$ does not have to be computed, since it will be zero for Gaussian input data.

From Eq. (5.183), it follows directly that the autospectrum

$$\begin{aligned} S_{vv}(f) &= \frac{1}{T} E[V^*(f)V(f)] \\ &= a_1^2 S_{xx}(f) + a_2^2 S_{x_2x_2}(f) + a_3^2 S_{x_3x_3}(f) + 2a_1 a_3 \operatorname{Real}[S_{xx_3}(f)] \end{aligned} \tag{5.192}$$

Thus, $S_{vv}(f)$ can be predicted from knowledge of a_1, a_2, and a_3 plus measurement of the input record $x(t)$ only.

When the total output record $y(t)$ in Figure 5.23 can be measured together with the input record $x(t)$, one can also compute its finite Fourier transform

$$Y(f) = \mathscr{F}[y(t)] \tag{5.193}$$

Together with $X(f)$ and $X_3(f)$, expected value operations over the n_d available pairs of records give the cross-spectral density functions

$$S_{xy}(f) = \frac{1}{T} E[X^*(f)Y(f)] \tag{5.194}$$

$$S_{x_2y}(f) = \frac{1}{T} E[X_2^*(f)Y(f)] \tag{5.195}$$

$$S_{x_3y}(f) = \frac{1}{T} E[X_3^*(f)Y(f)] \tag{5.196}$$

From Eq. (5.183), it follows that the cross-spectrum

$$S_{vy}(f) = \frac{1}{T} E[V^*(f)Y(f)] = a_1 S_{xy}(f) + a_2 S_{x_2y}(f) + a_3 S_{x_3y}(f) \tag{5.197}$$

Thus, $S_{vy}(f)$ can be predicted from knowledge of a_1, a_2 and a_3 plus simultaneous measurement of the input record $x(t)$ and the total output record $y(t)$.

The Fourier transform relation of Eq. (5.184) leads directly to the linear single-input/single-output spectral relation

$$S_{vy}(f) = A(f)S_{vv}(f) \tag{5.198}$$

where $S_{vn}(f)$ is zero from the assumptions that (1) $w(t)$ and $n(t)$ are uncorrelated, and (2) $w(t)$ is linearly related to $v(t)$. Hence, the linear frequency response function $A(f)$ for any $f \neq 0$ can be determined by the formula

$$A(f) = \frac{S_{vy}(f)}{S_{vv}(f)} \tag{5.199}$$

This formula shows how to identify $A(f)$ from knowledge of a_1, a_2, and a_3 plus measurement of $x(t)$ and $y(t)$.

From Eq. (5.184), since $w(t)$ and $n(t)$ are uncorrelated, the total measured output spectrum

$$S_{yy}(f) = S_{ww}(f) + S_{nn}(f) \tag{5.200}$$

where

$$S_{ww}(f) = |A(f)|^2 S_{vv}(f) \tag{5.201}$$

This result for $S_{ww}(f)$ can be predicted without measuring $w(t)$ by using the previous computations of $S_{vv}(f)$ from Eq. (5.192) and $A(f)$ from Eq. (5.199). If $S_{ww}(f)$ is close to $S_{yy}(f)$ at frequencies of interest where there are peaks in the output spectrum of $S_{yy}(f)$, then one can conclude that Figure 5.23 is a good model for the finite-memory nonlinear system that has been approximated.

Figure 5.23 is a Case 1 type of model where a linear system $A(f)$ follows the zero-memory nonlinear system. Similarly, one can consider a Case 2 type of model where a linear system $B(f)$ would precede the zero-memory nonlinear system. Techniques for analyzing Case 2 models, discussed in Section 5.3, are more complicated than the practical linear analysis procedure outlined here that requires computation only of ordinary autospectral and cross-spectral density functions.

5.5.2 Extended Nonlinear Models

A generalization of Figure 5.23 is drawn in Figure 5.24, consisting of a linear system $H(f)$ with an output $Y_h(f)$ in parallel with the finite-memory nonlinear system of Figure 5.23 producing the correlated output $Y_v(f)$. There is also uncorrelated output noise $N(f)$. This extended nonlinear model is suitable for physical problems where the total output $Y(f)$ is due to a linear transformation in parallel with any finite-memory nonlinear transformation that can be approximated by Figure 5.23. Thus, Figure 5.24 can cover many engineering applications.

By combining linear operations in Figure 5.24, it is clear that the next Figure 5.25 represents a Case 1 nonlinear model with correlated outputs $Y_1(f)$, $Y_2(f)$,

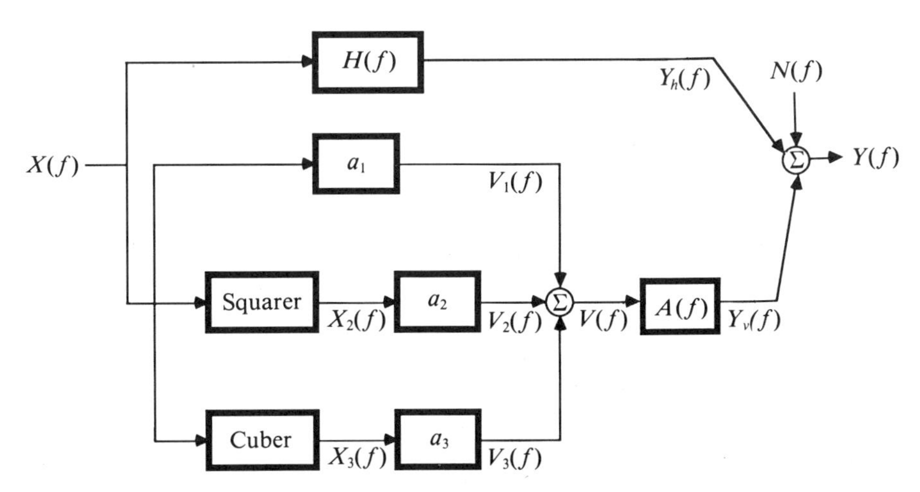

Figure 5.24 Linear system in parallel with Figure 5.23.

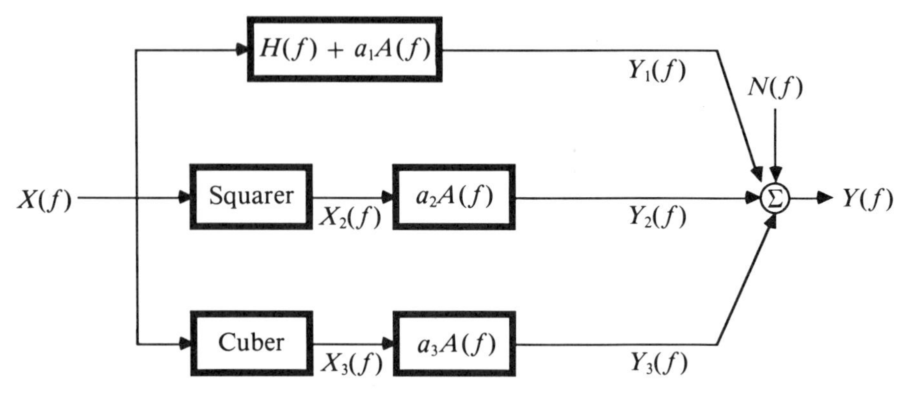

Figure 5.25 Case 1 nonlinear model with correlated outputs equivalent to Figure 5.24.

and $Y_3(f)$, where Figure 5.25 is equivalent to Figure 5.24. As previously done in deriving Figure 5.3 from Figure 5.1, Figure 5.25 can now be revised as shown in Figure 5.26 to yield a Case 1 nonlinear model with uncorrelated outputs $Y_a(f)$, $Y_b(f)$, and $Y_c(f)$, where Figure 5.26 is equivalent to Figure 5.25. Practical standard techniques for analyzing the response properties and identifying the system properties in Figures 5.25 and 5.26 are developed in this chapter. The overall optimum linear system $H_o(f) = [S_{xy}(f)/S_{xx}(f)]$ is the top path in Figure 5.26.

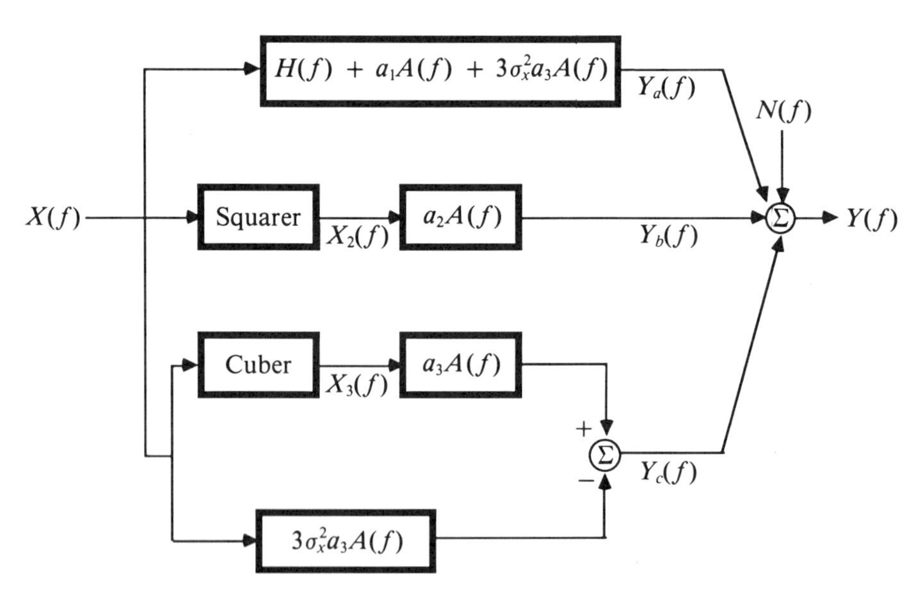

Figure 5.26 Revised Case 1 nonlinear model with uncorrelated outputs equivalent to Figure 5.25.

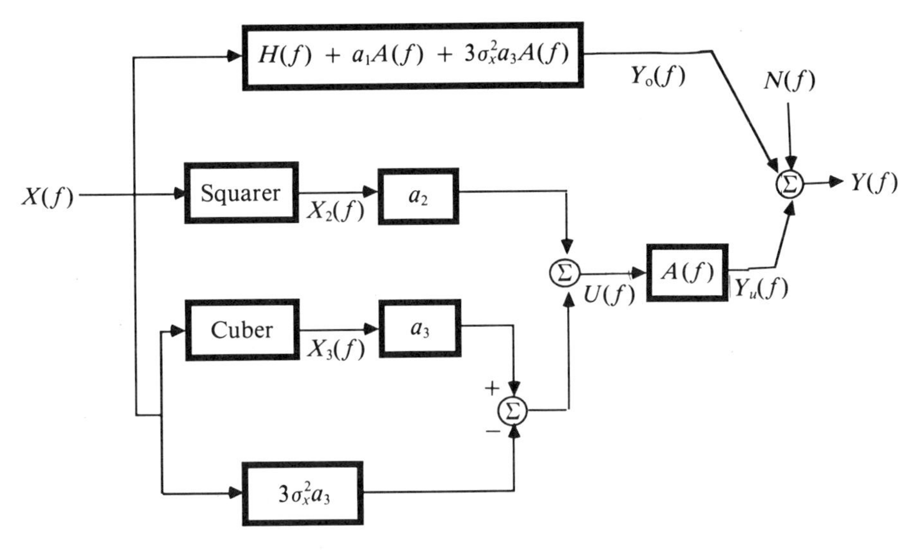

Figure 5.27 Revised Figure 5.24 with uncorrelated outputs.

Figure 5.26 can be redrawn as Figure 5.27 consisting of the optimum linear system $H_o(f)$ that produces the output $Y_o(f)$ in parallel with a revised finite-memory nonlinear system that produces the uncorrelated output $Y_u(f)$. The sum $[Y_h(f) + Y_v(f)]$ in Figure 5.24 is the same as the sum $[Y_o(f) + Y_u(f)]$ in Figure 5.27. Thus, Figures 5.24 and 5.27 are equivalent. Note also that Figure 5.25 can be replaced by the three-input/single-output linear model of Figure 5.2 with correlated inputs, and that Figure 5.26 can be replaced by the three-input/single-output linear model of Figure 5.4 with uncorrelated inputs.

Figures 5.24 and 5.27 represent nonlinear models of a linear system in parallel with a broad class of finite-memory nonlinear systems, namely, those finite-memory nonlinear systems that can be approximated by the finite-memory third-order-polynomial least-squares model of Figure 5.23. Still broader nonlinear problems of a linear system in parallel with *any* type of nonlinear system are considered in Chapter 7. A general analysis methodology is developed there to handle these nonlinear problems of widespread engineering importance.

This concludes Chapter 5.

6

STATISTICAL ERRORS IN NONLINEAR ESTIMATES

This chapter derives normalized random error formulas that are applicable to estimates from the nonlinear models discussed in Chapter 5. These random error formulas include estimates of the optimum linear system output spectrum, the linear coherence function, and the nonlinear system output spectra from square-law and cubic systems, together with estimates of their associated nonlinear coherence functions. Also derived are appropriate random error formulas for estimates of the special bispectral and trispectral density functions of a single frequency variable. These results are useful not only to evaluate measurements made from actual data but also to design experiments by assuming certain values for unknown terms and then solving for the number of independent averages to achieve a desired random error. Basic definitions, interpretations, and practical normalized random error formulas for statistical errors in estimates from linear models are in references 1, 2, and 8.

6.1 THIRD-ORDER NONLINEAR MODELS

The essence of the Case 1 and Case 2 single-input/single-output nonlinear models of Figures 5.3 and 5.6 is shown in Figure 6.1 where, for convenience only, the linear systems $A_2(f)$, $A_3(f)$, $B_2(f)$, and $B_3(f)$ are all set equal to unity at every value of f. From Eqs. (5.12) and (5.22), this makes $\bar{y}_2 = \sigma_x^2$ and $C(f) = 3\sigma_x^2$. Theoretical formulas will now be summarized for Figure 6.1 based upon previous relations in Chapter 5.

Input data $x(t)$ and output data $n(t)$ in Figure 6.1 are assumed to be zero mean value Gaussian stationary random processes. Output data $y_a(t)$, $y_b(t)$, and $y_c(t)$ are defined here to have zero mean values and be mutually uncorrelated. Thus,

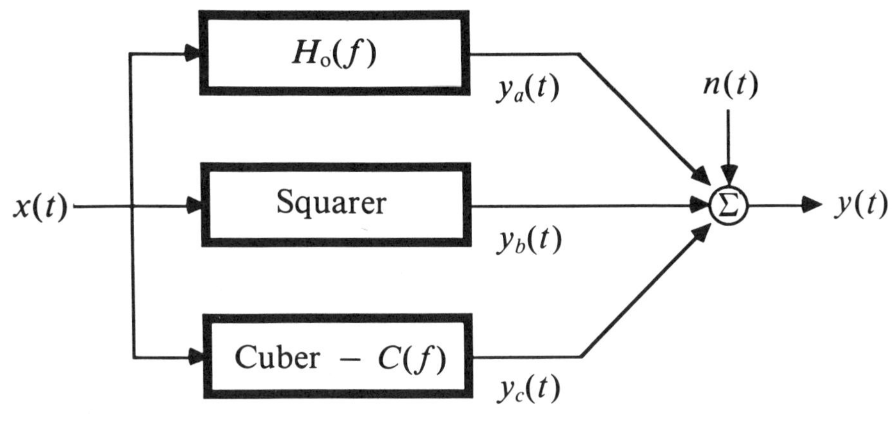

Figure 6.1 Third-order nonlinear model with uncorrelated outputs.

the total output data $y(t)$ in Figure 6.1 is

$$y(t) = y_a(t) + y_b(t) + y_c(t) + n(t) \tag{6.1}$$

where the first optimum linear output $y_a(t)$ satisfies $E[y_a(t)] = 0$.

The second squarer output

$$y_b(t) = x^2(t) - \sigma_x^2 \tag{6.2}$$

satisfies $E[y_b(t)] = 0$ with $E[y_a(t)y_b(t)] = 0$.

The third cuber output

$$y_c(t) = x^3(t) - 3\sigma_x^2 x(t) \tag{6.3}$$

satisfies $E[y_c(t)] = 0$ with $E[y_a(t)y_c(t)] = 0$ and $E[y_b(t)y_c(t)] = 0$.

Corresponding finite Fourier transforms are

$$Y(f) = Y_a(f) + Y_b(f) + Y_c(f) + N(f) \tag{6.4}$$

with

$$Y_a(f) = H_o(f)X(f) \tag{6.5}$$

$$Y_b(f) = \int X(u)X(f-u)\,du - \sigma_x^2\delta_1(f) \tag{6.6}$$

$$Y_c(f) = \iint X(u)X(v-u)X(f-v)\,du\,dv - 3\sigma_x^2 X(f) \tag{6.7}$$

The total output autospectral density function in Figure 6.1 satisfies

$$S_{yy}(f) = S_{y_a y_a}(f) + S_{y_b y_b}(f) + S_{y_c y_c}(f) + S_{nn}(f) \tag{6.8}$$

where, from Eqs. (5.29)–(5.31),

$$S_{y_a y_a}(f) = \gamma_{xy}^2(f) S_{yy}(f) \tag{6.9}$$

$$S_{y_b y_b}(f) = q_{xy_b}^2(f) S_{yy}(f) \tag{6.10}$$

$$S_{y_c y_c}(f) = q_{xy_c}^2(f) S_{yy}(f) \tag{6.11}$$

Hence, the output noise spectrum

$$S_{nn}(f) = [1 - \gamma_{xy}^2(f) - q_{xy_b}^2(f) - q_{xy_c}^2(f)] S_{yy}(f) \tag{6.12}$$

It is known also from Eqs. (5.24)–(5.26) that when $A_2(f) = A_3(f) = 1$, the terms

$$S_{y_a y_a}(f) = |H_o(f)|^2 S_{xx}(f) \tag{6.13}$$

$$S_{y_b y_b}(f) = 2 \int S_{xx}(u) S_{xx}(f - u)\, du \tag{6.14}$$

$$S_{y_c y_c}(f) = 6 \iint S_{xx}(u) S_{xx}(v - u) S_{xx}(f - v)\, du\, dv \tag{6.15}$$

Results from the last three equations give the coherence functions $\gamma_{xy}^2(f)$, $q_{xy_b}^2(f)$, and $q_{xy_c}^2(f)$, respectively, by dividing each of these results in turn by $S_{yy}(f)$. The optimum linear system $H_o(f)$ is the ratio of $S_{xy}(f)$ to $S_{xx}(f)$.

Consider now the individual output terms in Figure 6.1. The first output $y_a(t)$ is such that the ordinary cross-spectral density function

$$S_{xy}(f) = S_{xy_a}(f) = H_o(f) S_{xx}(f) \tag{6.16}$$

because

$$S_{xy_b}(f) = S_{xy_c}(f) = S_{xn}(f) = 0 \tag{6.17}$$

The second output $y_b(t)$ is such that, from Eq. (4.68), the special bispectral density function

$$S_{xxy}(f) = S_{xxy}(f, f) = S_{xxy_b}(f, f) = 2S_{xx}^2(f) \tag{6.18}$$

because

$$S_{xxy_a}(f, f) = S_{xxy_c}(f, f) = S_{xxn}(f, f) = 0 \tag{6.19}$$

The third output $y_c(t)$ is such that, from Eq. (4.69), the special trispectral density function

$$S_{xxxy}(f) = S_{xxxy}(f, f, f) = S_{xxxy_c}(f, f, f) = 6S_{xx}^3(f) \tag{6.20}$$

because

$$S_{xxxy_a}(f, f, f) = S_{xxxy_b}(f, f, f) = S_{xxxn}(f, f, f) = 0 \tag{6.21}$$

Calculation of general bispectral and trispectral density functions of two and three variables is not required to analyze or determine estimates from the nonlinear model of Figure 6.1.

Simpler, special cases of Figure 6.1 occur by assuming that the nonlinear model consists of parallel linear and square-law systems only (Figure 6.2) or that the nonlinear model consists of parallel linear and cubic systems only (Figure 6.3). Results will be stated that apply to these models.

Normalized random error formulas will now be derived for estimates of linear and nonlinear quantities in Figure 6.1. By definition, the *normalized random error* $\varepsilon[\hat{\phi}]$ for an estimate $\hat{\phi}$ of a true quantity $\phi \neq 0$ is the standard deviation of the estimate $\sigma[\hat{\phi}]$, divided by ϕ. For small values of ε, assuming negligible bias error, there is approximately 95% confidence that the true value of ϕ lies inside the interval $[1 \pm 2\varepsilon]\hat{\phi}$. To be specific, if $\varepsilon[\hat{\phi}] = 0.10$, there is approximately 95% confidence that the true value of ϕ lies inside the interval bounded by $0.80\hat{\phi}$ and $1.20\hat{\phi}$. This value of $\varepsilon[\hat{\phi}] = 0.10$ is commonly recommended for the design of physical experiments, keeping in mind that it represents only statistical errors. Design criteria of $\varepsilon[\hat{\phi}] < 0.05$ are often not justified because of other possible system and calibration errors.

Besides the error formulas in this chapter, Section 7.3 in Chapter 7 contains additional error formulas appropriate for parallel linear and nonlinear systems. These additional results are based upon two-input/single output linear formulas derived and discussed in references 1, 2, 5, and 8.

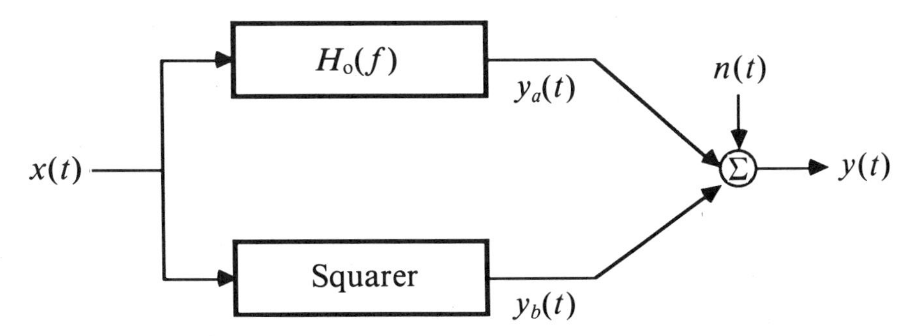

Figure 6.2 Parallel linear and square-law systems.

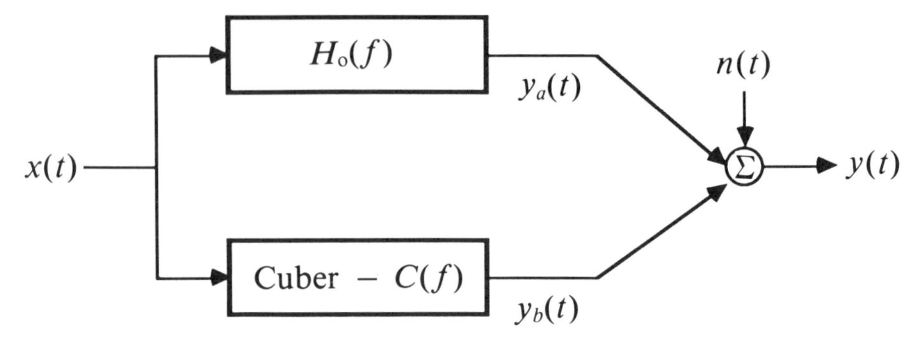

Figure 6.3 Parallel linear and cubic systems.

6.2 LINEAR SYSTEM ESTIMATES

6.2.1 Linear Output Spectrum Estimate

Consider, first of all, the linear system output spectrum estimate $\hat{S}_{y_a y_a}(f)$ in Figure 6.1, where the symbol ^ is used to indicate an estimate of the true value. Let the sum of the Gaussian output $y_a(t)$ with the Gaussian extraneous noise $n(t)$ be denoted by

$$y_0(t) = y_a(t) + n(t) \tag{6.22}$$

Then $y_0(t)$ is also Gaussian, and since $y_a(t)$ and $n(t)$ are uncorrelated, the spectral estimate is

$$\hat{S}_{yoyo}(f) = \hat{S}_{y_a y_a} + \hat{S}_{nn}(f) \tag{6.23}$$

From Eqs. (6.9) and (6.12), the true value is

$$S_{yoyo}(f) = [1 - q^2_{xy_b}(f) - q^2_{xy_c}(f)]S_{yy}(f) \tag{6.24}$$

Assuming $\hat{S}_{y_a y_a}(f)$ and $\hat{S}_{nn}(f)$ to be statistically independent, their variances satisfy

$$\sigma^2[\hat{S}_{yoyo}] = \sigma^2[\hat{S}_{y_a y_a}] + \sigma^2[\hat{S}_{nn}] \tag{6.25}$$

where the dependence upon f has been omitted to simplify the notation.

For Gaussian data, the variances of $\hat{S}_{yoyo}$ and $\hat{S}_{nn}$ from n_d averages are given by (reference 1)

$$\sigma^2[\hat{S}_{yoyo}] = (S^2_{yoyo}/n_d), \qquad \sigma^2[\hat{S}_{nn}] = (S^2_{nn}/n_d) \tag{6.26}$$

Hence,

$$\sigma^2[\hat{S}_{y_a y_a}] = \frac{S_{y_o y_o}^2 - S_{nn}^2}{n_d} \tag{6.27}$$

Substitution from Eqs. (6.12) and (6.24) shows

$$\sigma^2[\hat{S}_{y_a y_a}] = \frac{[2(1 - q_{xy_b}^2 - q_{xy_c}^2) - \gamma_{xy}^2]\gamma_{xy}^2 S_{yy}^2}{n_d} \tag{6.28}$$

Using Eq. (6.9) now proves

$$\varepsilon[\hat{S}_{y_a y_a}] = \frac{\sigma[\hat{S}_{y_a y_a}]}{S_{y_a y_a}} = \frac{[2(1 - q_{xy_b}^2 - q_{xy_c}^2) - \gamma_{xy}^2]^{1/2}}{|\gamma_{xy}|\sqrt{n_d}} \tag{6.29}$$

This is the normalized random error for the linear system output spectrum in Figure 6.1. The absolute value $|\gamma_{xy}|$ denotes the positive square root of γ_{xy}^2. Similar results occur later for other quantities.

For the model of Figure 6.2 where $q_{xy_c}^2 = 0$, one obtains

$$\varepsilon[\hat{S}_{y_a y_a}] = \frac{[2(1 - q_{xy_b}^2) - \gamma_{xy}^2]^{1/2}}{|\gamma_{xy}|\sqrt{n_d}} \tag{6.30}$$

For the model of Figure 6.3 where $q_{xy_b}^2 = 0$, one obtains

$$\varepsilon[\hat{S}_{y_a y_a}] = \frac{[2(1 - q_{xy_c}^2) - \gamma_{xy}^2]^{1/2}}{|\gamma_{xy}|\sqrt{n_d}} \tag{6.31}$$

Finally, one should note that if both $q_{xy_b}^2 = 0$ and $q_{xy_c}^2 = 0$, then

$$\varepsilon[\hat{S}_{y_a y_a}] = \frac{[2 - \gamma_{xy}^2]^{1/2}}{|\gamma_{xy}|\sqrt{n_d}} \tag{6.32}$$

In agreement with the known linear result for this special case (reference 1). In practice, unknown true quantities on the right-hand side of Eqs. (6.29)–(6.32) would be replaced by their estimated values. This is required also in later equations.

6.2.2 Linear Coherence Function Estimate

The linear coherence function estimate for the optimum linear system in Figure 6.1 is given by

$$\hat{\gamma}_{xy}^2(f) = \frac{\hat{S}_{y_a y_a}(f)}{\hat{S}_{yy}(f)} \tag{6.33}$$

Dropping the dependence upon f as before, the true value is

$$S_{y_a y_a} = \gamma_{xy}^2 S_{yy} \tag{6.34}$$

The finite differential operation gives

$$\Delta S_{y_a y_a} = \gamma_{xy}^2 \Delta S_{yy} + S_{yy} \Delta \gamma_{xy}^2$$

Hence,

$$S_{yy} \Delta \gamma_{xy}^2 = \Delta S_{y_a y_a} - \gamma_{xy}^2 \Delta S_{yy}$$

Squaring both sides,

$$S_{yy}^2 (\Delta \gamma_{xy}^2)^2 = (\Delta S_{y_a y_a})^2 - 2\gamma_{xy}^2 (\Delta S_{y_a y_a})(\Delta S_{yy}) + \gamma_{xy}^4 (\Delta S_{yy})^2$$

Taking expected values of both sides shows

$$S_{yy}^2 \sigma^2[\hat{\gamma}_{xy}^2] = \sigma^2[\hat{S}_{y_a y_a}] - 2\gamma_{xy}^2 \sigma^2[\hat{S}_{y_a y_a}] + \gamma_{xy}^4 \sigma^2[\hat{S}_{yy}] \tag{6.35}$$

A formula for the variance $\sigma^2[\hat{S}_{y_a y_a}]$ is given by Eq. (6.28). However, an exact formula for the variance $\sigma^2[\hat{S}_{yy}]$ is *not* known when the output $y(t)$ in Figure 6.1 is non-Gaussian data. If both $q_{xy_b}^2 = 0$ and $q_{xy_c}^2 = 0$, then $y(t)$ would be Gaussian, and the variance of $\hat{S}_{yy}$ from n_d averages would satisfy

$$n_d \sigma^2[\hat{S}_{yy}] = S_{yy}^2 \tag{6.36}$$

For this situation, Eq. (6.28) becomes

$$n_d \sigma^2[\hat{S}_{y_a y_a}] = [2 - \gamma_{xy}^2] \gamma_{xy}^2 S_{yy}^2 \tag{6.37}$$

Substitution of Eqs. (6.36) and (6.37) into Eq. (6.35) yields

$$n_d \sigma^2[\hat{\gamma}_{xy}^2] = 2\gamma_{xy}^2 [1 - 2\gamma_{xy}^2 + \gamma_{xy}^4] \tag{6.38}$$

Hence,

$$\varepsilon[\hat{\gamma}_{xy}^2] = \frac{\sigma[\hat{\gamma}_{xy}^2]}{\gamma_{xy}^2} = \frac{\sqrt{2}[1 - \gamma_{xy}^2]}{|\gamma_{xy}|\sqrt{n_d}} \tag{6.39}$$

in agreement with the known linear result for this special case (reference 1). Note that the normalized random error for the estimate requires knowledge of the true value. Similar results occur later.

For more general cases where $q_{xy_b}^2 \neq 0$ and $a_{xy_c}^2 \neq 0$, $y(t)$ would be non-Gaussian. Equation (6.36) can here be considered as only a first order of approximation. For n_d averages, the relation of Eq. (6.36) will be replaced by

$$n_d \sigma^2[\hat{S}_{yy}] \approx S_{yy}^2 \tag{6.40}$$

Substitution of Eqs. (6.28) and (6.40) into Eq. (6.35) now yields

$$n_d \sigma^2[\hat{\gamma}_{xy}^2] \approx 2\gamma_{xy}^2[(1 - 2\gamma_{xy}^2)(1 - q_{xy_b}^2 - q_{xy_c}^2) + \gamma_{xy}^4] \tag{6.41}$$

Then, in place of Eq. (6.39), one obtains

$$\varepsilon[\hat{\gamma}_{xy}^2] \approx \frac{\sqrt{2}[(1 - 2\gamma_{xy}^2)(1 - q_{xy_b}^2 - q_{xy_c}^2) + \gamma_{xy}^4]^{1/2}}{|\gamma_{xy}|\sqrt{n_d}} \tag{6.42}$$

This approximates the normalized random error for the linear coherence function estimate in Figure 6.1.

For the model of Figure 6.2 where $q_{xy_c}^2 = 0$,

$$\varepsilon[\hat{\gamma}_{xy}^2] \approx \frac{\sqrt{2}[(1 - 2\gamma_{xy}^2)(1 - q_{xy_b}^2) + \gamma_{xy}^4]^{1/2}}{|\gamma_{xy}|\sqrt{n_d}} \tag{6.43}$$

For the model of Figure 6.3 where $q_{xy_b}^2 = 0$,

$$\varepsilon[\hat{\gamma}_{xy}^2] \approx \frac{\sqrt{2}[(1 - 2\gamma_{xy}^2)(1 - q_{xy_c}^2) + \gamma_{xy}^4]^{1/2}}{|\gamma_{xy}|\sqrt{n_d}} \tag{6.44}$$

Finally, if both $q_{xy_b}^2 = 0$ and $q_{xy_c}^2 = 0$, then Eq. (6.42) becomes

$$\varepsilon[\hat{\gamma}_{xy}^2] \approx \frac{\sqrt{2}[1 - \gamma_{xy}^2]}{|\gamma_{xy}|\sqrt{n_d}} \tag{6.45}$$

which is the same as Eq. (6.39).

6.3 NONLINEAR SYSTEM ESTIMATES

6.3.1 Nonlinear Output Spectrum Estimate for Square-Law System

Consider now the square-law system output spectrum estimate $\hat{S}_{y_b y_b}(f)$ in Figure 6.1 where $y_b(t)$ is non-Gaussian. Let $y_p(t) = y_b(t) + n(t)$. Then $y_p(t)$ is also non-Gaussian, even when the extraneous noise $n(t)$ is Gaussian, and the estimated spectrum value is

$$\hat{S}_{y_p y_p}(f) = \hat{S}_{y_b y_b}(f) + \hat{S}_{nn}(f) \tag{6.46}$$

From Eqs. (6.10) and (6.12), the true value is

$$S_{y_py_p}(f) = [1 - \gamma_{xy}^2(f) - q_{xy_c}^2(f)]S_{yy}(f) \tag{6.47}$$

Assuming $\hat{S}_{y_by_b}(f)$ and $\hat{S}_{nn}(f)$ to be statistically independent, their variances satisfy

$$\sigma^2[\hat{S}_{y_py_p}] = \sigma^2[\hat{S}_{y_by_b}] + \sigma^2[\hat{S}_{nn}] \tag{6.48}$$

where the dependence upon f has been omitted to simplify the notation.

The variance of $\hat{S}_{nn}$ from n_d averages is given by Eq. (6.26) when $n(t)$ is Gaussian. However, a similar result for the variance of $\hat{S}_{y_py_p}$ requires $y_p(t)$ to be approximately Gaussian. Hence, as a first order of approximation only,

$$\sigma^2[\hat{S}_{nn}] \approx (S_{nn}^2/n_d), \qquad \sigma^2[\hat{S}_{y_py_p}] \approx (S_{y_py_p}^2/n_d) \tag{6.49}$$

Thus, Eq. (6.48) gives

$$\sigma^2[\hat{S}_{y_by_b}] \approx \frac{S_{y_py_p}^2 - S_{nn}^2}{n_d} \tag{6.50}$$

Substitution from Eqs. (6.12) and (6.47) shows

$$\sigma^2[\hat{S}_{y_by_b}] \approx \frac{[2(1 - \gamma_{xy}^2 - q_{xy_c}^2) - q_{xy_b}^2]q_{xy_b}^2 S_{yy}^2}{n_d} \tag{6.51}$$

Using Eq. (6.10) now proves

$$\varepsilon[\hat{S}_{y_by_b}] = \frac{\sigma[\hat{S}_{y_by_b}]}{S_{y_by_b}} \approx \frac{[2(1 - \gamma_{xy}^2 - q_{xy_c}^2) - q_{xy_b}^2]^{1/2}}{|q_{xy_b}|\sqrt{n_d}} \tag{6.52}$$

This approximates the normalized random error for the square-law system output spectrum estimate in Figure 6.1.

For the model of Figure 6.2 where $q_{xy_c}^2 = 0$,

$$\varepsilon[\hat{S}_{y_by_b}] \approx \frac{[2(1 - \gamma_{xy}^2) - q_{xy_b}^2]^{1/2}}{|q_{xy_b}|\sqrt{n_d}} \tag{6.53}$$

Finally, note that if also $\gamma_{xy}^2 = 0$, then

$$\varepsilon[\hat{S}_{y_by_b}] \approx \frac{[2 - q_{xy_b}^2]^{1/2}}{|q_{xy_b}|\sqrt{n_d}} \tag{6.54}$$

which has the same form as Eq. (6.32).

6.3.2 Nonlinear Output Spectrum Estimate for Cubic System

Consider next the cubic system output spectrum estimate $\hat{S}_{y_c y_c}(f)$ in Figure 6.1. The same steps carried out from Eqs. (6.46)–(6.52) with $\hat{S}_{y_c y_c}$ in place of $\hat{S}_{y_b y_b}$ indicate that the normalized random error

$$\varepsilon[\hat{S}_{y_c y_c}] \approx \frac{[2(1 - \gamma_{xy}^2 - q_{xy_b}^2) - q_{xy_c}^2]^{1/2}}{|q_{xy_c}|\sqrt{n_d}} \tag{6.55}$$

For the model of Figure 6.3 where $q_{xy_b}^2 = 0$, one obtains

$$\varepsilon[\hat{S}_{y_c y_c}] \approx \frac{[2(1 - \gamma_{xy}^2) - q_{xy_c}^2]^{1/2}}{|q_{xy_c}|\sqrt{n_d}} \tag{6.56}$$

If also $\gamma_{xy}^2 = 0$, then

$$\varepsilon[\hat{S}_{y_c y_c}] \approx \frac{[2 - q_{xy_c}^2]^{1/2}}{|q_{xy_c}|\sqrt{n_d}} \tag{6.57}$$

Note the similarity of this result for $\hat{S}_{y_c y_c}(f)$ to the previous results for $\hat{S}_{y_b y_b}(f)$ in Eq. (6.54) and $\hat{S}_{y_a y_a}(f)$ in Eq. (6.32).

6.3.3 Nonlinear Coherence Function Estimate for Square-Law System

The nonlinear coherence function estimate for the square-law system in Figure 6.1 is given by

$$\hat{q}_{xy_b}^2(f) = \frac{\hat{S}_{y_b y_b}(f)}{\hat{S}_{yy}(f)} \tag{6.58}$$

The same steps carried out previously to derive Eq. (6.45) can now be duplicated with $S_{y_a y_a}$ replaced by $S_{y_b y_b}$ and γ_{xy}^2 replaced by $q_{xy_b}^2$. This proves that the normalized random error is

$$\varepsilon[\hat{q}_{xy_b}^2] \approx \frac{\sqrt{2}[(1 - 2q_{xy_b}^2)(1 - \gamma_{xy}^2 - q_{xy_c}^2) + q_{xy_b}^4]^{1/2}}{|q_{xy_b}|\sqrt{n_d}} \tag{6.59}$$

For the model of Figure 6.2 where $q_{xy_c}^2 = 0$,

$$\varepsilon[\hat{q}_{xy_b}^2] \approx \frac{\sqrt{2}[(1 - 2q_{xy_b}^2)(1 - \gamma_{xy}^2) + q_{xy_b}^4]^{1/2}}{|q_{xy_b}|\sqrt{n_d}} \tag{6.60}$$

If also $\gamma_{xy}^2 = 0$, then

$$\varepsilon[\hat{q}_{xy_b}^2] \approx \frac{\sqrt{2}[1 - q_{xy_b}^2]}{|q_{xy_b}|\sqrt{n_d}} \tag{6.61}$$

which has the same form as Eq. (6.39).

6.3.4 Nonlinear Coherence Function Estimate for Cubic System

The nonlinear coherence function estimate for the cubic system in Figure 6.1 is given by

$$\hat{q}_{xy_c}^2(f) = \frac{\hat{S}_{y_cy_c}(f)}{\hat{S}_{yy}(f)} \tag{6.62}$$

Error analysis results are now the same as in Eqs. (6.59)–(6.61) by merely interchanging $q_{xy_b}^2$ and $q_{xy_c}^2$. Hence, for the model of Figure 6.1,

$$\varepsilon[\hat{q}_{xy_c}^2] \approx \frac{\sqrt{2}[(1 - 2q_{xy_c}^2)(1 - \gamma_{xy}^2 - q_{xy_b}^2) + q_{xy_c}^4]^{1/2}}{|q_{xy_c}|\sqrt{n_d}} \tag{6.63}$$

For the model of Figure 6.3 where $q_{xy_b}^2 = 0$, one obtains

$$\varepsilon[\hat{q}_{xy_c}^2] \approx \frac{\sqrt{2}[(1 - 2q_{xy_c}^2)(1 - \gamma_{xy}^2) + q_{xy_c}^4]^{1/2}}{|q_{xy_c}|\sqrt{n_d}} \tag{6.64}$$

Finally, if also $\gamma_{xy}^2 = 0$, then

$$\varepsilon[\hat{q}_{xy_c}^2] \approx \frac{\sqrt{2}[1 - q_{xy_c}^2]}{|q_{xy_c}|\sqrt{n_d}} \tag{6.65}$$

Note the similarity of this result for $\hat{q}_{xy_c}^2$ to the previous results for $\hat{q}_{xy_b}^2$ in Eq. (6.61) and $\hat{\gamma}_{xy}^2$ in Eq. (6.39).

6.3.5 Summary of Results

The main normalized random error formulas for the output spectrum estimates and coherence function estimates associated with the third-order nonlinear model of Figure 6.1 will now be compared side by side:

$$\varepsilon[\hat{S}_{y_ay_a}] \approx \frac{[2(1 - q_{xy_b}^2 - q_{xy_c}^2) - \gamma_{xy}^2]^{1/2}}{|\gamma_{xy}|\sqrt{n_d}} \tag{6.29}$$

$$\varepsilon[\hat{S}_{y_b y_b}] \approx \frac{[2(1 - \gamma_{xy}^2 - q_{xy_c}^2) - q_{xy_b}^2]^{1/2}}{|q_{xy_b}|\sqrt{n_d}} \tag{6.52}$$

$$\varepsilon[\hat{S}_{y_c y_c}] \approx \frac{[2(1 - \gamma_{xy}^2 - q_{xy_b}^2) - q_{xy_c}^2]^{1/2}}{|q_{xy_c}|\sqrt{n_d}} \tag{6.55}$$

$$\varepsilon[\hat{\gamma}_{xy}^2] \approx \frac{\sqrt{2}[(1 - 2\gamma_{xy}^2)(1 - q_{xy_b}^2 - q_{xy_c}^2) + \gamma_{xy}^4]^{1/2}}{|\gamma_{xy}|\sqrt{n_d}} \tag{6.42}$$

$$\varepsilon[\hat{q}_{xy_b}^2] \approx \frac{\sqrt{2}[(1 - 2q_{xy_b}^2)(1 - \gamma_{xy}^2 - q_{xy_c}^2) + q_{xy_b}^4]^{1/2}}{|q_{xy_b}|\sqrt{n_d}} \tag{6.59}$$

$$\varepsilon[\hat{q}_{xy_c}^2] \approx \frac{\sqrt{2}[(1 - 2q_{xy_c}^2)(1 - \gamma_{xy}^2 - q_{xy_b}^2) + q_{xy_c}^4]^{1/2}}{|q_{xy_c}|\sqrt{n_d}} \tag{6.63}$$

Various special cases of these formulas are shown in Sections 6.2.1–6.3.4. To apply these formulas to actual data, the unknown true linear and nonlinear coherence functions should be replaced by their measured values.

6.4 SPECIAL BISPECTRUM AND TRISPECTRUM ESTIMATES

Normalized random error formulas will now be derived for the special single-variable bispectrum and trispectrum estimates calculated using the $x(t)$ and $y(t)$ in the third-order nonlinear model of Figure 6.1. All results should be considered first-order approximations, since $y(t)$ is non-Gaussian.

6.4.1 Special Bispectral Density Function Estimate

The "raw" estimate of the special bispectral density function is obtained from Eq. (5.32) without taking expected values by

$$\tilde{S}_{xxy}(f) = \tilde{S}_{xxy}(f, f) = \frac{1}{T}[X^*(f)X^*(f)Y(2f)] \tag{6.66}$$

where $Y(2f)$ is given by Eq. (6.4) with f replaced by $2f$. For convenience, as per Figure 6.2, it will be assumed here that $Y_c(2f)$ is included with the uncorrelated noise term $N(2f)$ since it appears as noise to bispectral estimates. From Eq. (5.13) when $A_2(2f)$ is unity, the expected value

$$E[\tilde{S}_{xxy}(f)] = E[\tilde{S}_{xxy_b}(f)] = S_{xxy}(f) = 2S_{xx}^2(f) \tag{6.67}$$

Thus, $\tilde{S}_{xxy}(f)$ can be considered to be an unbiased estimate of $S_{xxy}(f)$ with $|\tilde{S}_{xxy}(f)|$ as an unbiased estimate of $|S_{xxy}(f)|$.

To obtain the variance of the "raw" estimate $|\tilde{S}_{xxy}(f)|$, one should compute $|\tilde{S}_{xxy}(f)|^2$ and take its expected value. This will require calculation of a six- and eight-fold Gaussian moment as indicated below, which can be evaluated by considering the products of all possible pairs of second-order moments. For $f \neq 0$, the quantity

$$E[|\tilde{S}_{xxy}(f)|^2] = E[|\tilde{S}_{xxy_b}(f)|^2] + E[|\tilde{S}_{xxn}(f)|^2]$$

$$= \frac{1}{T^2} \iint E[P(u, v, f)]\, du\, dv + E[|\tilde{S}_{xxn}(f)|^2] \tag{6.68}$$

plus other terms equal to zero. Here, the terms

$$P(u, v, f) = X^*(f)X^*(f)X(u)X(2f-u)X(f)X(f)X^*(v)X^*(2f-v) \tag{6.69}$$

and

$$\tilde{S}_{xxn}(f) = \frac{1}{T}[X^*(f)X^*(f)N(2f)] \tag{6.70}$$

For $f \neq 0$, evaluation of the eight-fold Gaussian moment in the first part of Eq. (6.68) followed by the double integration shows after a number of steps that

$$\frac{1}{T^2} \iint E[P(u, v, f)]\, du\, dv = 2TS_{xx}^2(f)S_{y_by_b}(2f) + 4S_{xx}^4(f) \tag{6.71}$$

For $f \neq 0$, evaluation of the six-fold Gaussian moment in the second part of Eq. (6.68) shows after several steps that

$$E[|\tilde{S}_{xxn}(f)|^2] = 2TS_{xx}^2(f)S_{nn}(2f) \tag{6.72}$$

Thus,

$$E[|\tilde{S}_{xxy}(f)|^2] = 2TS_{xx}^2(f)[S_{y_by_b}(2f) + S_{nn}(2f)] + 4S_{xx}^4(f) \tag{6.73}$$

The variance for the "raw" estimate is now given from Eqs. (6.67) and (6.73) by

$$\sigma^2[|\tilde{S}_{xxy}(f)|] = E[|\tilde{S}_{xxy}(f)|^2] - (E[|\tilde{S}_{xxy}(f)|])^2$$

$$= 2TS_{xx}^2(f)[S_{y_by_b}(2f) + S_{nn}(2f)] \tag{6.74}$$

"Smooth" estimates are obtained by averaging over n_d statistically independent "raw" estimates where $n_d > 1$. The associated variance for "smooth" estimates becomes

$$\sigma^2[|\hat{S}_{xxy}(f)|] = \frac{\sigma^2[|\tilde{S}_{xxy}(f)|]}{n_d} \tag{6.75}$$

This yields the normalized random error

$$\varepsilon[|\hat{S}_{xxy}(f)|] = \frac{\sigma[|\hat{S}_{xxy}(f)|]}{|S_{xxy}(f)|} = \frac{\sqrt{2T}S_{xx}(f)[S_{y_by_b}(2f) + S_{nn}(2f)]^{1/2}}{|S_{xxy}(f)|\sqrt{n_d}} \tag{6.76}$$

For $Y(2f) = Y_a(2f) + Y_b(2f) + N(2f)$, as per Figure 6.2, one obtains

$$S_{y_by_b}(2f) + S_{nn}(2f) \approx [1 - \gamma^2_{xy}(2f)]S_{yy}(2f) \tag{6.77}$$

From Eq. (6.67), the quantity $|S_{xxy}(f)|$ can be replaced by $2S^2_{xx}(f)$. Hence, as a first order of approximation for Figure 6.2, for $f \neq 0$,

$$\varepsilon[|\hat{S}_{xxy}(f)|] \approx \frac{\sqrt{T}\{[1 - \gamma^2_{xy}(2f)]S_{yy}(2f)\}^{1/2}}{S_{xx}(f)\sqrt{2n_d}} \tag{6.78}$$

Note that results at frequency $2f$ determine errors at frequency f. All of the terms on the right-hand are known from $x(t)$ and $y(t)$. In Eq. (6.78), for ε to be dimensionless, if $S_{xx}(f)$ has units of (volts)^2 per hertz, then $S_{yy}(2f)$ has units of (volts)^4 per hertz and T has units of seconds. Also, $|S_{xxy}(f)|$ has units of (volts)^4 (seconds)^2 or [(volts)^2 per hertz]^2. These units are consistent with the fact that this is a second-order system.

6.4.2 Special Trispectral Density Function Estimate

A similar procedure can be carried out for the special trispectral density function estimate. From Eq. (5.33), the "raw" estimate is defined without taking expected values by

$$\tilde{S}_{xxxy}(f) = \tilde{S}_{xxxy}(f, f, f) = \frac{1}{T}[X^*(f)X^*(f)X^*(f)Y(3f)] \tag{6.79}$$

where $Y(3f)$ is given by Eq. (6.4) with f replaced by $3f$. For convenience, as in Figure 6.3, it will be assumed here that $Y_b(3f)$ is included with the uncorrelated noise term $N(3f)$ since it appears as noise to trispectral estimates. From Eq. (5.14) when $A_3(3f)$ is unity, the expected value

$$E[\tilde{S}_{xxxy}(f)] = E[\tilde{S}_{xxxy_c}(f)] = S_{xxxy}(f) = 6S^3_{xx}(f) \tag{6.80}$$

Thus, $\tilde{S}_{xxxy}(f)$ can be considered to be an unbiased estimate of $S_{xxxy}(f)$ with $|\tilde{S}_{xxxy}(f)|$ as an unbiased estimate of $|S_{xxxy}(f)|$. One also obtains

$$E[\tilde{S}_{xxxy}(f)|^2] = E[|\tilde{S}_{xxxy_c}(f)|^2] + E[|\tilde{S}_{xxxn}(f)|^2] \tag{6.81}$$

plus other terms equal to zero. In the first part of Eq. (6.81), the quantity

$$\tilde{S}_{xxxy_c}(f) = \frac{1}{T}\,[X^*(f)X^*(f)X^*(f)Y_c(3f)] \tag{6.82}$$

where

$$Y_c(3f) = \iint X(u)X(v-u)X(3f-v)\,du\,dv - 3\sigma_x^2 X(3f) \tag{6.83}$$

Thus,

$$\tilde{S}_{xxxy_c}(f) = \frac{1}{T}\iint Q(u, v, f)\,du\,dv - \frac{3\sigma_x^2}{T}\,R(f) \tag{6.84}$$

where the terms

$$Q(u, v, f) = X^*(f)X^*(f)X^*(f)X(u)X(v-u)X(3f-v) \tag{6.85}$$

and

$$R(f) = X^*(f)X^*(f)X^*(f)X(3f) \tag{6.86}$$

In the second part of Eq. (6.81), the quantity

$$\tilde{S}_{xxxn}(f) = \frac{1}{T}\,[X^*(f)X^*(f)X^*(f)N(3f)] \tag{6.87}$$

For $f \neq 0$, evaluation of the eight-fold Gaussian moment in the second part of Eq. (6.81) shows after a number of steps that

$$E[|\tilde{S}_{xxxn}(f)|^2] = 6T^2 S_{xx}^3(f)S_{nn}(3f) \tag{6.88}$$

It is much more difficult to evaluate the first part of Eq. (6.81). One must compute the expected value of

$$|\tilde{S}_{xxxy_c}(f)|^2 = \frac{1}{T^2}\iiiint Q(u, v, f)Q^*(\alpha, \beta, f)\,d\alpha\,d\beta\,du\,dv$$
$$- \frac{3\sigma_x^2}{T^2}\iint [Q^*(u, v, f)R(f) + Q(u, v, f)R^*(f)]du\,dv + \frac{9\sigma_x^4}{T^2}\,|R(f)|^2 \tag{6.89}$$

This involves three terms, of which the hardest to evaluate is the first quadruple integral term where the integrand is a twelve-fold product of Gaussian terms. This can be reduced to certain products of meaningful different second-order moments, plus other terms that are zero or do not exist for $f \neq 0$. Finally, after many steps, one can derive the result

$$E[|\tilde{S}_{xxxy_c}(f)|^2] = 6T^2 S_{xx}^3(f) S_{y_c y_c}(3f) + 36 S_{xx}^6(f) \tag{6.90}$$

The variance for the "raw" estimate is now given by

$$\sigma^2[|\tilde{S}_{xxxy}(f)|] = 6T^2 S_{xx}^3(f)[S_{y_c y_c}(3f) + S_{nn}(3f)] \tag{6.91}$$

This result should be compared with Eq. (6.74).

"Smooth" estimates are obtained, as before, by averaging over n_d statistically independent "raw" estimates where $n_d > 1$. Hence,

$$\sigma^2[|\hat{S}_{xxxy}(f)|] = \frac{\sigma^2[|\tilde{S}_{xxxy}(f)|]}{n_d} \tag{6.92}$$

Thus, the normalized random error

$$\varepsilon[|\hat{S}_{xxxy}(f)|] = \frac{\sqrt{6}T[S_{xx}(f)]^{3/2}[S_{y_c y_c}(3f) + S_{nn}(3f)]^{1/2}}{|S_{xxxy}(f)|\sqrt{n_d}} \tag{6.93}$$

For $Y(3f) = Y_a(3f) + Y_c(3f) + N(3f)$, as in Figure 6.3, one obtains

$$S_{y_c y_c}(3f) + S_{nn}(3f) = [1 - \gamma_{xy}^2(3f)] S_{yy}(3f) \tag{6.94}$$

From Eq. (6.80), the quantity $|S_{xxxy}(f)|$ can be replaced by $6S_{xx}^3(f)$. Hence, as a first order of approximation for Figure 6.3,

$$\varepsilon[|\hat{S}_{xxxy}(f)|] \approx \frac{T\{[1 - \gamma_{xy}^2(3f)] S_{yy}(3f)\}^{1/2}}{[S_{xx}(f)]^{3/2}\sqrt{6n_d}} \tag{6.95}$$

Note that results at frequency $3f$ determine errors at frequency f. All of the terms on the right-hand side are known from $x(t)$ and $y(t)$. In Eq. (6.95), for ε to be dimensionless, if $S_{xx}(f)$ has units of (volts)^2 per hertz, then $S_{yy}(3f)$ has units of (volts)^6 per hertz and T has units of seconds. Also, $|S_{xxxy}(f)|$ has units of (volts)^6 (seconds)^3 or [(volts)^2 per hertz]^3. These units are consistent with the fact that this is a third-order system.

6.4.3 Summary of Results

Special Bispectrum Estimates

$$\tilde{S}_{xxy}(f) = \frac{1}{T}\,[X^*(f)X^*(f)Y(2f)] \tag{6.96}$$

$$\hat{S}_{xxy}(f) = \frac{1}{n_d T}\sum_{i=1}^{n_d} X_i^*(f)X_i^*(f)Y_i(2f) \tag{6.97}$$

$$S_{xxy}(f) = \frac{1}{T}\,E[X^*(f)X^*(f)Y(2f)] \tag{6.98}$$

$$E[\tilde{S}_{xxy}(f)] = E[\hat{S}_{xxy}(f)] = S_{xxy}(f) \tag{6.99}$$

$$\varepsilon[|\hat{S}_{xxy}(f)|] \approx \frac{\sqrt{T}\{[1-\gamma_{xy}^2(2f)]S_{yy}(2f)\}^{1/2}}{S_{xx}(f)\sqrt{2n_d}} \tag{6.100}$$

Equation (6.100) applies to Figure 6.1 and the nonlinear model of Figure 6.2 consisting of a linear system in parallel with a square-law system.

Special Trispectrum Estimates

$$\tilde{S}_{xxxy}(f) = \frac{1}{T}\,[X^*(f)X^*(f)X^*(f)Y(3f)] \tag{6.101}$$

$$\hat{S}_{xxxy}(f) = \frac{1}{n_d T}\sum_{i=1}^{n_d} X_i^*(f)X_i^*(f)X_i^*(f)Y_i(3f) \tag{6.102}$$

$$S_{xxxy}(f) = \frac{1}{T}\,E[X^*(f)X^*(f)X^*(f)Y(3f)] \tag{6.103}$$

$$E[\tilde{S}_{xxxy}(f)] = E[\hat{S}_{xxxy}(f)] = S_{xxxy}(f) \tag{6.104}$$

$$\varepsilon[|\hat{S}_{xxxy}(f)|] \approx \frac{T\{[1-\gamma_{xy}^2(3f)]S_{yy}(3f)\}^{1/2}}{[S_{xx}(f)]^{3/2}\sqrt{6n_d}} \tag{6.105}$$

Equation (6.105) applies to Figure 6.1 and the nonlinear model of Figure 6.3 consisting of a linear system in parallel with a cubic system.

6.5 FREQUENCY RESPONSE FUNCTION ESTIMATES

Normalized random error formulas will now be obtained for the gain factors of frequency response function estimates of $A_2(f)$ and $A_3(f)$ in the Case 1 nonlinear model of Figure 6.4 and of $B_2(f)$ and $B_3(f)$ in the Case 2 nonlinear model of Figure 6.5. These models are the same as Figures 5.1 and 5.5 in Chapter

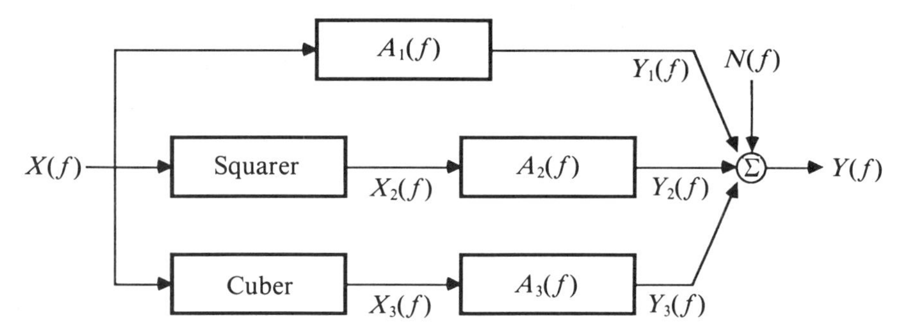

Figure 6.4 Case 1 nonlinear model.

5. The phase factors of frequency response function estimates will have small errors when associated gain factors have small errors. Phase factor error formulas are listed later in Section 7.3.

6.5.1 Case 1 Nonlinear Model

The Case 1 nonlinear model of Figure 6.4 contains the frequency response functions $A_2(f)$ and $A_3(f)$ that follow the squaring and cubing operations. For this model, from Eqs. (5.13) and (5.14), the special bispectrum $S_{xxy}(f)$ and the special trispectrum $S_{xxxy}(f)$ satisfy

$$S_{xxy}(f) = 2A_2(2f)S_{xx}^2(f) \tag{6.106}$$

$$S_{xxxy}(f) = 6A_3(3f)S_{xx}^3(f) \tag{6.107}$$

Assume that the input autospectral density function $S_{xx}(f)$ behaves like bandwidth-limited white noise. From Eqs. (6.106) and (6.107), normalized random errors will now be the same for $|\hat{S}_{xxy}(f)|$ and $|\hat{A}_2(2f)|$ and for $|\hat{S}_{xxxy}(f)|$

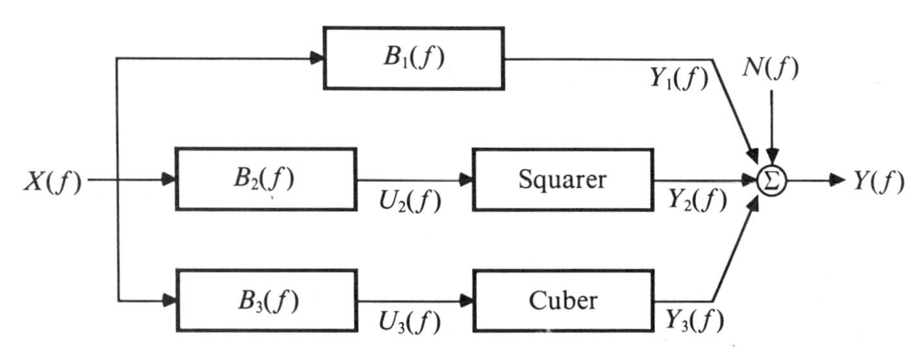

Figure 6.5 Case 2 nonlinear model.

and $|\hat{A}_3(3f)|$. This gives

$$\varepsilon[|\hat{A}_2(2f)|] = \varepsilon[|\hat{S}_{xxy}(f)|] \tag{6.108}$$

$$\varepsilon[|\hat{A}_3(3f)|] = \varepsilon[|\hat{S}_{xxxy}(f)|] \tag{6.109}$$

where the right-hand sides satisfy Eqs. (6.78) and (6.95). Results for $|\hat{A}_2(f)|$ are the same as for $|\hat{S}_{xxy}(f/2)|$, and results for $|\hat{A}_3(f)|$ are the same as for $|\hat{S}_{xxxy}(f/3)|$. Specifically,

$$\varepsilon[|\hat{A}_2(f)|] = \varepsilon[|\hat{S}_{xxy}(f/2)|] \tag{6.110}$$

$$\varepsilon[|\hat{A}_3(f)|] = \varepsilon[|\hat{S}_{xxxy}(f/3)|] \tag{6.111}$$

where $|\hat{A}_2(f)|$ is the gain factor estimate of $A_2(f)$ and $|\hat{A}_3(f)|$ is the gain factor estimate of $A_3(f)$.

6.5.2 Case 2 Nonlinear Model

The Case 2 nonlinear model of Figure 6.5 contains the frequency response functions $B_2(f)$ and $B_3(f)$ that precede the squaring and cubing operations. For this model, from Eqs. (5.86) and (5.87),

$$S_{xxy}(f) = 2B_2^2(f)S_{xx}^2(f) \tag{6.112}$$

$$S_{xxxy}(f) = 6B_3^3(f)S_{xx}^3(f) \tag{6.113}$$

Assume that the input autospectral density function $S_{xx}(f)$ behaves like bandwidth-limited white noise. From Eqs. (6.110) and (6.111), normalized random errors will now be the same for $|\hat{S}_{xxy}(f)|$ and $|\hat{B}_2^2(f)|$ and for $|\hat{S}_{xxxy}(f)|$ and $|\hat{B}_3^3(f)|$. This gives

$$\begin{aligned} \varepsilon[|\hat{B}_2^2(f)|] &= \varepsilon[|\hat{S}_{xxy}(f)|] \\ \varepsilon[|\hat{B}_3^3(f)|] &= \varepsilon[|\hat{S}_{xxxy}(f)|] \end{aligned} \tag{6.114}$$

where the right-hand sides satisfy Eqs. (6.78) and (6.95). For small ε, one obtains

$$\begin{aligned} \varepsilon[|\hat{B}_2(f)|] &\approx \tfrac{1}{2}\varepsilon[|\hat{S}_{xxy}(f)|] \\ \varepsilon[|\hat{B}_3(f)|] &\approx \tfrac{1}{3}\varepsilon[|\hat{S}_{xxxy}(f)|] \end{aligned} \tag{6.115}$$

where $|\hat{B}_2(f)|$ is the gain factor estimate of $B_2(f)$ and $|\hat{B}_3(f)|$ is the gain factor estimate of $B_3(f)$.

These results show that good estimates of the frequency response functions in Figures 6.4 and 6.5 will occur when there are good estimates of the special bispectrum $S_{xxy}(f)$ and the special trispectrum $S_{xxxy}(f)$.

6.6 NUMERICAL EXAMPLES

6.6.1 Linear and Nonlinear Output Spectrum Estimates

To apply the following formulas to actual data, the unknown true linear and nonlinear coherence functions should be replaced by their estimates. Assume that $n_d = 200$ and at frequency f_0 the terms

$$\hat{\gamma}_{xy}^2(f_0) \approx 0.50$$
$$\hat{q}_{xy_b}^2(f_0) \approx 0.20$$
$$\hat{q}_{xy_c}^2(f_0) \approx 0.10$$

Then, for the linear and nonlinear output spectra in Figure 6.1, from Eqs. (6.29), (6.52), and (6.55), respectively,

$$\varepsilon[\hat{S}_{y_a y_a}(f_0)] \approx 0.095$$
$$\varepsilon[\hat{S}_{y_b y_b}(f_0)] \approx 0.122$$
$$\varepsilon[\hat{S}_{y_c y_c}(f_0)] \approx 0.158$$

6.6.2 Special Bispectrum and Trispectrum Estimates

Assume in Figure 6.1 that the input autospectrum

$$S_{xx}(f) = \begin{cases} K & \text{for } |f| \leqslant F \\ 0 & \text{for } |f| > F \end{cases} \tag{6.116}$$

Then, from Eq. (6.67), the special bispectrum

$$S_{xxy}(f) = 2K^2 \qquad \text{only for } |f| \leqslant F \tag{6.117}$$

and from Eq. (6.80), the special trispectrum

$$S_{xxxy}(f) = 6K^3 \qquad \text{only for } |f| \leqslant F \tag{6.118}$$

Now, from Eq. (6.14), the autospectrum of the nonlinear squarer output $y_b(t)$ is

$$S_{y_b y_b}(f) = 2 \int S_{xx}(u) S_{xx}(f - u)\, du \tag{6.119}$$

where

$$S_{xx}(u) = K \qquad \text{only for } -F \leqslant u \leqslant F$$
$$S_{xx}(f - u) = K \qquad \text{only for } f - F \leqslant u \leqslant f + F$$

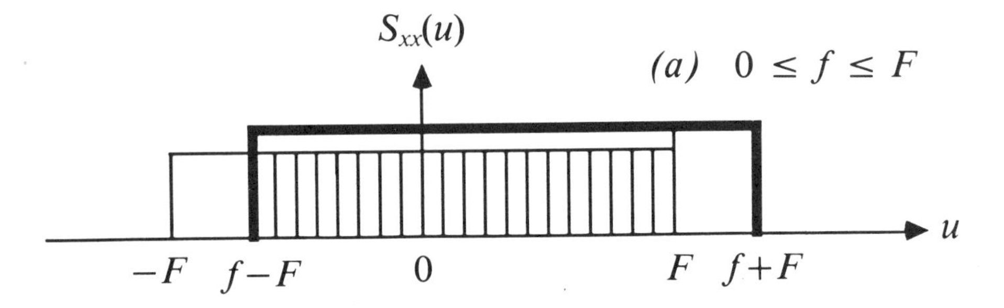

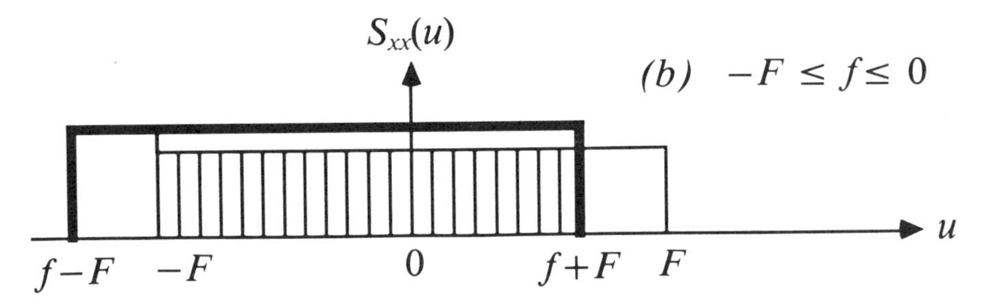

Figure 6.6 Plots of input spectra $S_{xx}(u)$ and $S_{xx}(f-u)$.

Thus, for $0 \leqslant f \leqslant F$, $S_{xx}(u)$ and $S_{xx}(f-u)$ have overlapping parts as sketched in Figure 6.6*a*. However, for $-F \leqslant f \leqslant 0$, $S_{xx}(u)$ and $S_{xx}(f-u)$ overlap differently as sketched in Figure 6.6*b*.

It follows that nonzero values of $S_{y_by_b}(f)$ occur only for

$$\begin{aligned} S_{y_by_b}(f) &= 2K^2 \int_{f-F}^{F} du = 2K^2(2F - f) \qquad \text{when } 0 \leqslant f \leqslant 2F \\ &= 2K^2 \int_{-F}^{f+F} du = 2K^2(2F + f) \qquad \text{when } -2F \leqslant f \leqslant 0 \end{aligned} \tag{6.120}$$

Equation (6.120) is plotted in Fig. 6.7. Note that $S_{y_by_b}(f)$ is an even function of f as required for autospectral density functions.

From Eqs. (6.14) and (6.15), the autospectrum of the nonlinear cuber output $y_c(t)$ is

$$\begin{aligned} S_{y_cy_c}(f) &= 6 \iint S_{xx}(u)S_{xx}(v-u)S_{xx}(f-v)\,du\,dv \\ &= 3 \int S_{y_by_b}(v)S_{xx}(f-v)\,dv \end{aligned} \tag{6.121}$$

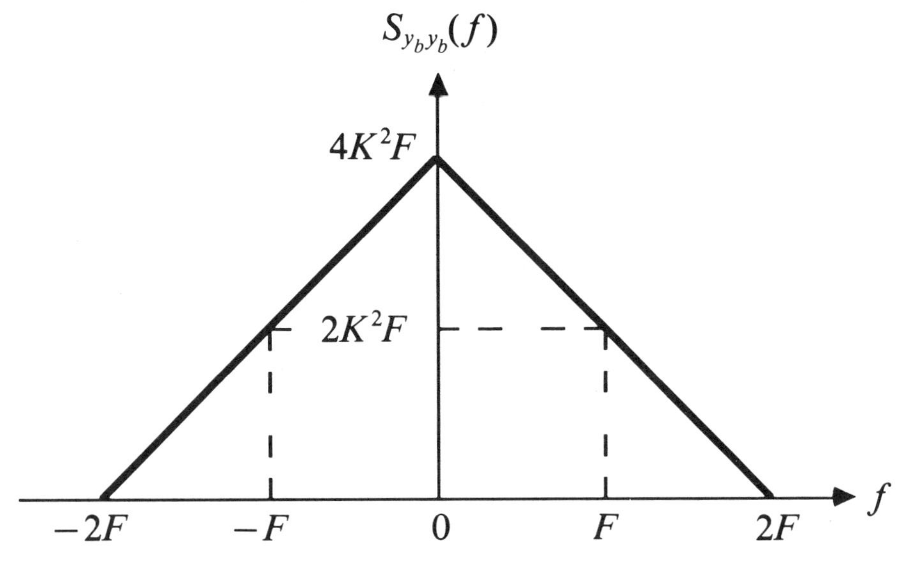

Figure 6.7 Plot of squarer output spectrum $S_{y_by_b}(f)$.

where the last equation occurs because $A_2(f) = A_3(f) = 1$. The autospectrum of the nonlinear squarer output $y_b(t)$ is

$$\begin{aligned} S_{y_by_b}(v) &= 2K^2(2F + v) \qquad \text{when } -2F \leqslant v \leqslant 0 \\ &= 2K^2(2F - v) \qquad \text{when } \quad 0 \leqslant v \leqslant 2F \end{aligned} \tag{6.122}$$

Also, the input autospectrum

$$S_{xx}(f - v) = K \qquad \text{only for } f - F \leqslant v \leqslant f + F \tag{6.123}$$

Hence, nonzero values of $S_{y_cy_c}(f)$ occur for $f \geqslant 0$ only when

$$S_{y_cy_c}(f) = 3K \int_{f-F}^{2F} S_{y_by_b}(v)\, dv, \qquad 0 \leqslant f \leqslant 3F \tag{6.124}$$

To evaluate Eq. (6.124), one must consider two separate ranges for values of f from $0 \leqslant f \leqslant F$ and $F \leqslant f \leqslant 3F$. Specifically, for $0 \leqslant f \leqslant F$, Eq. (6.124) becomes

$$\begin{aligned} S_{y_cy_c}(f) &= 6K^3 \int_{f-F}^{0} (2F + v)\, dv + 6K^3 \int_{0}^{2F} (2F - v)\, dv \\ &= 3K^3(7F^2 - 2Ff - f^2), \qquad 0 \leqslant f \leqslant F \end{aligned} \tag{6.125}$$

However, for $F \leqslant f \leqslant 3F$, one obtains

$$\begin{aligned} S_{y_c y_c}(f) &= 6K^3 \int_{f-F}^{2F} (2F - v)\, dv \\ &= 3K^3(3F - f)^2, \qquad F \leqslant f \leqslant 3F \end{aligned} \tag{6.126}$$

As a check, note that Eqs. (6.125) and (6.126) at $f = F$ give the same value

$$S_{y_c y_c}(F) = 12K^3F^2 \tag{6.127}$$

For negative values of f, since $S_{y_c y_c}(-f) = S_{y_c y_c}(f)$ one obtains

$$\begin{aligned} S_{y_c y_c}(f) &= 3K^3(3F + f)^2, & -3F \leqslant f \leqslant -F \\ &= 3K^3(7F^2 + 2Ff - f^2), & -F \leqslant f \leqslant 0 \end{aligned} \tag{6.128}$$

The results from Eqs. (6.125)–(6.128) are plotted in Figure 6.8.

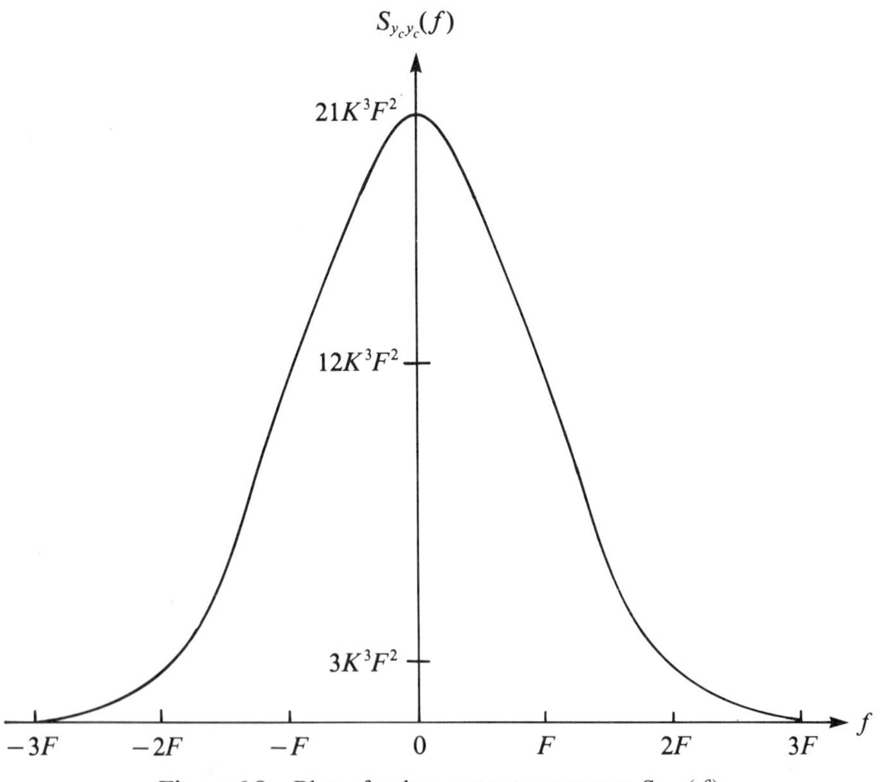

Figure 6.8 Plot of cuber output spectrum $S_{y_c y_c}(f)$.

For a numerical example, consider the frequency $f = (2F/3)$. Here, the autospectrum of Eq. (6.120) is

$$S_{y_b y_b}(2F/3) = (8/3)K^2 F \tag{6.129}$$

Assume that $S_{nn}(2F/3) \ll S_{y_b y_b}(2F/3)$ so that $S_{y_b y_b}(2F/3)$ represents the right-hand side of Eq. (6.77). Now, by substituting Eqs. (6.116) and (6.129) into Eq. (6.78), one obtains

$$\varepsilon[|\hat{S}_{xxy}(F/3)|] \approx \left(\frac{2}{\sqrt{3n_{\mathrm{d}}}}\right)\sqrt{FT} \tag{6.130}$$

This gives the normalized random error for the special bispectral density function estimate at $f = (F/3)$. To be definite, suppose $F = 2\,Hz$, $T = 8$ sec, and $n_{\mathrm{d}} = 1000$. Then $\varepsilon \approx 0.15$ at $f = 0.667$ Hz here.

Consider next the frequency $f = F$. Here, the autospectrum of Eq. (6.126) is

$$S_{y_c y_c}(F) = 12K^3 F^2 \tag{6.131}$$

Assume that $S_{nn}(F) \ll S_{y_c y_c}(F)$ so that $S_{y_c y_c}(F)$ represents the right-hand side of Eq. (6.94). Now, by substituting Eqs. (6.116) and (6.131) into Eq. (6.95), one obtains

$$\varepsilon[|\hat{S}_{xxxy}(F/3)|] \approx \left(\sqrt{\frac{2}{n_{\mathrm{d}}}}\right) FT \tag{6.132}$$

This gives the normalized random error for the special trispectral density function estimate at $f = (F/3)$. To be definite, suppose $F = 2$ Hz, $T = 8$ sec, and $n_d = 10{,}000$. Then $\varepsilon \approx 0.23$ at $f = 0.667$ Hz here. A value of $n_{\mathrm{d}} = 1000$ as before gives a meaningless result here.

6.6.3 Record Length Requirements for Nonlinear Estimates

For ordinary autospectrum estimates such as $\hat{S}_{xx}(f)$, the normalized random error formula at any frequency is given by the simple equation (reference 1)

$$\varepsilon[\hat{S}_{xx}] \approx \frac{1}{\sqrt{n_{\mathrm{d}}}} \tag{6.133}$$

Replacing n_{d} here by n_1 and solving for n_1 yields

$$n_1 \approx \frac{1}{\varepsilon^2} \tag{6.134}$$

For special bispectrum estimates such as $\hat{S}_{xxy}(f)$, the normalized random error at $f = (F/3)$ is given by Eq. (6.130). Replacing n_d in Eq. (6.130) by n_2 and solving for n_2 yields

$$n_2 \approx \frac{1.33(FT)}{\varepsilon^2} \tag{6.135}$$

For special trispectrum estimates such as $\hat{S}_{xxxy}(f)$, the normalized random error at $f = (F/3)$ is given by Eq. (6.132). Replacing n_d in Eq. (6.132) by n_3 and solving for n_3 yields

$$n_3 \approx \frac{2(FT)^2}{\varepsilon^2} \tag{6.136}$$

Relations will now be found between n_1, n_2, and n_3 for this example where

n_1 = number of averages for autospectrum estimates

n_2 = number of averages for bispectrum estimates

n_3 = number of averages for trispectrum estimates

For the same value of ε, these relations are

$$n_2 \approx 1.33(FT)n_1 \tag{6.137}$$

$$n_3 \approx 1.50(FT)n_2 \tag{6.138}$$

$$n_3 \approx 2.00(FT)^2 n_1 \tag{6.139}$$

Since $(FT) \gg 1$, in general, one sees that

$$n_3 > 10n_2 > 100n_1 \tag{6.140}$$

For example, if $F = 2\,\text{Hz}$ and $T = 8\,\text{sec}$, then $FT = 16$. Here, $n_2 \approx 21n_1$, $n_3 \approx 24n_2$, and $n_3 \approx 512n_1$. Thus, $\varepsilon = 0.10$ requires $n_1 = 100$, $n_2 = 2100$, and $n_3 = 51{,}200$. These requirements apply only to this example, but are indicative of the increased data needed when computing special bispectrum and special trispectrum estimates compared to computing ordinary autospectrum estimates.

This concludes Chapter 6.

7

PARALLEL LINEAR AND NONLINEAR SYSTEMS

This chapter presents a recommended methodology for analyzing the spectral properties of stationary random data passing through a single-input/single-output nonlinear model with parallel linear and nonlinear systems of arbitrary type. From measurements of the input data and total output data, this general methodology shows how to decompose the total output spectral density function into its uncorrelated linear and nonlinear components. A large class of single-input/single-output nonlinear models is made to be equivalent to two-input/single-output linear models. Procedures discussed are how to identify system properties and how to evaluate statistical errors in estimates. Two engineering applications detailed are (a) nonlinear models for wave forces on structures, and (b) nonlinear models for slowly varying drift forces on structures. The chapter concludes with connections between these new techniques for nonlinear system identification from random data and the determination of physical parameters in nonlinear differential equations of motion.

7.1 ANALYSIS METHODOLOGY

A general physical situation is pictured in Figure 7.1 of a single-input/single-output nonlinear model with parallel linear and nonlinear systems of arbitrary type where the outputs can be correlated. As shown, the total output $y(t)$ due to the input $x(t)$ is the sum of a linear output $y_h(t)$, a nonlinear output $y_v(t)$, and an extraneous noise term $n(t)$ that will be assumed uncorrelated with $y_h(t)$ and $y_v(t)$. The input data can be Gaussian or non-Gaussian. The nonlinear system in Figure 7.1 can be of broader scope than previous bilinear and trilinear systems from Chapter 4 as special cases in addition to covering other possible nonlinear

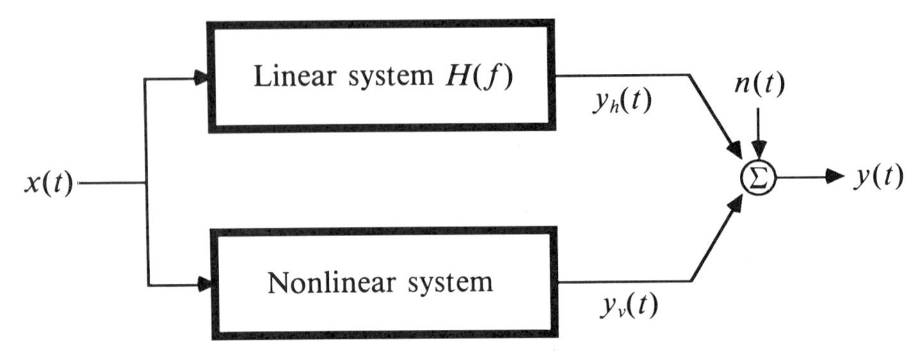

Figure 7.1 Single-input/single-output nonlinear model with parallel linear and nonlinear systems where outputs can be correlated.

systems such as those in Chapter 5. Consequently, the nonlinear output $y_v(t)$ can here be any desired function of $x(t)$. Also, there is no requirement that $y_h(t)$ and $y_v(t)$ be uncorrelated. Thus,

$$y(t) = y_h(t) + y_v(t) + n(t) \tag{7.1}$$

and the output autospectral density function becomes

$$S_{yy}(f) = S_{y_h y_h}(f) + S_{y_v y_v}(f) + S_{y_h y_v}(f) + S^*_{y_h y_v}(f) + S_{nn}(f) \tag{7.2}$$

where, in general, the cross-spectrum term

$$S_{y_h y_v}(f) \neq 0 \tag{7.3}$$

From measurements only of $x(t)$ and $y(t)$, in practice, it is difficult to decompose $S_{yy}(f)$ into its linear and nonlinear components because $S_{y_h y_v}(f) \neq 0$. A recommended analysis methodology will now be outlined to help solve such decomposition problems by showing how to change this nonlinear model of Figure 7.1 so that $x(t)$ will give a different optimum linear output $y_o(t)$ and a different uncorrelated nonlinear output $y_u(t)$ as pictured in Figure 7.2. Specifically, it is required that

$$y(t) = y_o(t) + y_u(t) + n(t) \tag{7.4}$$

with the cross-spectrum term

$$S_{y_o y_u}(f) = 0 \tag{7.5}$$

For this new model, it will be much simpler to decompose $S_{yy}(f)$ into its linear and nonlinear components. The desired $S_{yy}(f)$ will be

$$S_{yy}(f) = S_{y_o y_o}(f) + S_{y_u y_u}(f) + S_{nn}(f) \tag{7.6}$$

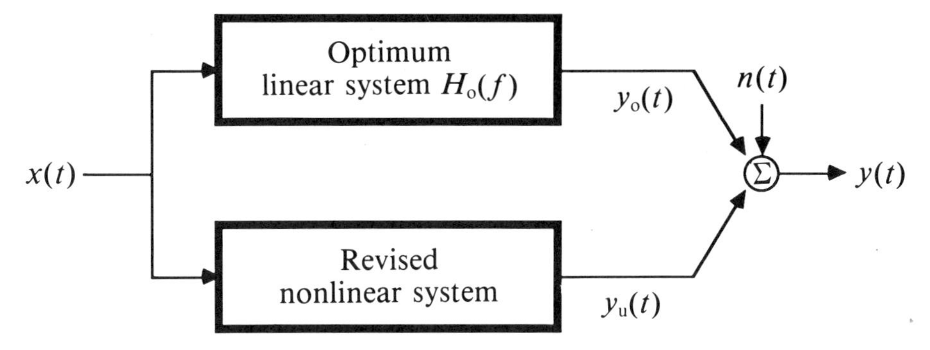

Figure 7.2 Nonlinear model with parallel optimum linear system and revised nonlinear system where outputs are uncorrelated.

Equations (7.1) and (7.4) have the same $y(t)$ and $n(t)$. Hence, it is required that the two models satisfy the time and Fourier transform equations

$$y_h(t) + y_v(t) = y_o(t) + y_u(t) \tag{7.7}$$

$$Y_h(f) + Y_v(f) = Y_o(f) + Y_u(f) \tag{7.8}$$

The optimum linear system $H_o(f)$ for Figure 7.2 may or may not be the same as the linear system $H(f)$ shown in Figure 7.1, depending upon the correlation between $x(t)$ and $y_v(t)$. To be specific, by taking appropriate Fourier transforms of Eq. (7.1), one obtains

$$Y(f) = Y_h(f) + Y_v(f) + N(f) \tag{7.9}$$

where

$$Y_h(f) = H(f)X(f) \tag{7.10}$$

and $Y_v(f)$ is the Fourier transform of the nonlinear output $y_v(t)$. Now, assuming $S_{xn}(f) = 0$, the input/output cross-spectrum $S_{xy}(f)$ between $x(t)$ and the total output $y(t)$ is

$$S_{xy}(f) = S_{xy_h}(f) + S_{xy_v}(f) \tag{7.11}$$

Here, $S_{xy_h}(f)$ is the cross-spectrum between $x(t)$ and the linear output $y_h(t)$ given by

$$S_{xy_h}(f) = H(f)S_{xx}(f) \tag{7.12}$$

and $S_{xy_v}(f)$ is the cross-spectrum between $x(t)$ and the nonlinear output $y_v(t)$.

The optimum linear system $H_o(f)$ is known from the usual formula

$$H_o(f) = \frac{S_{xy}(f)}{S_{xx}(f)} \tag{7.13}$$

Hence, Eqs. (7.11)–(7.13) prove that

$$H_o(f) = H(f) + \frac{S_{xy_v}(f)}{S_{xx}(f)} \tag{7.14}$$

Thus, $H_o(f) = H(f)$ if and only if $S_{xy_v}(f) = 0$. In general, $S_{xy_v}(f) \neq 0$ so that $H_o(f) \neq H(f)$.

If $H(f)$ is replaced by $H_o(f)$ as shown in Figure 7.2, then the output $Y(f)$ may be separated into a new optimum linear component $Y_o(f)$ and a new nonlinear component $Y_u(f)$ where Fourier transforms of Eq. (7.4) give

$$Y(f) = Y_o(f) + Y_u(f) + N(f) \tag{7.15}$$

with

$$Y_o(f) = H_o(f)X(f) \tag{7.16}$$

From Eq. (7.8), $Y_u(f)$ must satisfy

$$Y_u(f) = Y_h(f) + Y_v(f) - Y_o(f) \tag{7.17}$$

Hence, from Eqs. (7.10), (7.14), and (7.16),

$$Y_o(f) = Y_h(f) + \left[\frac{S_{xy_v}(f)}{S_{xx}(f)}\right] X(f) \tag{7.18}$$

and from Eq. (7.17),

$$Y_u(f) = Y_v(f) - \left[\frac{S_{xy_v}(f)}{S_{xx}(f)}\right] X(f) \tag{7.19}$$

It follows immediately that

$$S_{xy_u}(f) = \frac{1}{T} E[X^*(f)Y_u(f)] = 0 \tag{7.20}$$

where expected value operations are over an ensemble of records of length T. Thus, the input $x(t)$ is uncorrelated with the output $y_u(t)$ in Figure 7.2. Equation (7.16) then shows that

$$S_{y_o y_u} = \frac{1}{T} E[Y_o^*(f)Y_u(f)] = H_o^*(f)S_{xy_u}(f) = 0 \tag{7.21}$$

This proves that the outputs $y_o(t)$ and $y_u(t)$ in Figure 7.2 are uncorrelated as stated in Eq. (7.5). From Eq. (7.15), one now obtains the desired result of Eq.

(7.6), namely,

$$S_{yy}(f) = S_{y_o y_o}(f) + S_{y_u y_u}(f) + S_{nn}(f) \tag{7.22}$$

with

$$S_{y_o y_o}(f) = |H_o(f)|^2 S_{xx}(f) = \frac{|S_{xy}(f)|^2}{S_{xx}(f)} \tag{7.23}$$

The formula for $S_{y_u y_u}(f)$ depends upon the particular type of nonlinear operations that may be involved. Equation (7.19) shows that $S_{y_u y_u}(f)$ is a function of $S_{y_v y_v}(f)$ and $S_{xy_v}(f)$ using the $y_v(t)$ of Figure 7.1. Specifically,

$$S_{y_u y_u}(f) = \frac{1}{T} E[Y_u^*(f) Y_u(f)] = S_{y_v y_v}(f) - \frac{|S_{xy_v}(f)|^2}{S_{xx}(f)} \tag{7.24}$$

Note that $S_{y_u y_u}(f) = S_{y_v y_v}(f)$ if and only if $S_{xy_v}(f) = 0$. In general, $S_{xy_v}(f) \neq 0$ so that $S_{y_u y_u}(f) \neq S_{y_v y_v}(f)$.

The ordinary linear coherence function $\gamma_{xy}^2(f)$ from the optimum linear output $y_o(t)$ is

$$\gamma_{xy}^2(f) = \frac{S_{y_o y_o}(f)}{S_{yy}(f)} = \frac{|S_{xy}(f)|^2}{S_{xx}(f) S_{yy}(f)} \tag{7.25}$$

The associated nonlinear coherence function $q_{xy}^2(f)$ from the uncorrelated nonlinear output $y_u(t)$ is

$$q_{xy}^2(f) = \frac{S_{y_u y_u}(f)}{S_{yy}(f)} \tag{7.26}$$

where $S_{y_u y_u}(f)$ is computed by Eq. (7.24). In terms of these coherence functions, the output noise spectrum $S_{nn}(f)$ becomes

$$S_{nn}(f) = [1 - \gamma_{xy}^2(f) - q_{xy}^2(f)] S_{yy}(f) \tag{7.27}$$

This shows the importance of obtaining $q_{xy}^2(f)$ to reduce $S_{nn}(f)$. For good models, $S_{nn}(f)$ should be small compared to $S_{yy}(f)$ at selected frequencies of interest. These frequencies are where the peak values occur in the output spectrum of $S_{yy}(f)$.

7.2 EQUIVALENT TWO-INPUT/SINGLE-OUTPUT LINEAR MODELS

Suppose the nonlinear system in Figure 7.1 is a general finite-memory nonlinear system consisting of a zero-memory nonlinear system with output $v(t)$, that is followed by a linear system $A(f)$ with output $y_v(t)$. Figure 7.1 then becomes

Figure 7.3, where the linear output $y_h(t)$ and the nonlinear output $y_v(t)$ can be correlated. By applying the methodology in Section 7.1, Figure 7.3 can be replaced by Figure 7.4 consisting of a revised zero-memory nonlinear system with uncorrelated output $u(t)$, that is followed by the same linear system $A(f)$ as before but with output $y_u(t)$. The previous linear system $H(f)$ is replaced by the optimum linear system $H_o(f)$. Now, in Figure 7.4, the linear output $y_o(t)$ and the nonlinear output $y_u(t)$ are uncorrelated. Note that Figures 7.3 and 7.4 are Case 1 type nonlinear models analogous to Figures 5.1 and 5.3.

Relations between quantities shown in Figures 7.3 and 7.4 can be stated as special cases of results in Section 7.1. These relations can also be derived independently and include the following formulas.

$$C(f) = [S_{xy_v}(f)/S_{xx}(f)] = [S_{xv}(f)/S_{xx}(f)]A(f) \tag{7.28}$$

$$H_o(f) = [S_{xy}(f)/S_{xx}(f)] = H(f) + C(f) \tag{7.29}$$

$$Y_o(f) = Y_h(f) + C(f)X(f) \tag{7.30}$$

$$Y_u(f) = Y_v(f) - C(f)X(f) \tag{7.31}$$

$$Y_o(f) = H_o(f)X(f), \qquad Y_h(f) = H(f)X(f) \tag{7.32}$$

$$Y_u(f) = A(f)U(f), \qquad Y_v(f) = A(f)V(f) \tag{7.33}$$

$$U(f) = V(f) - [S_{xv}(f)/S_{xx}(f)]X(f) \tag{7.34}$$

$$S_{xu}(f) = 0 \tag{7.35}$$

$$S_{uu}(f) = S_{vv}(f)[1 - \gamma_{xv}^2(f)] \tag{7.36}$$

$$S_{uy}(f) = S_{vy}(f) - [S_{vx}(f)/S_{xx}(f)]S_{xy}(f) \tag{7.37}$$

For the single-input/single-output nonlinear models of Figure 7.3 and 7.4, assume that

1. $x(t)$ and $y(t)$ can be measured, and
2. $v(t)$ can be calculated from $x(t)$.

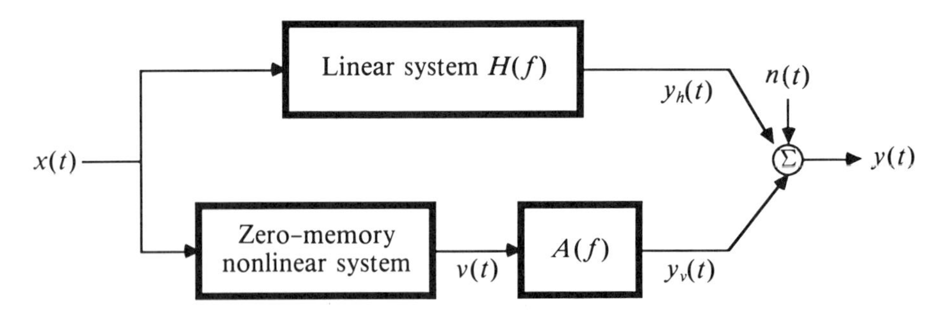

Figure 7.3 Single-input/single-output nonlinear model with parallel linear and zero-memory nonlinear systems where outputs can be correlated.

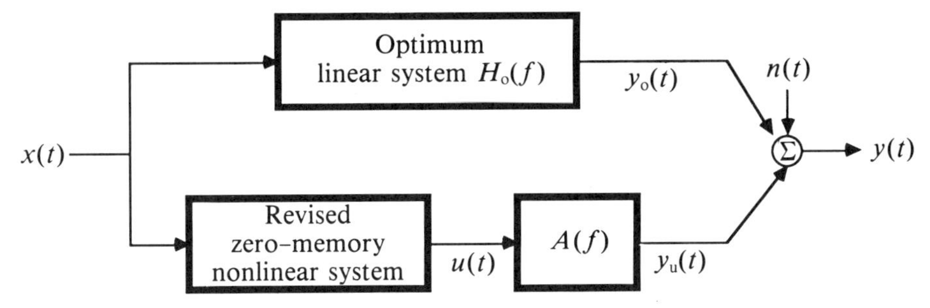

Figure 7.4 Nonlinear model with parallel optimum linear system and revised zero-memory nonlinear system where outputs are uncorrelated.

The single-input/single-output nonlinear models of Figures 7.3 and 7.4 can then be replaced by the equivalent two-input/single-output linear models of Figures 7.5 and 7.6, respectively. From knowledge of $x(t)$, $y(t)$, and $v(t)$, one can compute their individual autospectral density functions as well as their various cross-spectral density functions. One can then compute $S_{uu}(f)$ and $S_{uy}(f)$ from Equations (7.36) and (7.37), respectively. Note that Figures 7.5 and 7.6 are analogous to Figures 5.2 and 5.4.

It is straightforward to compute various desired quantities in Figures 7.5 and 7.6. When $C(f) \neq 0$, Eqs. (7.28) and (7.29) show that $H_o(f) \neq H(f)$ and is dependent upon the particular input excitation. Thus, assuming a conventional single-input/single-output linear model in place of the correct nonlinear model of Figure 7.1 and computing only $H_o(f)$ leads to an erroneous estimate of $H(f)$ and no estimate of $A(f)$. Using the notation in reference 1, the two correlated inputs in Figure 7.5 are $X_1(f) = X(f)$ and $X_2(f) = V(f)$. The two uncorrelated inputs in Figure 7.6 are $X_1(f) = X(f)$ and $X_{2\cdot 1}(f) = U(f)$, where $U(f)$ is computed by Eq. (7.34).

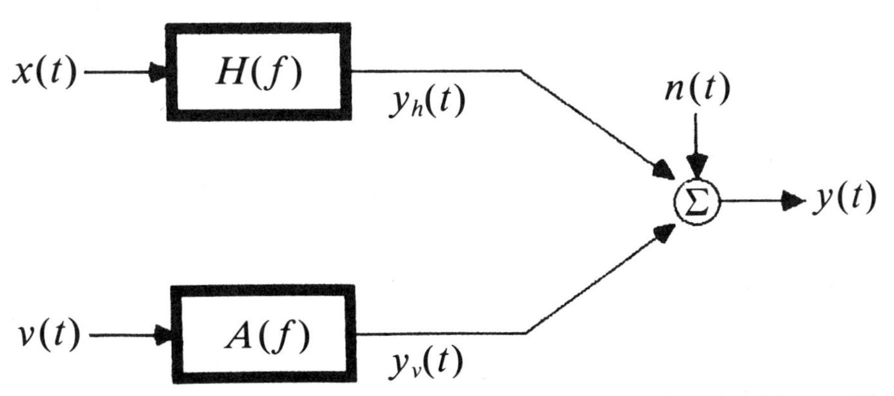

Figure 7.5 Two-input/single-output linear model equivalent to Figure 7.3 where inputs can be correlated.

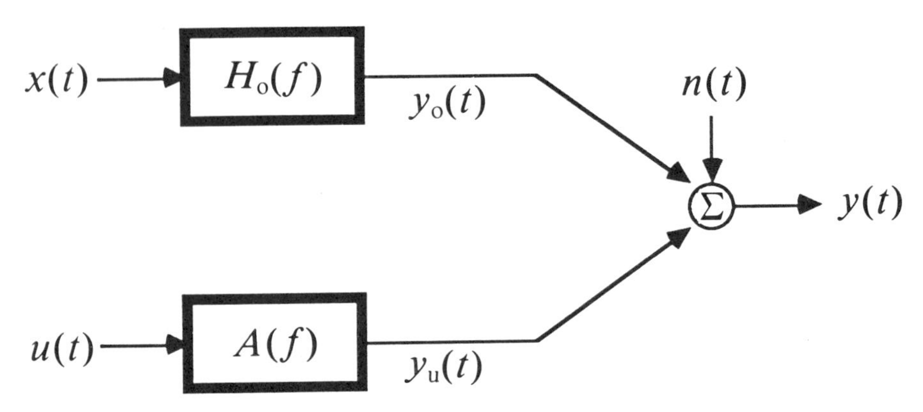

Figure 7.6 Two-input/single-output linear model equivalent to Figure 7.4 where inputs are uncorrelated.

Some important equations for Figures 7.4–7.6 are

$$S_{yy}(f) = S_{y_o y_o}(f) + S_{y_u y_u}(f) + S_{nn}(f) \tag{7.38}$$

$$H_o(f) = \frac{S_{xy}(f)}{S_{xx}(f)} = |H_o(f)|e^{-j\phi_o(f)} \tag{7.39}$$

$$S_{y_o y_o}(f) = |H_o(f)|^2 S_{xx}(f) = \gamma_{xy}^2(f) S_{yy}(f) \tag{7.40}$$

$$A(f) = \frac{S_{uy}(f)}{S_{uu}(f)} = |A(f)|d^{-j\phi_a(f)} \tag{7.41}$$

$$S_{y_u y_u}(f) = |A(f)|^2 S_{uu}(f) = q_{xy}^2(f) S_{yy}(f) \tag{7.42}$$

The meaning of these quantities is clear from previous discussions. Note that only first-order autospectral and cross-spectral density functions are required in these equations, and the results apply to Gaussian or non-Gaussian input data.

7.3 STATISTICAL ERRORS IN ESTIMATES

Statistical normalized random error and bias error formulas will now be stated for estimates of quantities in Figures 7.4 and 7.6. These formulas are derived in reference 1 and summarized in reference 2. The basis for these results is that $x(t)$ and $u(t)$ are uncorrelated inputs to $H_o(f)$ and $A(f)$, respectively. For these error formulas to be strictly valid, both $x(t)$ and $u(t)$ should be Gaussian. This can often be assumed for the input data $x(t)$ but will not be true for $u(t)$, since $u(t)$ comes from a nonlinear operation on $x(t)$. Thus, error formulas for $A(f)$, $S_{y_u y_u}(f)$, and $q_{xy}^2(f)$ should be regarded only as guidelines.

The normalized random errors for the two output spectra estimates $S_{y_o y_o}(f)$

and $S_{y_u y_u}(f)$ are approximated by

$$\varepsilon[\hat{S}_{y_o y_o}(f)] \approx \frac{[2 - \hat{\gamma}^2_{xy}(f)]^{1/2}}{|\hat{\gamma}_{xy}(f)|\sqrt{n_d}} \tag{7.43}$$

$$\varepsilon[\hat{S}_{y_u y_u}(f)] \approx \frac{[2 - \hat{q}^2_{xy}(f)]^{1/2}}{|\hat{q}_{xy}(f)|\sqrt{n_d - 1}} \tag{7.44}$$

where n_d is the number of independent averages used to compute the original spectral density quantities. Required values of n_d to achieve desired values of $\varepsilon[\hat{S}_{y_o y_o}(f)]$ or $\varepsilon[\hat{S}_{y_u y_u}(f)]$ are obtained by solving these equations for n_d. Bias errors for these spectra estimates will be minimized when there is good bandwidth resolution.

The normalized random errors for the two coherence function estimates $\hat{\gamma}^2_{xy}(f)$ and $\hat{q}^2_{xy}(f)$ are approximated by

$$\varepsilon[\hat{\gamma}^2_{xy}(f)] \approx \frac{\sqrt{2}[1 - \hat{\gamma}^2_{xy}(f)]}{|\hat{\gamma}_{xy}(f)|\sqrt{n_d}} \tag{7.45}$$

$$\varepsilon[\hat{q}^2_{xy}(f)] \approx \frac{\sqrt{2}[1 - \hat{q}^2_{xy}(f)]}{|\hat{q}_{xy}(f)|\sqrt{n_d - 1}} \tag{7.46}$$

Bias errors for these coherence estimates are

$$b[\hat{\gamma}^2_{xy}(f)] \approx \frac{[1 - \hat{\gamma}^2_{xy}(f)]^2}{n_d} \tag{7.47}$$

$$b[\hat{q}^2_{xy}(f)] \approx \frac{[1 - \hat{q}^2_{xy}(f)]^2}{n_d - 1} \tag{7.48}$$

Results in Eqs. (7.43)–(7.48) apply to Figure 7.2 as well as to Figures 7.4 and 7.6 when $\hat{\gamma}^2_{xy}(f)$ and $\hat{q}^2_{xy}(f)$ can be computed.

Normalized random errors for the frequency response function estimates $\hat{H}_o(f)$ and $\hat{A}(f)$ are required for both gain factor and phase factor estimates. For gain factor estimates $|\hat{H}_o(f)|$ and $|\hat{A}(f)|$, the normalized random errors are approximated by

$$\varepsilon[|\hat{H}_o(f)|] \approx \frac{[1 - \hat{\gamma}^2_{xy}(f)]^{1/2}}{|\hat{\gamma}_{xy}(f)|\sqrt{2n_d}} \tag{7.49}$$

$$\varepsilon[|\hat{A}(f)|] \approx \frac{[1 - \hat{q}^2_{xy}(f)]^{1/2}}{|\hat{q}_{xy}(f)|\sqrt{2(n_d - 1)}} \tag{7.50}$$

The standard deviations (in radians) of the two associated phase factor estimates $\hat{\phi}_o(f)$ and $\hat{\phi}_a(f)$ (in radians) are approximated by

$$\sigma[\hat{\phi}_o(f)] \approx \sin^{-1}\{\varepsilon[|\hat{H}_o(f)|]\} \tag{7.51}$$

$$\sigma[\hat{\phi}_a(f)] \approx \sin^{-1}\{\varepsilon[|\hat{A}(f)|]\} \tag{7.52}$$

For small values of $\varepsilon[|\hat{H}_o|]$ and $\varepsilon[|\hat{A}|]$, these equations can be further simplified to

$$\sigma[\hat{\phi}_o(f)] \approx \varepsilon[|\hat{H}_o(f)|] \tag{7.53}$$

$$\sigma[\hat{\phi}_a(f)] \approx \varepsilon[|\hat{A}(f)|] \tag{7.54}$$

Various bias errors in frequency response function estimates that should be minimized as much as possible in practice are also covered in reference 1.

These random error formulas from Section 7.3 are suitable for parameter estimates from nonlinear models of wave forces and drift forces in Sections 7.4 and 7.5. The same random error formulas apply to theoretical two-sided spectral quantities involving $S(f)$ functions and to measurable one-sided spectral quantities involving $G(f)$ functions as listed in Table 1.2. The sensitivity of the wave force results in Section 7.4 to the Gaussian assumption on input data is discussed in reference 4.

Example of Experimental Design

Suppose that stationary random data $x(t)$ and $y(t)$ are to be measured for the nonlinear model of Figure 7.2 or 7.4. Suppose that the desired bandwidth resolution is $\Delta f = 0.10$ Hz and that the desired normalized random errors in estimates of $\hat{G}_{y_o y_o}(f)$ and $\hat{G}_{y_u y_u}(f)$ are 10% each. What should be the total record length T_{total} for the experiment when the expected coherence function estimates are $\hat{\gamma}^2_{xy}(f) = 0.60$ and $\hat{q}^2_{xy}(f) = 0.20$?

The total record length $T_{\text{total}} = n_d T$ where $T = (1/\Delta f)$ and n_d is the number of distinct subrecords of length T each. From $\Delta f = 0.10\,Hz$, it follows that each subrecord should be of length

$$T = (1/\Delta f) = 10\ \text{sec}$$

From Eq. (7.43), the solution for n_d to estimate $G_{y_o y_o}(f)$ is

$$n_d \approx \frac{[2 - \hat{\gamma}^2_{xy}(f)]}{\hat{\gamma}^2_{xy}(f)\varepsilon^2[\hat{G}_{y_o y_o}(f)]}$$

For $\hat{\gamma}^2_{xy}(f) = 0.60$ and $\varepsilon = 0.10$, this gives

$$n_d \approx 234$$

Hence, for estimates of $G_{y_o y_o}$, the experiment should be designed so that the total record length is

$$T_{\text{total}} = n_d T \approx 39 \text{ min}$$

From Eq. (7.44), the solution for n_d to estimate $G_{y_u y_u}(f)$ is

$$n_d \approx \frac{[2 - \hat{q}^2_{xy}(f)]}{\hat{q}^2_{xy}(f)\varepsilon^2[\hat{G}_{y_u y_u}(f)]}$$

For $\hat{q}^2_{xy}(f) = 0.20$ and $\varepsilon = 0.10$, this gives

$$n_d \approx 900$$

Hence, for estimates of $G_{y_u y_u}(f)$, the experiment should be designed so that the total record length is

$$T_{\text{total}} = n_d T \approx 150 \text{ min}$$

This numerical example indicates the nature of increased numbers of averages and longer total record lengths to obtain equally valid estimates of nonlinear output spectra compared to linear output spectra for situations where $\hat{q}^2_{xy}(f) < \hat{\gamma}^2_{xy}(f)$. Similar requirements apply to other parallel linear and nonlinear systems where the nonlinear coherence functions are smaller than the linear coherence functions.

7.4 NONLINEAR WAVE FORCE MODELS

Two important engineering applications will now be discussed that involve procedures in this chapter for analyzing parallel linear and nonlinear systems. These applications concern nonlinear models for

(a) wave forces on structures, and
(b) slowly varying drift forces on structures.

Oceanographic research and development work consider the structures to be systems such as vertical piles, ships, or offshore oil and gas platforms. Through changes in terminology and interpretation, similar models and procedures are applicable to the analysis of linear and nonlinear properties for electrical, mechanical, and propulsion systems in airplanes, aerospace vehicles, automotive vehicles, power plants, bridges, tall buildings, and other engineering fields. Formulas in this section and in Section 7.5 will be stated in terms of measurable one-sided spectral quantities $G(f)$.

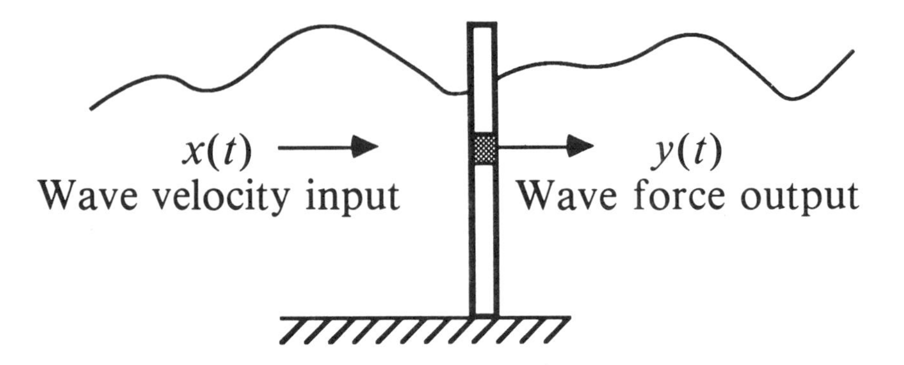

Figure 7.7 Illustration of wave force problem.

The general problem to be analyzed for wave forces on small-diameter fixed structures is shown in Figure 7.7. The structure is assumed to be subject to Gaussian stationary random ocean waves represented by the wave velocity input $x(t)$ at a specified depth. This input record produces a wave force output at the same depth denoted by the output record $y(t)$.

Morison's equation is widely used to study such wave forces on structures (references 4, 28, 39, and 44) where the output wave force $y(t)$ due to an input wave velocity $x(t)$ consists of two main components:

1. A *linear* inertial force $m(t)$ proportional to the derivative $\dot{x}(t)$ of $x(t)$.
2. A *nonlinear* drag force $d(t)$ proportional to $x(t)|x(t)|$, a squarer with sign.

Morison's equation is thus of the form

$$y(t) = m(t) + d(t) = C_1\dot{x}(t) + C_2 x(t)|x(t)| \tag{7.55}$$

where the coefficients C_1 and C_2 are assumed to be constants, independent of frequency. Experimental work shows that these coefficients can vary with frequency so that Morison's equation needs to be extended.

7.4.1 Case 1 and Case 2 Nonlinear Models

Generalizations of Morison's equation are drawn in the Case 1 and Case 2 nonlinear wave force models of Figures 7.8 and 7.9, respectively, where C_1 and C_2 are replaced by linear frequency response functions $H(f)$ and $A(f)$ or $H(f)$ and $B(f)$. The problem of concern is to determine the spectral properties of the linear inertial component $m(t)$ and the nonlinear drag component $d(t)$ from measurements of the input wave velocity $x(t)$ and the total wave force output $y(t)$. Assuming the input $x(t)$ to behave like Gaussian random data, this problem

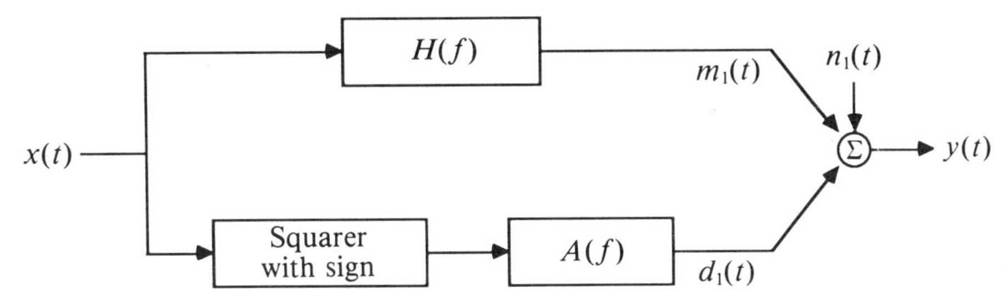

Figure 7.8 Case 1 nonlinear wave force model with parallel linear and nonlinear systems.

is difficult for two main reasons:

1. $d(t)$ has a nonlinear relationship involving a squarer with sign.
2. The parallel outputs $m(t)$ *and* $d(t)$ are correlated.

Note also that Figures 7.8 and 7.9 show different possible values for outputs $m_1(t)$ versus $m_2(t)$ and $d_1(t)$ versus $d_2(t)$, as well as for the noise terms $n_1(t)$ versus $n_2(t)$, so that it is not obvious in advance which of these models will best fit the measured data.

The basis for analyzing the nonlinear wave force models of Figures 7.8 and 7.9 is as follows:

1. Replace the squarer with sign, $x|x|$, by its third-order polynomial least-squares expansion of Eq. (2.196). This gives an approximation to $x|x|$ where a linear operation is in parallel with a cubic operation as per Figures 5.19–5.21.

2. Determine for both Figures 7.8 and 7.9 an overall optimum linear system and a parallel revised nonlinear system where outputs will be uncorrelated. This gives equivalent new models to Figures 7.8 and 7.9.

3. For the resulting new Figure 7.8, where a linear system $A(f)$ follows the revised nonlinear system, use the Case 1 procedures in Section 5.2. Apply the

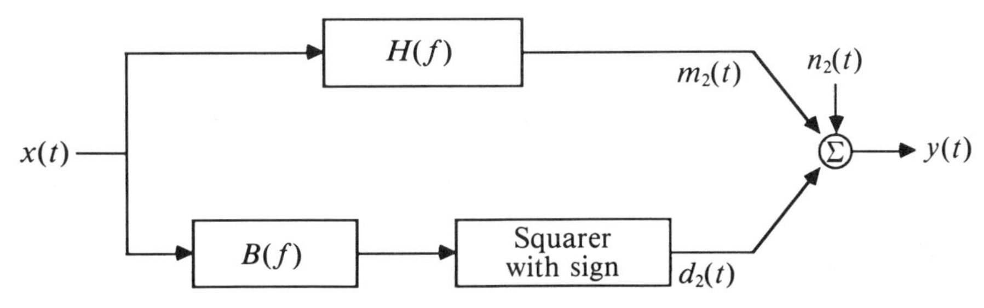

Figure 7.9 Case 2 nonlinear wave force model with parallel linear and nonlinear systems.

practical linear system identification techniques from Sections 5.2.4 and 5.5 that require the computation of only ordinary spectral density functions.

4. For the resulting new Figure 7.9, where a linear system $B(f)$ precedes the revised nonlinear system, use the Case 2 procedures in Section 5.3. This requires the computation of the special bispectral and trispectral density functions in Sections 5.2.3 and 5.3.3.

5. Case 1 models are preferred over Case 2 models when they give smaller values for the output noise spectrum, and conversely.

The two models of Figures 7.8 and 7.9 were studied extensively in past work (references 4 and 44). Techniques from Chapter 5 were employed to analyze the data resulting from model basin experiments, where design of the experiments was based upon statistical error analysis formulas stated in Chapter 6 and Section 7.3. Measured results from these model basin experiments show that for vertical piles in random seas, the model in Figure 7.8 (where the linear system $A(f)$ follows the nonlinear operation) is superior to the model in Figure 7.9 (where the linear system $B(f)$ precedes the nonlinear operation) in giving smaller values for the spectral values of the output noise $G_{n_1n_1}(f)$ versus $G_{n_2n_2}(f)$. For other physical problems, one can envision situations where the model of Figure 7.9 might be preferred over Figure 7.8 so that work to analyze Figure 7.9 is justified. Where appropriate, to simplify the analysis, one should choose a model where a linear system follows the nonlinear operation over an alternative model where a linear system precedes the nonlinear operation.

Further discussion will now be devoted only to the Case 1 nonlinear wave force model of Figure 7.8. The Case 2 nonlinear wave force model of Figure 7.9 can be similarly developed. Two problems will be treated:

1. *Spectral Decomposition Problem.* Given $H(f)$ and $A(f)$ plus measurement only of $x(t)$, determine the spectral properties of $m(t)$ and $d(t)$. If $y(t)$ is measured as well as $x(t)$, determine also the spectral properties of $n(t)$.

2. *System Identification Problem.* From simultaneous measurements of both $x(t)$ and $y(t)$, identify the optimum frequency response function properties of $H(f)$ and $A(f)$ to minimize the autospectrum of $n(t)$.

7.4.2 Spectral Decomposition Problem

To solve the spectral decomposition problem for the wave force model in Figure 7.8, an approximation is required for the nonlinear operation $x|x|$. Assuming $x(t)$ to be a sample time-history of a Gaussian stationary random process with zero mean value, the zero-memory third-order polynomial least-squares approximation is given from Eq. (2.196) by $v(t) \simeq x(t)|x(t)|$ where

$$v(t) = 3k\sigma_x^2 x(t) + kx^3(t), \qquad k = \sqrt{(2/\pi)}/3\sigma_x \tag{7.56}$$

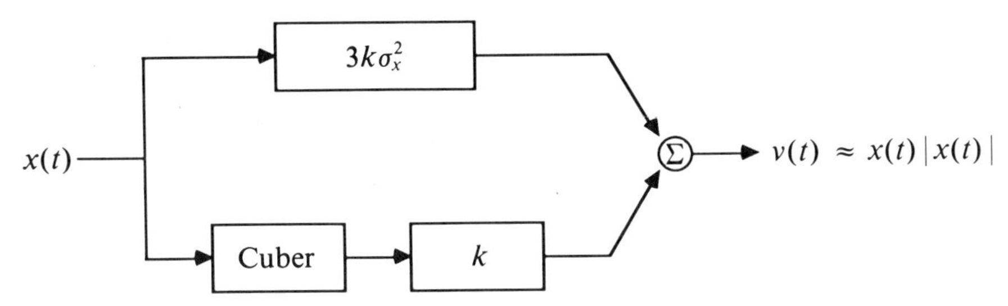

Figure 7.10 Zero-memory third-order polynomial nonlinear model for squarer with sign.

The factor k is the same as a_3 in Eq. (2.195), and σ_x is the standard deviation of $x(t)$. In words, $x|x|$, the squarer with sign, can be replaced by the sum of a linear operation plus a cubic operation as diagrammed in Figure 7.10. This is the same as Figure 5.19.

Substitution of Figure 7.10 into Figure 7.8 yields the nonlinear wave force model shown in Figure 7.11 which, henceforth, is used to approximate the Case 1 model in Figure 7.8. Formulas will now be stated to compute the autospectral density functions of $m(t)$ and $d(t)$ from knowledge of $H(f)$ and $A(f)$, using measurement of only $x(t)$. Note that Figure 7.11 is a special case of Figure 5.15 and, hence, can be analyzed by the formulas listed in Section 5.4.2. These formulas can also be derived directly from basic frequency-domain averaging operations without using the results in Chapter 5. Figure 7.11 is equivalent to a two-input/single-output linear model with correlated inputs $x(t)$ and $v(t)$.

The following equations apply to the proposed nonlinear wave force model of Figure 7.11, where the notation and quantities should be clear from previous

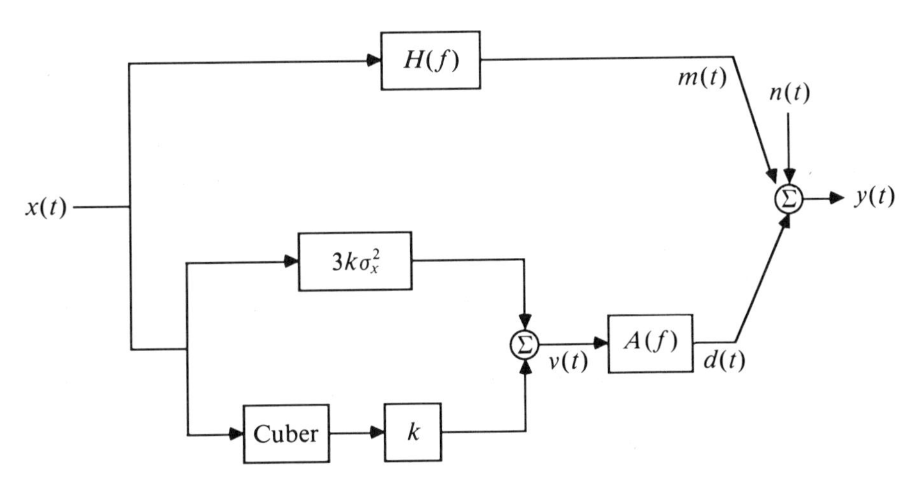

Figure 7.11 Nonlinear wave force model approximating Figure 7.8.

material. Fourier transform relations are

$$Y(f) = M(f) + D(f) + N(f) \tag{7.57}$$

$$M(f) = H(f)X(f) \tag{7.58}$$

$$D(f) = A(f)V(f) \tag{7.59}$$

Here,

$$V(f) = k[X_3(f) + 3\sigma_x^2 X(f)] \tag{7.60}$$

where $X_3(f)$ is the Fourier transform of $x_3(t) = x^3(t)$ denoted by

$$X_3(f) = \mathscr{F}[x_3(t)] = \mathscr{F}[x^3(t)] \tag{7.61}$$

The measurable one-sided autospectral density function of $x_3(t) = x^3(t)$ can be computed by

$$G_{x_3x_3}(f) = \frac{2}{T} E[X_3^*(f)X_3(f)] \tag{7.62}$$

where expected value operations are over an ensemble of records of length T. Cross-spectral density function relations are

$$G_{xx_3}(f) = 3\sigma_x^2 G_{xx}(f) \tag{7.63}$$

$$G_{xv}(f) = 6k\sigma_x^2 G_{xx}(f) \tag{7.64}$$

These two results show that $x(t)$ and $x_3(t)$, as well as $x(t)$ and $v(t)$, are correlated. The autospectral density function of $v(t)$ is given by

$$G_{vv}(f) = k^2[27\sigma_x^4 G_{xx}(f) + G_{x_3x_3}(f)] \tag{7.65}$$

The autospectral density function of the inertial term $m(t)$ in Figure 7.11 is

$$G_{mm}(f) = |H(f)|^2 G_{xx}(f) \tag{7.66}$$

The autospectral density function of the drag term $d(t)$ in Figure 7.11 is

$$G_{dd}(f) = |A(f)|^2 G_{vv}(f) \tag{7.67}$$

The cross-spectral density function between $m(t)$ and $d(t)$ is given by

$$G_{md}(f) = 6k\sigma_x^2 H^*(f)A(f)G_{xx}(f) \tag{7.68}$$

Thus, $m(t)$ and $d(t)$ are correlated so that the total output autospectral density function $G_{yy}(f)$ in Figure 7.11, assuming $n(t)$ to be uncorrelated with both $m(t)$ and $d(t)$, is given by

$$G_{yy}(f) = G_{mm}(f) + G_{dd}(f) + G_{md}(f) + G_{md}^*(f) + G_{nn}(f) \tag{7.69}$$

This last result can be used to solve for $G_{nn}(f)$ from knowledge of the other quantities when both $x(t)$ and $y(t)$ are measured. These formulas solve the spectral decomposition problem.

7.4.3 System Identification Problem

The starting point for this system identification problem is the model in Figure 7.11, where now the properties of $H(f)$ and $A(f)$ are *not* known. Optimum properties are to be determined from simultaneous measurements of $x(t)$ and $y(t)$, based upon minimizing the autospectral density function $G_{nn}(f)$ of $n(t)$ over all possible choices of linear systems to predict $y(t)$ from $x(t)$.

As before, it is assumed that the Gaussian input $x(t)$ has a zero mean value. Some extensions of these results are required for nonzero mean value inputs, which change Figure 7.11 into parallel linear, square-law, and cubic systems. Only zero mean value inputs are treated here.

An equivalent model to Figure 7.11 with correlated outputs $y_1(t)$ and $y_3(t)$ is shown in Figure 7.12. An equivalent model to Figure 7.11 with uncorrelated outputs $y_o(t)$ and $y_u(t)$ is shown in Figure 7.13. Note that Figure 7.13 can also be analyzed as a two-input/single-output linear model with uncorrelated inputs $x(t)$ and $u(t)$. The following equations apply to Figures 7.12 and 7.13.

Fourier transform relations for Figure 7.12 are

$$Y(f) = Y_1(f) + Y_3(f) + N(f) \tag{7.70}$$

$$Y_1(f) = [H(f) + 3k\sigma_x^2 A(f)]X(f) \tag{7.71}$$

$$Y_3(f) = kA(f)X_3(f) \tag{7.72}$$

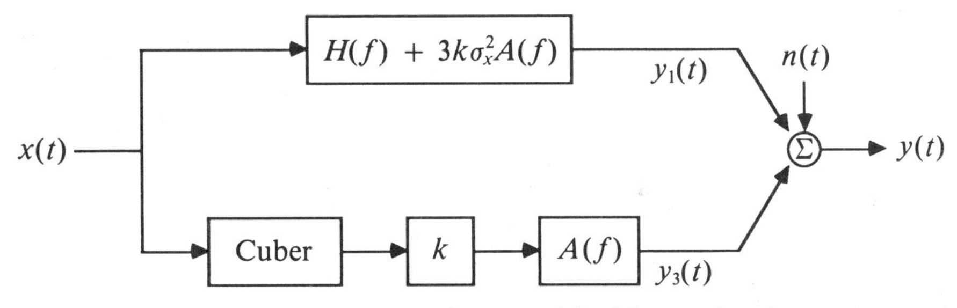

Figure 7.12 Nonlinear wave force model with correlated outputs.

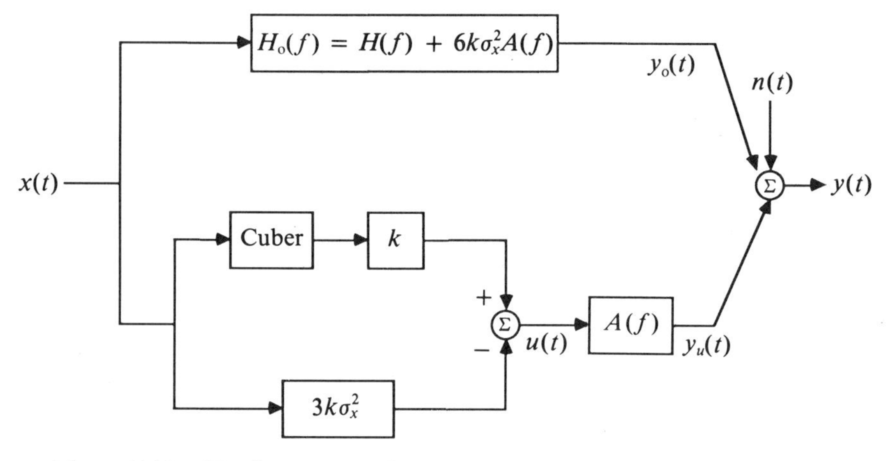

Figure 7.13 Nonlinear wave force model with uncorrelated outputs.

where $X_3(f)$ is the Fourier transform of $x^3(t)$. The cross-spectral density function from Eqs. (7.70)–(7.72) and (7.63) is

$$G_{xy}(f) = [H(f) + 6k\sigma_x^2 A(f)]G_{xx}(f) \tag{7.73}$$

Hence, the optimum linear system $H_o(f)$ is given by

$$H_o(f) = \frac{G_{xy}(f)}{G_{xx}(f)} = H(f) + 6k\sigma_x^2 A(f) \tag{7.74}$$

Determination of $H(f)$ requires separate knowledge of $A(f)$.

Fourier transform relations for Figure 7.13 are

$$Y(f) = Y_o(f) + Y_u(f) + N(f) \tag{7.75}$$

$$Y_o(f) = H_o(f)X(f) \tag{7.76}$$

$$Y_u(f) = A(f)U(f) \tag{7.77}$$

$$U(f) = k[X_3(f) - 3\sigma_x^2 X(f)] \tag{7.78}$$

Here, $G_{xu}(f) = 0$ so that $x(t)$ and $u(t)$ are uncorrelated. The autospectral density function of $u(t)$ is given by

$$G_{uu}(f) = k^2[G_{x_3x_3}(f) - 9\sigma_x^4 G_{xx}(f)] \tag{7.79}$$

The cross-spectral density function between $u(t)$ and $y(t)$ is

$$G_{uy}(f) = k[G_{x_3y}(f) - 3\sigma_x^2 G_{xy}(f)] \tag{7.80}$$

where

$$G_{x_3y}(f) = \frac{2}{T} E[X_3^*(f)Y(f)] \tag{7.81}$$

can be computed by taking expected value operations over an ensemble of records of length T. Also, in Figure 7.13, since $u(t)$ is uncorrelated with $y_o(t)$ and $n(t)$,

$$G_{uy}(f) = G_{uy_u}(f) \tag{7.82}$$

Hence, the frequency response function $A(f)$ is

$$A(f) = \frac{G_{uy_u}(f)}{G_{uu}(f)} = \frac{G_{uy}(f)}{G_{uu}(f)} \tag{7.83}$$

where Eq. (7.80) is divided by Eq. (7.79). The system $H(f)$ follows from Eqs. (7.74) and (7.83). These formulas solve the system identification problem.

Equations (7.66) and (7.67) can now be applied using the estimates of $H(f)$ and $A(f)$ to compute the spectral properties of the linear inertial force $m(t)$ and the nonlinear drag force $d(t)$. Other analyses can also be conducted to provide further physical insight into the linear and uncorrelated nonlinear parts of the spectral properties of $d(t)$. Also, appropriate linear and nonlinear coherence functions can be computed.

For the nonlinear wave force model in Figure 7.13, the total output autospectral density function is given by the formula

$$G_{yy}(f) = G_{y_oy_o}(f) + G_{y_uy_u}(f) + G_{nn}(f) \tag{7.84}$$

where

$$G_{y_oy_o}(f) = |H_o(f)|^2 G_{xx}(f) \tag{7.85}$$

$$G_{y_uy_u}(f) = |A(f)|^2 G_{uu}(f) \tag{7.86}$$

with $G_{uu}(f)$ computed by Eq. (7.79). The linear coherence function between the input $x(t)$ and the total output $y(t)$ is defined by

$$\gamma_{xy}^2(f) = \frac{G_{y_oy_o}(f)}{G_{yy}(f)} = \frac{|G_{xy}(f)|^2}{G_{xx}(f)G_{yy}(f)} \tag{7.87}$$

The nonlinear coherence function between $x(t)$ and $y(t)$ is defined by

$$q_{xy}^2(f) = \frac{G_{y_uy_u}(f)}{G_{yy}(f)} \tag{7.88}$$

where $G_{y_u y_u}(f)$ is computed by Eq. (7.86). This completes the treatment here of the nonlinear wave force model.

7.5 NONLINEAR DRIFT FORCE MODELS

The general problem to be analyzed for slowly varying drift forces on structures is shown in Figure 7.14. Here, $x(t)$ and $y(t)$ have different interpretations than the $x(t)$ and $y(t)$ in Figure 7.7. The input $x(t)$ represents in Figure 7.14 the wave elevation (height) rather than the wave velocity of Figure 7.7. This input record produces a force on the ship that causes the ship to move. The output $y(t)$ represents in Figure 7.14 the ship motion output rather than the wave force output of Figure 7.7.

In Figure 7.14, the force acting on the ship is assumed to consist of two components:

1. A *linear* term proportional to $x(t)$.
2. A *nonlinear* term proportional to the squared envelope signal $u(t)$ of $x(t)$.

This nonlinear term is called the *slowly varying drift force*. The resulting ship motion output $y(t)$ thus takes the form

$$y(t) = k_1 x(t) + k_2 u(t) \tag{7.89}$$

where the coefficients k_1 and k_2 are assumed to be constants, independent of frequency. The squared envelope signal $u(t)$ is a zero-memory nonlinear function of $x(t)$ that is poorly represented by a third-order-polynomial least-squares approximation. Models like Figure 7.3 are not appropriate.

A generalization of Eq. (7.79) is drawn in Figure 7.15, where the coefficients k_1 and k_2 are replaced by linear frequency response functions $H_1(f)$ and $H_2(f)$. The problem of concern is to determine the spectral properties of the linear component $y_1(t)$ and the nonlinear component $y_2(t)$ from measurements of the

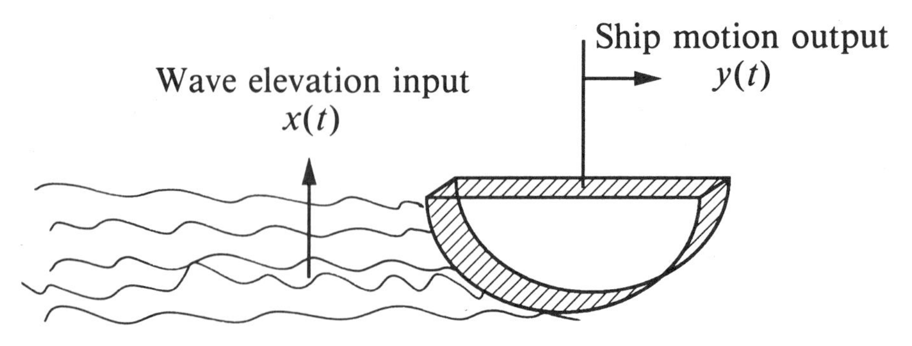

Figure 7.14 Illustration of drift force problem.

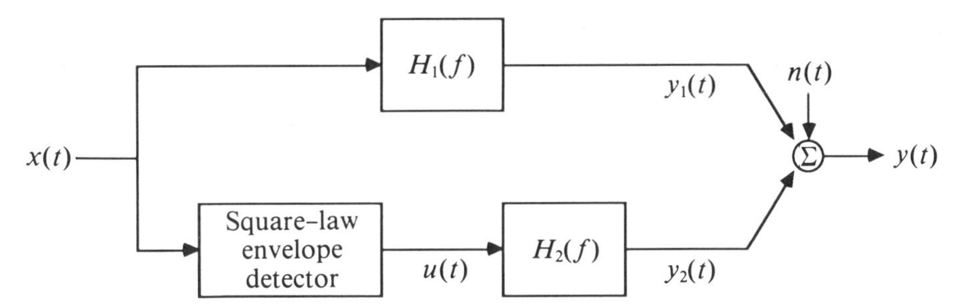

Figure 7.15 Nonlinear drift force model with parallel linear and square-law envelope detector systems.

input wave elevation input $x(t)$ and the total ship motion output $y(t)$. For Gaussian input data, the outputs $y_1(t)$ and $y_2(t)$ will be uncorrelated. This makes Figure 7.15 a special case of Figure 7.2 in Section 7.1, where $y_1(t)$ is the same as $y_o(t)$ and $y_2(t)$ is the same as $y_u(t)$. It is not necessary here to replace Figure 7.1 by Figure 7.2. Note also that Figure 7.15 is equivalent to a two-input/single-output linear model with uncorrelated inputs $x(t)$ and $u(t)$.

7.5.1 Previous Bilinear Models

Previous nonlinear drift force models, such as those assumed in references 14, 24 and 36 are more complicated than the proposed model of Figure 7.15 because they employ a bilinear weighting function $h_2(\tau_1, \tau_2)$ and a bilinear frequency response function $H_2(f, g)$ to represent the nonlinear system. Specifically, Figure 7.15 is replaced by the more general Figure 7.16, where $y_1(t)$ is the same but $y_2(t)$ is different. In Figure 7.16, from Eq. (3.34), the bilinear output $y_2(t)$ is given by

$$y_2(t) = \iint h_2(\tau_1, \tau_2)x(t - \tau_1)x(t - \tau_2)\,d\tau_1\,d\tau_2 \tag{7.90}$$

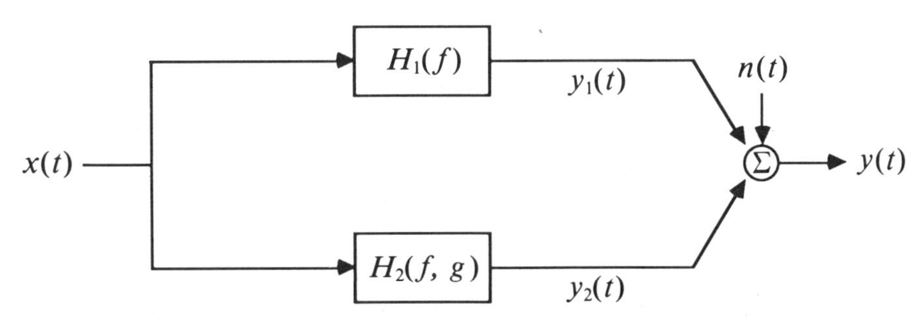

Figure 7.16 Nonlinear drift force model with parallel linear and bilinear systems.

From Eq. (3.55), the quantity $Y_2(f)$ is given at any $f \neq 0$ by

$$Y_2(f) = \int H_2(\alpha, f - \alpha) X(\alpha) X(f - \alpha)\, d\alpha \tag{7.91}$$

where $X(f)$ and $Y_2(f)$ are Fourier transforms of $x(t)$ and $y_2(t)$, respectively.

The bilinear frequency response function $H_2(f, g)$ from Eq. (3.45) is the double Fourier transform of the bilinear weighting function $h_2(\tau_1, \tau_2)$, namely,

$$H_2(f, g) = \iint h_2(\tau_1, \tau_2) e^{-j2\pi(f\tau_1 + g\tau_2)}\, d\tau_1\, d\tau_2 \tag{7.92}$$

This quantity $H_2(f, g)$ in two frequency variables is much more difficult to compute and interpret than the simpler $H_2(f)$ in Figure 7.15. The optimum solution for $H_2(f, g)$ is stated in Eq. (4.126). This complicated model will no longer be considered.

7.5.2 Basic Formulas for Proposed Model

The following basic formulas apply to the proposed nonlinear drift force model of Figure 7.15. In the time domain, the formulas are

$$y(t) = y_1(t) + y_2(t) + n(t) \tag{7.93}$$

where the linear output term $y_1(t)$ and the nonlinear output term $y_2(t)$ are

$$y_1(t) = \int_{-\infty}^{\infty} h_1(\tau) x(t - \tau)\, d\tau \tag{7.94}$$

$$y_2(t) = \int_{-\infty}^{\infty} h_2(\tau) u(t - \tau)\, d\tau \tag{7.95}$$

As proved in reference 1, the output $u(t)$ of the square-law envelope detector is

$$u(t) = x^2(t) + \tilde{x}^2(t) \tag{7.96}$$

where

$$\tilde{x}(t) = \text{Hilbert transform of } x(t) \tag{7.97}$$

The symbol ˜ used here should not be confused with "raw" estimates. When $x(t)$ has a zero mean value, then $\tilde{x}(t)$ will also have a zero mean value. However, the mean value of $u(t)$, denoted by $E[u(t)]$, is

$$E[u(t)] = E[x^2(t)] + E[\tilde{x}^2(t)] = 2\sigma_x^2 \tag{7.98}$$

where σ_x^2 is the variance of $x(t)$.

In the frequency domain, with mean values removed prior to taking Fourier transforms, the formulas are

$$Y(f) = Y_1(f) + Y_2(f) + N(f) \tag{7.99}$$

where

$$Y_1(f) = H_1(f)X(f) \tag{7.100}$$

$$Y_2(f) = H_2(f)U(f) \tag{7.101}$$

The Fourier transform of Eq. (7.96) yields

$$U(f) = \int_{-\infty}^{\infty} [X(\alpha)X(f-\alpha) + \tilde{X}(\alpha)\tilde{X}(f-\alpha)]\, d\alpha \tag{7.102}$$

where the quantity $\tilde{X}(f)$ is defined by

$$\tilde{X}(f) = B(f)X(f) \tag{7.103}$$

using

$$B(f) = -j \operatorname{sgn} f = \begin{cases} -j, & f > 0 \\ 0, & f = 0 \\ j, & f < 0 \end{cases} \tag{7.104}$$

Hence,

$$U(f) = \int_{-\infty}^{\infty} [1 + B(\alpha)B(f-\alpha)]X(\alpha)X(f-\alpha)\, d\alpha \tag{7.105}$$

For any $f > 0$, the product quantity

$$B(\alpha)B(f-\alpha) = \begin{cases} 1, & \alpha < 0 \\ -1, & 0 < \alpha < f \\ 1, & \alpha > f \end{cases} \tag{7.106}$$

Thus, for any $f > 0$,

$$U(f) = 2\int_{-\infty}^{0} X(\alpha)X(f-\alpha)\, d\alpha + 2\int_{f}^{\infty} X(\alpha)X(f-\alpha)\, d\alpha \tag{7.107}$$

Since $X(-\alpha) = X^*(\alpha)$, this is the same as

$$U(f) = 2\int_{0}^{\infty} X^*(\alpha)X(f+\alpha)\, d\alpha + 2\int_{f}^{\infty} X(\alpha)X(f-\alpha)\, d\alpha \tag{7.108}$$

Equation (7.108) shows how to compute $U(f)$ from $X(f)$ by integrating over positive frequencies only.

Two problems will now be treated using the nonlinear drift force model of Figure 7.15.

1. *Spectral Decomposition Problem.* Given $H_1(f)$ and $H_2(f)$ plus measurement only of $x(t)$, determine the spectral properties of $y_1(t)$ and $y_2(t)$. If $y(t)$ is measured as well as $x(t)$, determine also the spectral properties of $n(t)$.
2. *System Identification Problem.* From simultaneous measurements of both $x(t)$ and $y(t)$, identify the optimum frequency response function properties of $H_1(f)$ and $H_2(f)$ to minimize the autospectrum of $n(t)$.

These two problems are the same type of problems previously solved for wave force problems in Section 7.4.

7.5.3 Spectral Decomposition Problem

In Figure 7.15, for Gaussian input data, the linear output term $y_1(t)$ and the nonlinear output term $y_2(t)$ will be uncorrelated. This is because $y_1(t)$ is uncorrelated with both $x^2(t)$ and $\tilde{x}^2(t)$, the two parts of the square-law envelope detector output $u(t)$. Hence, the total output autospectral density function $G_{yy}(f)$ is given by the formula

$$G_{yy}(f) = G_{y_1y_1}(f) + G_{y_2y_2}(f) + G_{nn}(f) \tag{7.109}$$

where

$$G_{y_1y_1}(f) = |H_1(f)|^2 G_{xx}(f) \tag{7.110}$$

$$G_{y_2y_2}(f) = |H_2(f)|^2 G_{uu}(f) \tag{7.111}$$

These are results for two separate single-input/single-output linear systems.

The quantity $G_{uu}(f)$ can be computed directly from $U(f)$ for an ensemble of stationary random records of length T by the expected value operation

$$G_{uu}(f) = \frac{2}{T} E[U^*(f)U(f)] \tag{7.112}$$

Since $U(f)$ is known from $X(f)$ by Eq. (7.108), this result shows that $G_{uu}(f)$ is a function of $X(f)$. Another useful theoretical formula to compute $G_{uu}(f)$ from knowledge of the input autospectral density function $G_{xx}(f)$ is derived in reference 1. For any $f > 0$, this formula is

$$G_{uu}(f) = 2\int_{-\infty}^{\infty} [1 + B(\alpha)B(f-\alpha)]G_{xx}(\alpha)G_{xx}(f-\alpha)\,d\alpha \tag{7.113}$$

Thus, from measurement only of $x(t)$, one can compute both $G_{xx}(f)$ and $G_{uu}(f)$.

From knowledge of $H_1(f)$ and $G_{xx}(f)$, Eq. (7.110) computes the spectral properties of the linear component $y_1(t)$. From knowledge of $H_2(f)$ and $G_{uu}(f)$, Eq. (7.111) computes the spectral properties of the nonlinear component $y_2(t)$. To obtain the spectral properties of $n(t)$, the total output $y(t)$ must also be measured. Then $G_{yy}(f)$ can be computed. Equation (7.109) now gives the spectral properties of $n(t)$ by

$$G_{nn}(f) = G_{yy}(f) - G_{y_1y_1}(f) - G_{y_2y_2}(f) \tag{7.114}$$

These formulas solve the spectral decomposition problem.

The linear coherence function between the input $x(t)$ and the total output $y(t)$ is defined here by

$$\gamma_{xy}^2(f) = \frac{G_{y_1y_1}(f)}{G_{yy}(f)} = \frac{|G_{xy}(f)|^2}{G_{xx}(f)G_{yy}(f)} \tag{7.115}$$

The nonlinear coherence function between $x(t)$ and $y(t)$ is defined by

$$q_{xy}^2(f) = \frac{G_{y_2y_2}(f)}{G_{yy}(f)} \tag{7.116}$$

where $G_{y_2y_2}(f)$ is computed by Eq. (7.111). In terms of these linear and nonlinear coherence functions, the output noise spectrum is

$$G_{nn}(f) = [1 - \gamma_{xy}^2(f) - q_{xy}^2(f)]G_{yy}(f) \tag{7.117}$$

7.5.4 System Identification Problem

Assume now that the properties of $H_1(f)$ and $H_2(f)$ are *not* known in the nonlinear drift force model of Figure 7.15. Optimum properties are to be determined from simultaneous measurements of $x(t)$ and $y(t)$, based upon minimizing the autospectral density function $G_{nn}(f)$ of $n(t)$ over all possible choices of linear systems to predict $y(t)$ from $x(t)$. It is assumed that the input $x(t)$ follows a Gaussian distribution with zero mean value.

From Eqs. (7.99) and (7.100), the cross-spectral density function $G_{xy}(f)$ between $x(t)$ and $y(t)$ is given by

$$G_{xy}(f) = G_{xy_1}(f) = H_1(f)G_{xx}(f) \tag{7.118}$$

because

$$G_{xu}(f) = 0 \tag{7.119}$$

and

$$G_{xn}(f) = 0 \tag{7.120}$$

Equation (7.119) occurs because $x(t)$ and $u(t)$ are uncorrelated when $x(t)$ is Gaussian. Equation (7.120) occurs because $H_1(f)$ is the same as the optimum linear system $H_o(f)$ and the optimum linear system automatically makes $n(t)$ uncorrelated with $x(t)$.

The optimum linear system $H_o(f)$ is given by the formula

$$H_o(f) = \frac{G_{xy}(f)}{G_{xx}(f)} = H_1(f) \tag{7.121}$$

and is the same as $H_1(f)$ by Eq. (7.118). Thus, $H_1(f)$ can be identified using only $x(t)$ and $y(t)$.

From Eqs. (7.99) and (7.101), the cross-spectral density function $G_{uy}(f)$ between $u(t)$ and $y(t)$ is

$$G_{uy}(f) = H_2(f)G_{uu}(f) \tag{7.122}$$

because of Eqs. (7.119) and (7.120). It follows that

$$H_2(f) = \frac{G_{uy}(f)}{G_{uu}(f)} \tag{7.123}$$

The autospectral density function $G_{uu}(f)$ can be computed directly by Eq. (7.112), and the cross-spectral density function $G_{uy}(f)$ can be computed directly by

$$G_{uy}(f) = \frac{2}{T} E[U^*(f)Y(f)] \tag{7.124}$$

where $U(f)$ is known from $X(f)$ by Eq. (7.108). Thus, $H_2(f)$ can be identified using only $x(t)$ and $y(t)$. These formulas solve the system identification problem.

7.6 NONLINEAR DIFFERENTIAL EQUATIONS OF MOTION

Random data analysis of the nonlinear systems in Section 7.2 converts a single-input/single-output nonlinear model with parallel linear and finite-memory nonlinear systems into an equivalent two-input/single-output linear model. Similarly, as discussed in Section 5.2, a more general single-input/single-output nonlinear model consisting of a linear system in parallel with an arbitrary number of different finite-memory nonlinear systems can be converted into an equivalent multiple-input/single-output linear model. Conversion of these Case 1 type single-input/single-output nonlinear models allows them to be analyzed by known practical linear techniques from references 1 and 2.

Important connections exist between this new approach for random data analysis of nonlinear systems and nonlinear differential equations of motion. The connections will be shown here for nonlinear differential equations of motion representing single-degree-of-freedom (SDOF) systems with nonlinear stiffness terms, where the physical parameters can be determined from input/output measurements in an experimental program.

These techniques also apply to determining nonlinear damping terms or other types of nonlinear properties in proposed nonlinear differential equations of motion, when experimental measurements are made only of the input data and the output data from the system under investigation.

Reference 21 compares complicated bispectral analysis procedures to the alternative practical two-input/single-output linear approach in reference 3 and demonstrates the superiority of this alternative approach in an example. References 32 and 33 discuss extensions of the ideas that can replace reverse single-input/single-output nonlinear models by multiple-input/single-output linear models, and illustrate the results with three computer simulated engineering examples of single-degree-of-freedom systems with nonlinear terms where (1) a turbulent square-law buffeting force occurs, (2) the spring force involves a cubic nonlinearity (Duffing equation), and (3) the damping force involves a squarer with sign (Morison equation).

7.6.1 Single-Degree-of-Freedom Linear Systems

If $x(t)$ is a force input to a typical SDOF linear system with mass m, damping coefficient c, and stiffness coefficient k, and if $y(t)$ is the resulting displacement output, then the linear differential equation of motion in the time domain is

$$m\ddot{y}(t) + c\dot{y}(t) + ky(t) = x(t) \tag{7.125}$$

Fourier transforms of both sides give the associated frequency-domain relation

$$[k - (2\pi f)^2 m + j(2\pi f)c]Y(f) = X(f) \tag{7.126}$$

where $X(f)$ and $Y(f)$ are Fourier transforms of $x(t)$ and $y(t)$, respectively. Equation (7.126) is the same as

$$Y(f) = H(f)X(f) \tag{7.127}$$

where $H(f)$ is the linear frequency response function given by

$$H(f) = [k - (2\pi f)^2 m + j(2\pi f)c]^{-1} \tag{7.128}$$

Equation (7.128) can also be expressed as in Eq. (3.83) by

$$H(f) = (1/k)[1 - (f/f_n)^2 + j2\zeta(f/f_n)]^{-1} \tag{7.129}$$

where the undamped natural frequency f_n and the damping ratio ζ are defined by

$$f_n = (1/2\pi)\sqrt{k/m} \tag{7.130}$$

$$\zeta = c/2\sqrt{km} \tag{7.131}$$

Assume that input data are stationary random data. Then output data will also be stationary random data. From Eq. (7.127), it follows by spectral analysis that the desired linear frequency response function can be identified by the familiar formula

$$H(f) = \frac{S_{xy}(f)}{S_{xx}(f)} \tag{7.132}$$

Here, $S_{xy}(f)$ is the cross-spectral density function between the force input $x(t)$ and the displacement output $y(t)$, and $S_{xx}(f)$ is the autospectral density function of the force input $x(t)$. Computation of $H(f)$ from measured $x(t)$ and $y(t)$ by means of Eq. (7.132) enables one to determine the physical parameters m, c, k, f_n, and ζ that appear in Eqs. (7.130) and (7.131). This then gives the associated linear differential equation of motion (7.125). Here, Gaussian force input data produces Gaussian displacement output data.

The connections between the measured $H(f)$ and the physical parameters can be shown in the following way. From measurement of $H(f)$, one can find the frequency f_n where the peak value of $|H(f)|$ occurs and one can measure the values of $H(0)$ and $|H(f_n)|$. See Figure 3.10. From Equation (7.129) these values will be

$$H(0) = (1/k) \tag{7.133}$$

$$|H(f_n)| = (1/2\zeta k) \tag{7.134}$$

This knowledge of f_n, $H(0)$, and $|H(f_n)|$ will then give the physical parameters k, m, ζ, and c by using the relations of Eqs. (7.130)–(7.134) to obtain

$$k = 1/H(0) \tag{7.135}$$

$$m = k/(2\pi f_n)^2 = 1/(2\pi f_n)^2 H(0) \tag{7.136}$$

$$\zeta = 1/2k|H(f_n)| = H(0)/2|H(f_n)| \tag{7.137}$$

$$c = 2\zeta\sqrt{km} = 1/(2\pi f_n)|H(f_n)| \tag{7.138}$$

These formulas are well-known results that show the connection between measurement of $H(f)$ and the physical parameters in its associated SDOF linear differential equation of motion.

7.6.2 Reversal of Input and Output

Consider an experimental program where the force and displacement are measured. To apply the new approach for random data analysis of nonlinear systems to nonlinear differential equations of motion, as in Section 2.8, the roles of the force and displacement should be reversed so that the displacement becomes the input $x(t)$ and the force becomes the output $y(t)$. The linear differential equation of motion in the time domain instead of Eq. (7.125) is now

$$y(t) = m\ddot{x}(t) + c\dot{x}(t) + kx(t) \tag{7.139}$$

where $x(t)$ is the displacement input and $y(t)$ is the force output. The associated Fourier transform relation for Eq. (7.139) in the frequency domain is

$$Y(f) = A_1(f)X(f) \tag{7.140}$$

where

$$A_1(f) = [k - (2\pi f)^2 m + j(2\pi f)c] \tag{7.141}$$

Note that $A_1(f)$ is the inverse of the $H(f)$ of Eq. (7.128), namely,

$$A_1(f) = [H(f)]^{-1} \tag{7.142}$$

From Eq. (7.140), by ordinary spectral analysis of stationary random data, the frequency response function $A_1(f)$ can be identified by the formula

$$A_1(f) = \frac{S_{xy}(f)}{S_{xx}(f)} \tag{7.143}$$

Here, $S_{xy}(f)$ is now the cross-spectral density function between the displacement input $x(t)$ and the force output $y(t)$, and $S_{xx}(f)$ is the autospectral density function of the displacement input $x(t)$. Computation of $A_1(f)$ from measured $x(t)$ and $y(t)$ by means of Eq. (7.143) enables one to determine the parameters m, c, and k in Eq. (7.141), which then gives the reverse linear differential equation of motion (7.139). Instead of Eqs. (7.133) and (7.134), one will now obtain the results

$$A_1(0) = k \tag{7.144}$$

$$|A_1(f_n)| = 2\zeta k \tag{7.145}$$

Then

$$m = k/(2\pi f_n)^2 = A_1(0)/(2\pi f_n)^2 \tag{7.146}$$

$$c = 2\zeta\sqrt{km} = |A_1(f_n)|/(2\pi f_n) \tag{7.147}$$

These computed parameters apply to Eq. (7.125).

7.6.3 Nonlinear Stiffness Terms

When nonlinear stiffness terms are involved with SDOF systems, a nonlinear differential equation of motion is typically

$$m\ddot{y}(t) + c\dot{y}(t) + ky(t) + k_2y^2(t) + k_3y^3(t) = x(t) \tag{7.148}$$

where $x(t)$ is the force input and $y(t)$ is the resulting displacement output. Here $y^2(t)$ and $y^3(t)$ represent nonlinear stiffness terms multiplied by constants k_2 and k_3, respectively. The associated Fourier transform relation using the $H(f)$ of Eq. (7.128) is

$$[H(f)]^{-1}Y(f) + k_2Y_2(f) + k_3Y_3(f) = X(f) \tag{7.149}$$

where $Y_2(f)$ and $Y_3(f)$ are Fourier transforms of $y^2(t)$ and $y^3(t)$, respectively. Equation (7.149) is the same as

$$Y(f) = H(f)[X(f) - k_2Y_2(f) - k_3Y_3(f)] \tag{7.150}$$

This result represents a feedback system. From knowledge of the physical parameters m, c, and k in $H(f)$, and the physical parameters k_2 and k_3, simulation studies with stationary random data can determine the displacement output $Y(f)$ and $y(t)$ as a function of the force input $X(f)$ and $x(t)$. Here, Gaussian force input data produces non-Gaussian displacement output data.

With measured experimental data, none of this simulation work is required and none of the physical parameters need to be specified. Instead, the physical parameters can be determined by applying the new nonlinear system analysis techniques discussed in this book. The roles of $x(t)$ and $y(t)$ should be reversed so that Eq. (7.148) becomes the reverse relation

$$y(t) = m\ddot{x}(t) + c\dot{x}(t) + kx(t) + k_2x^2(t) + k_3x^3(t) \tag{7.151}$$

Here, $x(t)$ is now the measured non-Gaussian displacement input and $y(t)$ is the measured force output. The associated Fourier transform relation to Eq. (7.151), using the $A_1(f)$ of Eq. (7.141), is

$$Y(f) = A_1(f)X(f) + k_2X_2(f) + k_3X_3(f) \tag{7.152}$$

where $X_2(f)$ and $X_3(f)$, the Fourier transforms of $x^2(t)$ and $x^3(t)$, respectively, can be computed from the measured $x(t)$.

Equation (7.152) is a special case of the Case 1 nonlinear model displayed in Figure 5.1 where $A_2(f) = k_2$ and $A_3(f) = k_3$. For non-Gaussian displacement input data, the conditioned spectral density techniques from references 1 and 2 can be used to identify the systems $A_1(f)$, $A_2(f)$, and $A_3(f)$ because Figure 5.1 is always equivalent to a three-input/single-output linear model with correlated inputs $X(f)$, $X_2(f)$, and $X_3(f)$, as shown in Figure 5.2. A different Figure 5.4 follows for non-Gaussian input data using conditioned records $X_{2\cdot 1}(f)$ and $X_{3\cdot 2!}(f)$, with different formulas than those listed in Section 5.2.2 that apply to Gaussian displacement input data.

When the constants k_2 and k_3 in Eq. (7.152) are replaced by more general linear systems described by $A_2(f)$ and $A_3(f)$, respectively, Eq. (7.152) takes the form

$$Y(f) = A_1(f)X(f) + A_2(f)X_2(f) + A_3(f)X_3(f) + N(f) \tag{7.153}$$

where $N(f)$ is the Fourier transform of extraneous uncorrelated output noise $n(t)$ representing all possible deviations from an ideal model. More complicated nonlinear stiffness and damping terms will be obtained if these new procedures show that the linear systems $A_2(f)$ and $A_3(f)$ are not constants but instead are functions of frequency.

For example, suppose one finds that $A_1(f)$ satisfies Eq. (7.141), $N(f) = 0$, and

$$A_2(f) = 0 \tag{7.154}$$

$$A_3(f) = k_3 + j(2\pi f)c_3 \tag{7.155}$$

where k_3 and c_3 are constants. The corresponding reverse nonlinear differential equation of motion in place of Eq. (7.151) will then be

$$\begin{aligned} y(t) &= m\ddot{x}(t) + c\dot{x}(t) + kx(t) + k_3x^3(t) + c_3\frac{d}{dt}[x^3(t)] \\ &= m\ddot{x}(t) + c\dot{x}(t) + kx(t) + k_3x^3(t) + 3c_3x^2(t)\dot{x}(t) \end{aligned} \tag{7.156}$$

Special cases of this equation should be noted. When $c_3 = 0$, Eq. (7.156) takes the form of Duffing's equation. When $k_3 = 0$, Eq. (7.156) takes the form of Van der Pol's equation. These particular nonlinear equations are important in the field of nonlinear vibrations in electrical and mechanical systems.

7.6.4 Extended Nonlinear Terms

A SDOF nonlinear differential equation of motion with extended nonlinear terms can be expressed as

$$m\ddot{y}(t) + c\dot{y}(t) + ky(t) + g[y(t)] = x(t) \tag{7.157}$$

where

$$x(t) = \text{force input}$$
$$y(t) = \text{displacement output}$$
$$g[y(t)] = \text{arbitrary nonlinear transformation of } y(t)$$

Assume that the quantity $g[y(t)]$ is a sum of $(q - 1)$ different zero-memory nonlinear functions $g_i[y(t)]$ with different constant coefficients a_i such that

$$g[y(t)] = \sum_{i=2}^{q} a_i g_i[y(t)] \tag{7.158}$$

Then, the proposed nonlinear differential equation of motion is

$$m\ddot{y}(t) + c\dot{y}(t) + ky(t) + \sum_{i=2}^{q} a_i g_i[y(t)] = x(t) \tag{7.159}$$

Verification of terms in Eq. (7.159) can be obtained from measured experimental data by applying the techniques discussed here where (1) the roles of $x(t)$ and $y(t)$ are reversed, and (2) the resulting single-input/single-output nonlinear model is replaced by an equivalent multiple-input/single-output linear model. Specifically, by reversing the meaning of $x(t)$ and $y(t)$, Eq. (7.159) becomes the reverse relation

$$y(t) = m\ddot{x}(t) + c\dot{x}(t) + kx(t) + \sum_{i=2}^{q} a_i g_i[x(t)] \tag{7.160}$$

where

$$x(t) = \text{displacement input}$$
$$y(t) = \text{force output}$$

Then, from Eq. (7.160), Fourier transforms of both sides give the associated frequency-domain formula

$$Y(f) = A_1(f)X(f) + \sum_{i=2}^{q} a_i X_i(f) \tag{7.161}$$

where $A_1(f)$ is given by Eq. (7.141) and where the terms $X_i(f)$ are the Fourier transforms of $g_i[x(t)]$ for $i = 2, 3, \ldots, q$. A further extension of Eq. (7.161) occurs by replacing the constants a_i with linear frequency response functions $A_i(f)$ where $i = 2, 3, \ldots, q$. Thus, as per Eq. (7.153), Eq. (7.161) with the addition of extraneous uncorrelated output noise becomes a Case 1 type nonlinear model

$$Y(f) = A_1(f)X(f) + \sum_{i=2}^{q} A_i(f)X_i(f) + N(f) \tag{7.162}$$

From measurement of $x(t)$ and $y(t)$, one can compute all of the proposed $g_i[x(t)]$ to obtain the various $X_i(f)$ terms in Eq. (7.162). One can also compute $X(f)$ and $Y(f)$. The solution of the $A_i(f)$ in Eq. (7.162) is now a straightforward matter using the multiple-input/single-output linear techniques in references 1 and 2. The particular reverse nonlinear differential equation of motion of Eq. (7.160) will be verified if $A_1(f)$ agrees with Eq. (7.141), if the $A_i(f)$ for $i = 2, 3, \ldots, q$ are constants, and if the output noise $N(f)$ is negligible. Thus, physical parameters in proposed SDOF nonlinear differential equations of motion such as Eq. (7.159) can be determined from measured random data.

If the $A_i(f)$ for $i = 2, 3, \ldots, q$ vary with frequency, one obtains more general SDOF nonlinear differential equations of motion than Eq. (7.159). The $g_i[x(t)]$ for $i = 2, 3, \ldots, q$ identify the different proposed nonlinear system amplitude-domain properties and the linear frequency response functions $A_i(f)$ identify the proposed nonlinear system frequency-domain properties.

This concludes Chapter 7.

REFERENCES

1. Bendat, J. S., and Piersol, A. G., *Random Data: Analysis and Measurement Procedures*, 2nd ed., Wiley-Interscience, New York, 1986.
2. Bendat, J. S., and Piersol, A. G., *Engineering Applications of Correlation and Spectral Analysis*, Wiley-Interscience, New York, 1980.
3. Bendat, J. S., and Piersol, A. G., Spectral analysis of nonlinear systems involving square-law operations, *Journal of Sound and Vibration*, **81**(2), 199 (1982).
4. Bendat, J. S., and Piersol, A. G., Decomposition of wave forces into linear and nonlinear components, *Journal of Sound and Vibration*, **106**(3), 391 (1986).
5. Bendat, J. S., Statistical errors for nonlinear system measurements involving square-law operations, *Journal of Sound and Vibration*, **90**(2), 275 (1983).
6. Bendat, J. S., Solutions for the multiple input/output problem, *Journal of Sound and Vibration*, **44**(3), 311 (1976).
7. Bendat, J. S., System identification from multiple input/output data, *Journal of Sound and Vibration*, **49**(3), 293 (1976).
8. Bendat, J. S., Statistical errors in measurement of coherence functions and input/output quantities, *Journal of Sound and Vibration*, **59**(3), 405 (1978).
9. Bendat, J. S., Modern analysis procedures for multiple input/output problems, *Journal Acoustical Society of America*, **68**(2), 498 (1980).
10. Bendat, J. S., *The Hilbert Transform and Applications to Correlation Measurements*, Bruel & Kjaer, Naerum, Denmark, 1986.
11. Braun, S., *Mechanical Signature Analysis*, Academic Press, New York, 1986.
12. Broch, J. T., *Nonlinear Systems and Random Vibration*, Bruel & Kjaer, Naerum, Denmark, 1975.

13. Bussgang, J. J., Cross-correlation functions of amplitude distorted Gaussian inputs, in *Nonlinear Systems*, A. H. Haddad, Ed., Dowden, Hutchinson & Ross, Stroudsburg, Pennsylvania, 1975.
14. Choi, D., Miksad, R. W., and Powers, E. J., Application of digital cross-bispectral analysis techniques to model the nonlinear response of a moored vessel system in random seas, *Journal of Sound and Vibration*, **99**(3), 309 (1985).
15. Crandall, S. H., and Mark, W. D., *Random Vibration in Mechanical Systems*, Academic Press, New York, 1963.
16. Dalzell, J. F., Cross-bispectral analysis: application to ship resistance in waves, *Journal of Ship Research*, **18**, 62 (1974).
17. Deutsch, R., *Nonlinear Transformations of Random Processes*, Prentice-Hall, Englewood Cliffs, New Jersey, 1962.
18. Dodds, C. J., and Robson, J. D., Partial coherence in multivariate random processes, *Journal of Sound and Vibration*, **42**(2), 243 (1975).
19. Ewins, D. J., *Modal Testing: Theory and Practice*. Wiley, New York, 1985.
20. Fitzpatrick, J. A., and Rice, H. J., Simplified coherence functions for multiple input/output analysis, *Journal of Sound and Vibration*, **122**(1), 171 (1988).
21. Fitzpatrick, J. A., and Rice, H. J., An alternative to the bispectral approach for the analysis of flow induced vibration problems, *Proc. International Symposium on Flow Induced Vibration and Noise*, American Society of Mechanical Engineers, Chicago, **2**, 197 (1988).
22. French, A. S., Practical nonlinear system analysis by Wiener kernal estimation in the frequency domain, *Biological Cybernetics*, **24**, 111 (1976).
23. Hasselman, K., Munk, W., and MacDonald, G., Bispectra of Ocean Waves, in *Time Series Analysis*, W. Rosenblatt, Ed., Wiley, New York, 1963.
24. Kim, C. H., and Breslin, J. P., Prediction of slow drift oscillations of a moored ship in head seas, *Proc. Behavior of Offshore Structures*, BOSS '76, Trondheim, Norway, 320 (1976).
25. Kim, K. S., and Shin, Y. S., Application of a new Hilbert transform method to nonlinearity identification, *Proc. 7th International Modal Analysis Conference*, Las Vegas, Nevada, 1386 (1989).
26. Marmelis, P. Z., and Marmelis, V. Z., *Analysis of Physiological Systems: The White-Noise Approach*, Plenum Press, New York, 1978.
27. Matsuoka, T., and Ulrych, T. J., Phase estimation using the bispectrum, *Proc. of the IEEE*, **72**(10), 1403 (1988).
28. Morison, J. R., et al, The force exerted by surface waves on piles, *Journal of Petroleum Techniques*, **189**, 149 (1950).
29. Newland, D. E., *Random Vibrations and Spectral Analysis*, 2nd ed., Longman, New York, 1984.
30. Nikias, C. L., and Raghuveer, M. R., Bispectrum estimation: A digital signal processing framework, *Proc. of the IEEE*, **75**(7), 869 (1987).

31. Price, R., A useful theorem for nonlinear devices having Gaussian inputs, in *Nonlinear Systems*, A. H. Haddad, Ed., Dowden, Hutchinson & Ross, Stroudsburg, Pennsylvania, 1975.
32. Rice, H. J., and Fitzpatrick, J. A., A generalized technique for spectral analysis of nonlinear systems, *Mechanical Systems and Signal Processing*, **2**(2), 195 (1988).
33. Rice, H. J., Esmonde, H., and Fitzpatrick, J. A., A spectral method for identifying quadratic damping in single degree of freedom systems, *Proc. International Symposium on Flow Induced Vibration and Noise*, American Society of Mechanical Engineers, Chicago, **3**, 269 (1988).
34. Rice, S. O., Mathematical analysis of random noise, in *Selected Papers on Noise and Stochastic Processes*, N. Wax, Ed., Dover, New York, 1954.
35. Rice, S. O., and Bedrosian, E., The output properties of Volterra systems (Nonlinear systems with memory) driven by harmonic and Gaussian inputs, *Proc. of the IEEE*, **59**(12) 1688 (1971).
36. Roberts, J. B., Nonlinear analysis of slow drift oscillations of moored vessels in random seas, *Journal of Ship Research*, **25**, 130 (1981).
37. Rozario, N. and Papoulis, A., The identification of certain nonlinear systems by only observing the output, *Workshop on Higher-Order Spectral Analysis*, Office of Naval Research and National Science Foundation, Vail, Colorado, 78 (1989).
38. Rugh, W. J., *Nonlinear System Theory: The Volterra/Wiener Approach*, Johns Hopkins University Press, Baltimore, Maryland, 1981.
39. Sarpkaya, T., and Isaacson, M., *Mechanics of Wave Forces on Offshore Structures*, Van Nostrand Reinhold, New York, 1981.
40. Schetzen, M., *The Volterra and Wiener Theories of Nonlinear Systems*, Wiley-Interscience, New York, 1980.
41. Schueller, G. I., and Bucher, C. G., Non-Gaussian response of systems under dynamic excitation, *Stochastic Structural Dynamics*, S. T. Ariaratnam, G. I. Schueller, and I. Elishakoff, Eds., Elsevier Applied Science, New York, 1988.
42. Tick, L. J., The estimation of transfer functions of quadratic systems, *Technometrics*, **3**, 563 (1961).
43. Vinh, T., Liu, H., and Djouder, M., Second order transfer function: computation and physical interpretation, *Proc. 5th International Modal Analysis Conference*, London, 587 (1987).
44. Vugts, J. H., and Bouquet, A. G., A nonlinear frequency domain description of wave forces on an element of a vertical pile, *Proc. Behavior of Offshore Structures*, BOSS '85, Delft, Netherlands, 239 (1985).
45. Yang, C. Y., *Random Vibration of Structures*, Wiley-Interscience, New York, 1986.

INDEX

GLOSSARY OF SYMBOLS

Symbol	Meaning
a, b	Arbitrary constants
$A(f), B(f)$	Linear system frequency response functions
$E[\ ]$	Expected value of [ ]
f	Cyclical frequency
$\mathscr{F}[\ ]$	Fourier transform of [ ]
$G_{xx}(f)$	Autospectral density function (one-sided)
$G_{xy}(f)$	Cross-spectral density function (one-sided)
$\mathscr{G}(f)$	Energy spectral density function (one-sided)
$h(\tau)$	Linear weighting function (first-order)
$h(\tau, t)$	Time-varying linear weighting function
$h(\tau_1, \tau_2)$	Bilinear weighting function (second-order)
$h(\tau_1, \tau_2, \tau_3)$	Trilinear weighting function (third-order)
$H(f)$	Linear frequency response function (first-order)
$H_0(f)$	Optimum linear frequency response function
$\|H(f)\|$	System gain factor
$H(f_1, f_2)$	Bilinear frequency response function (second-order)
$H(f_1, f_2, f_3)$	Trilinear frequency response function (third-order)
j	$\sqrt{-1}$
$J(f, g)$	Double Fourier transform of $h(\tau, t)$
$L_1(x)$	Linear system operator
$L_2(x)$	Special bilinear system operator
$L_3(x)$	Special trilinear system operator
$L_2(x, y)$	Bilinear system operator
$L_3(x, y, z)$	Trilinear system operator
n_d	Number of distinct averages
$p(x)$	Probability density function (first-order)
$p_2(y)$	Nonlinear system output probability density function
$p(x_1, x_2)$	Joint probability density function (second-order)
Prob[ ]	Probability of [ ]
$q^2_{xy}(f)$	Nonlinear coherence function
$R_{xx}(\tau)$	Autocorrelation function (first-order)
$R_{xy}(\tau)$	Cross-correlation function (first-order)
$R_{xxx}(\tau_1, \tau_2)$	Second-order autocorrelation function
$R_{xxy}(\tau_1, \tau_2)$	Second-order cross-correlation function